D1547883

FOOD MICROBIOLOGY:
ADVANCES AND PROSPECTS

A complete list of titles in the
Society for Applied Bacteriology Symposium Series
appears at the end of this volume

THE SOCIETY FOR APPLIED BACTERIOLOGY
SYMPOSIUM SERIES NO. 11

FOOD MICROBIOLOGY: ADVANCES AND PROSPECTS

Edited by

T. A. ROBERTS

AND

F. A. SKINNER

1983

ACADEMIC PRESS

A Subsidiary of Harcourt Brace Jovanovich, Publishers

LONDON · NEW YORK
PARIS · SAN DIEGO · SAN FRANCISCO · SÃO PAULO
SYDNEY · TOKYO · TORONTO

ACADEMIC PRESS INC. (LONDON) LTD.
24–28 OVAL ROAD
LONDON NW1 7DX

U.S. Edition published by
ACADEMIC PRESS INC.
111 FIFTH AVENUE
NEW YORK, NEW YORK 10003

British Library Cataloguing in Publication Data
Food microbiology.—(Society for Applied Bacteriology
symposium ISSN 0300-9610; 11)
1. Food—Microbiology—Congresses
I. Roberts, T. A.   II. Skinner, F. A.
III. Series
641.3′015′76   QR115

ISBN 0-12-589670-0

Photoset by Paston Press, Norwich
Printed in Great Britain by
T. J. Press (Padstow) Ltd., Padstow, Cornwall

Contributors

A. C. BAIRD-PARKER, *Unilever Research, Colworth Laboratory, Colworth House, Sharnbrook, Bedford MK44 1LQ, UK*

P. G. BEAN, *Metal Box p.l.c., Research and Development Division, Denchworth Road, Wantage OX12 9BP, UK*

F. W. BEECH, *Department of Agriculture and Horticulture, Long Ashton Research Station, University of Bristol, Long Ashton, Bristol BS18 9AF, UK*

M. H. BROWN, *Unilever Research, Colworth Laboratory, Colworth House, Sharnbrook, Bedford MK44 1LQ, UK*

D. C. BULL, *RHM Research Ltd., The Lord Rank Research Centre, Lincoln Road, High Wycombe, Bucks HP12 3QR, UK*

R. H. DAINTY, *Agricultural Research Council, Meat Research Institute, Langford, Bristol BS18 7DY, UK*

R. R. DAVENPORT, *Department of Agriculture and Horticulture, Long Ashton Research Station, University of Bristol, Long Ashton, Bristol BS18 9AF, UK*

J. DE MAAKER, *Sprenger Instituut, P.O. Box 17, 6700 AA Wageningen, Haagsteeg 6, Wageningen, The Netherlands*

Y. DE WITTE, *Sprenger Instituut, P.O. Box 17, 6700 AA Wageningen, Haagsteeg 6, Wageningen, The Netherlands*

INGER ERICHSEN, *Swedish Meat Research Institute, POB 504, 244 00 Kävlinge, Sweden*

B. C. FLETCHER, *Unilever Research, Colworth Laboratory, Colworth House, Sharnbrook, Bedford MK44 1LQ, UK*

B. A. FRIEND, *Department of Food Science and Technology, University of Nebraska, Lincoln, NE 68583, USA*

G. A. GARDNER, *Ulster Curer's Association, 2 Greenwood Avenue, Belfast BT4 3JL, Northern Ireland, UK*

R. J. GILBERT, *Food Hygiene Laboratory, Central Public Health Laboratory, 175 Colindale Avenue, London NW9 5HT, UK*

G. W. GOULD, *Unilever Research, Colworth Laboratory, Colworth House, Sharnbrook, Bedford MK44 1LQ, UK*

G. HOBBS, *Ministry of Agriculture, Fisheries and Food, Torry Research Station, P.O. Box 31, 135 Abbey Road, Aberdeen AB9 8DG, Scotland, UK*

B. JARVIS, *The British Food Manufacturing Industries Research Association, Randalls Road, Leatherhead, Surrey KT22 7RY, UK*

D. C. KILSBY, *Unilever Research, Colworth Laboratory, Colworth House, Sharnbrook, Bedford MK44 1LQ, UK*

P. C. KOEK, *Sprenger Instituut, P.O. Box 17, 6700 AA Wageningen, Haagsteeg 6, Wageningen, The Netherlands*

B. A. LAW, *Microbiology Department, National Institute for Research in Dairying, Shinfield, Reading, Berks RG2 9AT, UK*

F. J. LEY, *Irradiated Products Ltd., Moray Road, Elgin Industrial Estate, Swindon, Wiltshire SN2 6DU, UK*

L. A. MABBITT, *Microbiology Department, National Institute for Research in Dairying, Shinfield, Reading, Berks RG2 9AT, UK*

G. C. MEAD, *Agricultural Research Council, Food Research Institute, Colney Lane, Norwich, Norfolk NR4 7UA, UK*

D. A. A. MOSSEL, *Laboratory of Microbiology, Department of the Science of Food of Animal Origin, Faculty of Veterinary Medicine, The University of Utrecht, Bilstraat 172, 3572 BP Utrecht, The Netherlands*

J. R. PELLON, *Department of Nutrition and Food Science, Massachusetts Institute of Technology, Cambridge, Massachusetts 02139, USA*

T. A. ROBERTS, *Agricultural Research Council, Meat Research Institute, Langford, Bristol BS18 7DY, UK*

B. SIMONSEN, *Danish Meat Products Laboratory, Ministry of Agriculture, Howitzvej 13, 2000, Copenhagen F, Denmark*

M. SHAHANI, *Department of Food Science and Technology, University of Nebraska, Lincoln, NE 68583, USA*

B. G. SHAW, *Agricultural Research Council, Meat Research Institute, Langford, Bristol BS18 7DY, UK*

A. J. SINSKEY, *Department of Nutrition and Food Science, Massachusetts Institute of Technology, Cambridge, Massachusetts 02139, USA*

G. L. SOLOMONS, *RHM Research Ltd., The Lord Rank Research Centre, Lincoln Road, High Wycombe, Bucks HP12 3QR, UK*

Preface

A symposium was held in July 1981 at the University of Bristol to mark the 50th Anniversary of the Society for Applied Bacteriology. It was appropriate that the topic was on aspects of food microbiology because the society had its origins in the dairy sector of that industry, and a majority of members remain associated with the many facets of microbiological aspects of food. Ten years ago the society held a symposium, also at the University of Bristol, on 'Microbial changes in foods'. In 1981 leading specialists were asked to review progress in their field since that symposium, and, where possible, to gaze into the future and attempt to predict trends under the title 'Advances and prospects in food microbiology'.

An introductory chapter on the microbial ecology of foods is followed by considerations of the microbiology of milk, red meats, fish, poultry and their products, and beverages. Cultured dairy foods and fermented meat and fish products are considered in additional chapters. Pasteurized cured meats are described as an example where it may be possible to develop a mathematical model to predict the microbiological response upon storage. Developments in preservation by heat and the renewed interest in food irradiation are reviewed. Trends in food-borne illness are described and the impact of novel microbiological methods and of genetic engineering are considered. Finally, the need for and the value of microbiological specifications are considered in relation to different sampling plans, and a consensus of opinion of the future by a panel of microbiologists is given in a Delphi-type forecast.

T. A. ROBERTS
Agricultural Research Council
Meat Research Institute,
Langford, Bristol BS18 7DY

F. A. SKINNER
Formerly of the
Soil Microbiology Department,
Rothamsted Experimental Station,
Harpenden,
Hertfordshire AL5 2JQ

Contents

ix

Predictive Modelling of Food Safety with Particular Reference to *Clostridium botulinum* **in Model Cured Meat Systems**
T. A. ROBERTS AND B. JARVIS

Developments in Heat Treatment Processes for Shelf-stable Products
P. G. BEAN

New Interest in the Use of Irradiation in the Food Industry
F. J. LEY

New Methods for Controlling the Spoilage of Milk and Milk Products
B. A. LAW AND L. A. MABBITT

Microbial and Chemical Changes in Chill-stored Red Meats
R. H. DAINTY, B. G. SHAW AND T. A. ROBERTS

Microbial Spoilage of Cured Meats
G. A. GARDNER

Effect of Packaging and Gaseous Environment on the Microbiology and Shelf Life of Processed Poultry Products
G. C. MEAD

Microbial Spoilage of Fish
G. HOBBS

Food Microbiology into the Twenty-first Century—a Delphi Forecast

B. JARVIS

Essentials and Perspectives of the Microbial Ecology of Foods

D. A. A. MOSSEL

*Laboratory of Microbiology, Department of the Science of
Food of Animal Origin, Faculty of Veterinary Medicine,
The University of Utrecht, Utrecht, The Netherlands*

Contents

1. Character, Establishment and Metabolic Activity of the Specific Microbial Community Structure of Foods at the Stage of Distribution

A. Historical introduction

STUDIES ON the microbial ecology of foods have long been marred by the approach that serves merely to demonstrate the presence of particular organisms, as revealed by isolation procedures, rather than to determine numbers of colony forming units of important micro-organisms and their changes as affected by time and temperature. This lack of quantification has yielded records of the presence of almost all types of organisms in virtually all foods. None the less, in fact, the microbial population of every food item at the moment of sale has a specific character.

This situation results from the effect of various selective influences on the initial microbial population of foods. This primary microflora originates mostly from one or more of the principal habitats: soil, surface water, dust,

FOOD MICROBIOLOGY
ISBN 0 12 589670 0

the gastro-intestinal and respiratory tracts and skin of animals and man, and the environment wherein food is manufactured or prepared, and it has a transient character (Mossel & Westerdijk 1949; Mossel 1971). Various ecological determinants (Ingram 1948; Clayson 1955; Clayson & Blood 1957; Post *et al.* 1961; Stadhouders 1975; Gill & Newton 1978; Hurst & Collins-Thompson 1979; McMeekin & Thomas 1979; Grau 1981; Newton & Gill 1981) soon exert a selective effect by favouring the proliferation of certain organisms that will subsequently outnumber the other components of the initial microflora. This dynamic initial state (Odum 1969) leads to the ultimate colonization of every food by a more permanent community of particular groups of organisms (Alexander 1971), named the microbial *association* of a given food (Ingram 1955; Mossel & Ingram 1955; Dowdell & Board 1971; Halls & Board 1973; McMeekin 1975, 1977; Daud *et al.* 1979; Grau 1979; Skovgaard 1979; Tolle 1979).

The term 'microbial association' was suggested for use in food micro-biology by the Dutch mycologist Johanna Westerdijk (1949). It followed the introduction of the general concept 'association' by Alexander von Humboldt about 1805, more than fifty years before Ernst Haeckel founded the branch of biology which was to become known as ecology. The specific character of spoilage associations of commodities like beer, wine and milk had been empirically recognized, *avant la lettre* by Pasteur (1858). Studies by Beijerinck (1908, 1913) and by Baas Becking (1934) on the selective pressures that determined the character of the microbial ecology of soil and water led Westerdijk to an understanding of the establishment of food associations.

The operative selective influences had been indicated as 'le terrain' (the soil) by Pasteur. Westerdijk (1949) grouped them according to (i) their character, i.e. abiotic or biotic; and (ii) whether they reside in the food itself ('intrinsic' parameters), in its processing, or in the environment in which the food is stored or transported ('extrinsic' factors). After different responses to the same parameter were demonstrated, in terms of growth of and metabolism by micro-organisms (Mossel 1971), the following grouping of determinants was introduced.

B. Parameters affecting colonization of foods

Many parameters have been identified as affecting the fate of the initial microbial community structure of foods. They determine whether any naturally occurring group of organisms will remain dormant, proliferate or die off, and hence completely control the composition of the microbial association at the stage of sale (Table 1). These parameters are generally divided into four groups.

(i) *Intrinsic parameters*

First and foremost, intrinsic chemical, physical and structural factors affect the fate of the original microbial population of a food. Intrinsic parameters include, in declining order or selective pressure exerted: water activity, pH and buffering power, the presence of naturally occurring or added antimicrobial components, E_h and redox poising capacity and nutrient composition (Mossel 1982). It is noteworthy that different values of these determinants may exist in the same solid food; in deep tissue of livers, e.g. the association is composed of facultative organisms, whereas the surface community structure is dominated by strict aerobes (Gill & de Lacy 1982).

As shown in Section C, a decisive intrinsic influence upon the type of microbial colonization of foods and its biochemical inference is also often exerted by adhesion of organisms to surfaces (Berkeley *et al*. 1980; Bitton & Marshall 1980), leading to spatial heterogeneity. This is a situation similar to that observed in pathology (Beachy 1981; Elliott *et al*. 1981), with respect to the effects of the colonization of: the intestinal lumen by *Salmonella* (Jones *et al*. 1981), *Vibrio* spp. (Levett & Daniel 1981), *Yersinia enterocolitica* (Vesikari *et al*. 1981) and *Escherichia coli* (Dickinson *et al*. 1980; Cravioto *et al*. 1982); the urinary tract by *E. coli* (Schaeffer *et al*. 1981) or *Candida albicans* (Sobel *et al*. 1981); and the oral cavity, not only by streptococci (Reed *et al*. 1980), but also by *Neisseria* spp. (Salit & Morton 1981), *Eikenella* (Yamazaki *et al*. 1981), *Bacteroides* (Slots & Gibbons 1978) and actinomycetes (Qureshi & Gibbons 1981).

(ii) *Processing factors*

Various determinants of the mode of processing applied to a food may exert a marked influence on the primary community structure as it was after selection by intrinsic factors.

Most processing procedures lead to the elimination of organisms as is the case with currently applied heat treatments. It may occur in the future following the use of gamma irradiation and innocuous chemical sterilizing agents.

However, food processing may also favour colonization. Microbial proliferation may occur when the product is held under temperature conditions that allow growth and when intrinsic attributes permit. Sometimes the acquisition of organisms from apparatus or added ingredients is hard to avoid.

TABLE 1

A review of some, mostly more recent, investigations wherein the fate of particular micro-organisms in foods and the genesis of specific microbial community structures (associations) were studied

Food studied	Organism monitored	Reference
Baked products	*Bacillus* spp.	Farmiloe *et al*. (1954)
	Moulds	Spicher (1980)
	Yeasts	Seiler (1980
Beef	*Cl. perfringens*	Thompson *et al*. (1979)
	Spoilage association	Pierson *et al*. (1970); Patterson & Gibbs (1978)
Beef, vacuum packaged	*Salmonella* spp.	Goodfellow & Brown (1978); Kennedy *et al*. (1980)
	Staph. aureus	Kennedy *et al*. (1980)
	Spoilage association	Pierson *et al*. (1970); Hanna *et al*. (1977); Newton & Gill (1978); Hanna *et al*. (1982*a*)
Cereals	Moulds	Christensen & Kaufmann (1969); Milton & Jarrett (1969, 1970); Lillehoj *et al*. (1976)
Cheese, hard	*Salmonella* spp.	Park *et al*. (1969)
	Enteropathogenic *E. coli*	Kornacki & Marth (1982)
soft	Enterobacteriaceae	Mourgues *et al*. (1979)
spread	*Cl. botulinum*	Tanaka *et al*. (1979); Kautter *et al*. (1981)
Chocolate	*Salmonella* spp.	Tamminga *et al*. (1976, 1977)
Condensed milk	*Staph. aureus*, *B. cereus*	Ducha Sardana (1976)
Edible offal	Spoilage association	Patterson & Gibbs (1979)
	Enterobacteriaceae	Bijker (1981); Hanna *et al*. (1982*b*)
Cured meats	*Bacillus* spp.	Mol & Timmers (1970); Bell & Gill (1982)
	Cl. perfringens	Roberts & Smart (1976)
	Cl. botulinum	Roberts & Smart (1976); Roberts *et al*. (1981)
	Lactic acid bacteria	Cavett (1962); Sinell & Levetzow (1967); Mol & Timmers (1970); Egan *et al*. (1980); Bell & Gill (1982)
	Salmonella spp., *Staph. aureus*	Stiles & Ng (1979)
	B. cereus, *Cl. perfringens*	Steele & Stiles (1981)
Dried foods		
Milk	*E. coli*	Chopin *et al*. (1977); Thompson *et al*. (1978)
	Salmonella spp.	Licari & Potter (1970); Ray *et al*. (1971)
	Staph. aureus	Galesloot & Stadhouders (1968); Chopin *et al*. (1978)
Vegetables	Enterobacteriaceae, group D streptococci, micrococci	Holtzapffel & Coutinho (1969)
Various	*Staph. aureus*, *Salm. newport*	Christian & Stewart (1973)
Fish	Microbial community structure	Lerke *et al*. (1965); Lee & Harrison (1968); Herbert *et al*. (1971); van Spreekens (1977); Banks *et al*. (1980)

(continued)

TABLE 1 (continued)

Food studied	Organism monitored	Reference
Fruits and juices	Yeasts	Dennis & Buhagiar (1980); Suresh et al. (1982)
	Moulds	Dennis & Davis (1977)
	Salmonella spp., Shigella spp.	Mossel & de Bruin (1960)
Frozen foods	Salmonella spp.	Wallace & Park (1933); Georgala & Hurst (1963)
	Staph. aureus	Dack & Lippitz (1962)
	Spoilage association	Berry (1933); Michener (1979)
Lamb	Spoilage association	Gill (1976); Gill & Newton (1977)
Lamb, vacuum-packed	Spoilage association	Newton et al. (1977)
Margarine	Enterobacteriaceae	Mossel (1970)
	Yeasts and moulds	Tuynenburg Muys (1971)
Meat, minced and	Enterobacteriaceae	Kleeberger & Busse (1975)
loaves	Pseudomonadaceae	Kleeberger & Busse (1975)
	Salmonella spp.	Tamminga et al. (1982)
	Staph. aureus	Bunck et al. (1977); Dahl et al. (1980)
	Spoilage association	Harrison et al. (1981)
Meat and chicken	Salmonella spp.	Winter et al. (1953)
salads	Staph. aureus	Holtzapffel & Mossel (1968); Doyle et al. (1982)
Non-alcoholic	Yeasts	Mossel & Scholts (1964); Sand &
beverages		van Grinsven (1976)
Oysters	Various enteric pathogens	Son & Fleet (1980)
	Spoilage association	Qadri et al. (1976)
Pasta	Salmonella spp., group D streptococci	Rayman et al. (1979)
Pies	B. cereus, Salmonella spp., Staph. aureus	Wyatt & Guy (1981a, b)
Pork	Spoilage association	Gill & Newton (1977)
Potatoes	Spoilage association	Ingram (1934); Buttiaux & Catsaras (1964)
Potatoes, baked	Cl. botulinum	Sugiyama et al. (1981)
Poultry	Salmonella spp.	Mulder (1982)
	Spoilage association	McMeekin et al. (1975, 1977); Gill & Newton (1977); Barnes et al. (1979)
Salmon, smoked	Micrococci	Lee & Pfeifer (1973
Sandwiches, packed	Salmonella spp.	Swaminathan et al. (1981)
	Staph. aureus	Bennet & Amos (1982)
	Cl. botulinum	Kautter et al. (1981a)
Shrimps	Spoilage association	Harrison & Lee (1969); van Spreekens (1977)
Tomato juice	Salmonella spp., Shigella spp.	Mossel & de Bruin (1960)
	Spoilage association	Juven (1979)
Vegetables	Salmonella spp.	Splitstoesser & Segen (1970)
	Enterobacteriaceae	Wright et al. (1976)
	Ps. aeruginosa	King et al. (1976)
	Spoilage bacteria	Splitstoesser (1970); Brocklehurst & Lund (1981); Lund (1982)
	Moulds	Thorne (1972); Geeson (1979)
	Yeasts	Dennis & Buhagiar (1980)

(iii) *Extrinsic parameters*

External selective influences of the greatest relevance under practical conditions include temperature, relative humidity and the composition of the gaseous phase prevailing during transport, distribution and storage. The microbial association of chilled foods, for example, consists, in essence, of psychrotrophic organisms. However, whereas slime-forming, aerobic Gram negative rods will predominate on refrigerated meat carcasses, vacuum-packed meat cuts will be colonized by facultatively anaerobic, mostly Gram positive bacteria, leading to souring (Fig. 1).

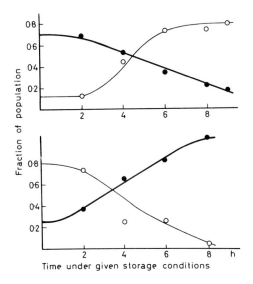

Fig. 1. The fate of the initial community structure of fresh meat during chilled storage under aerobic conditions (top) and in anaerobic packs (bottom) leading to markedly different final microbial associations. ● G +ve fraction; ○ G −ve fraction.

(iv) *Implicit factors*

Intrinsic biotic parameters playing an important role in the genesis of food associations are often indicated as *implicit* factors. This is consonant with the same term in mathematics because their influence cannot be divorced from that of the parameters mentioned above. Implicit factors in food spoilage include microbial antagonism, synergism and syntrophism.

Synergism and antagonism result mainly from the formation of stimulatory and inhibitory substances, respectively; in the latter case bacteriocins are sometimes involved (Reeves 1979; Means & Olson 1981; Nieves *et al.* 1981; Sahl & Brandis 1981). Syntrophism is the interesting phenomenon whereby each of the two components of a microbial association permits development of the other by forming a nutrient essential for its proliferation (Nurmikko 1956; Yeoh *et al.* 1968; McInerney & Bryant 1981; McInerney *et al.* 1981; Bochem *et al.* 1982).

Synergistic phenomena may markedly affect the microbial association of foods (Duitschaever & Irvine 1971; Megee *et al.* 1972; Moon & Reinbold 1976; Board *et al.* 1979; Treleaven *et al.* 1980; Yildiz & Westhoff 1981; Aries *et al.* 1982; Montville 1982; Bell & Gill 1982). None the less the practical impact of antagonistic interactions observed for the first time by Eijkman in 1904, is much more important (Kraft *et al.* 1976; Smither 1978; Raccach *et al.* 1979; Gibbs *et al.* 1980). As illustrated in Table 2, outbreaks of food-transmitted disease or spoilage phenomena may not occur despite the presence of the causative organisms, because growth and metabolism of the latter may be controlled by antagonistic interactions.

(v) *Net effects*

It is essential to note that the effect of a given parameter of intrinsic, extrinsic or implicit nature is, as a rule, also strongly determined by the values of other factors affecting microbial growth mentioned earlier (Genigeorgis *et al.* 1969, 1971; Boylan *et al.* 1976; Northolt *et al.* 1979; Beuchat 1981*a,b*; La Rocco & Martin 1981; Roberts *et al.* 1981; Tesone *et al.* 1981). This is an illustration of Blackman's (1905) law of superimposed limiting factors (Heath 1969).

The combined influence of the many parameters determing the fate of micro-organisms in foods, but particularly that of a_w, pH and antimicrobial constituents, results in the recognition of three broad classes of foods. These include perishable commodities such as fresh meats; weakly preserved foodstuffs, e.g. jams, condensed milk and fermented sausages; and the fully shelf-stable commodities such as dried foods. In ecological terms these foods represent niches of increasing resistance to colonization. The effects of ecological determinants on the origin of the typical associations of the foods of major importance and their keeping qualities are presented in Table 3.

TABLE 2

Examples of antagonistic interactions of practical significance in foods

Affected organism	Effector	Reference
B. cereus	Group D streptococci	Kafel & Ayres (1969)
Cl. botulinum	Bacillus spp.	Mossel (1969); Kwan & Lee (1974);
	Brevibacterium linens	Smith (1975); Graham (1978)
	Cl. perfringens	
	Cl. sporogenes	
	Cocci	
	Enterobacteriaceae	
	Lactobacillaceae	
	Moraxella sp.	
	Ps. aeruginosa	
	Strep. lactis	
	Saprophytic flora—not identified	
Cl. perfringens	Cl. sporogenes	Sinell & Levetzow (1967); Mossel
	Lactobacillaceae	(1969); Jayne-Williams (1973);
	Group D streptococci	Tabatai & Walker (1974);
		Gilliland & Speck (1977)
E. coli, incl.	Lactobacillaceae	Frank & Marth (1977)
enteropathogenic types		
Salmonella	E. coli	Mossel (1969); Oblinger & Kraft
	Lactobacillaceae	(1970); Sorrells & Speck (1970);
	Pseudomonas spp.	Gilliland & Speck (1977)
	Saprophytic flora—not identified	
Staph. aureus	Aeromonas	Mossel (1969); Oblinger & Kraft
	Bacillus spp.	(1970); Haines & Harmon
	Enterobacteriaceae	(1973); Gilliland & Speck (1977);
	Lactobacillaceae	Miller & Ledford (1977);
	Pseudomonas/Acinetobacter	Stadhouders et al. (1978)
	Staph. epidermidis	
	Streptococci	
	Saprophytic flora—not identified	
Yersinia enterocolitica	Psychrotrophs	Hanna et al. (1977a); Stern et al.
		(1980)
Brochothrix	Betabacterium breve,	Roth & Clark (1975); Collins-
	Lactobacillus plantarum	Thompson & Rodriguez Lopez
		(1980)
Pseudomonas/	Lactobacillaceae	Price & Lee (1970); Dubois et al.
Acinetobacter		(1979)
Yeasts	Aeromonas spp.	Buck et al. (1962)
	Alcaligenes spp.	
	Flavobacterium spp.	
	Pseudomonas spp.	
	Vibrio spp.	

C. Factors influencing the metabolic activity of microbial associations of foods

The previous considerations apply primarily to assimilative properties of micro-organisms and hence to formation of biomass. However, as stressed particularly by Sharpe (1980), population studies alone have only a limited bearing on the main interest of the food trade, which is the character of the microbial dissimilation leading to deterioration of foods. Hence, as illustrated in Fig. 2, a clear distinction has to be drawn between microbial proliferation in a food and the metabolic activities of the ultimate community structure. Like the colonization of foods, however, spoilage of every particular commodity will follow a similar course despite considerable initial differences in the composition of the microbial community structure.

The biochemical activities of food associations, like microbial proliferation, are determined by many of the intrinsic and extrinsic ecological determinants discussed earlier. These include temperature, pH, a_w and, especially, their combined effects (Barnes & Melton 1971; Ingram & Dainty 1971; Levin 1972; Dainty et al. 1975; Daud et al. 1978; Dainty & Hibbard 1980; Mabbitt 1981). Furthermore the availability of essential nutrients, such as amino acids and sugars (Gill 1976) and certain lipids (Bours & Mossel 1973; Delarras 1982) may affect the spoilage pattern. In addition, synergism (Bell & Gill 1982), but particularly antagonism amongst components of the association may exert a marked influence on metabolic processes occurring in foods (cf. Table 2 and the references presented there).

Some members of the microbial community structure of a food, although not dormant in the sense of growth curve D in Fig. 2, may none the less remain metabolically inactive (Daud et al. 1979) and will therefore take no part in any dissimilative microbial activity occurring in the commodity. When both growth (curve P, Fig. 2) and metabolism do occur, a given combination of parameters may exert different influences on proliferation and dissimilation (Troller & Stinson 1978; Mildenhall et al. 1981), or not affect different populations in the same way (Egan & Shay 1982). In such cases population distributions are clearly not, or not always, reflected quantitatively in deterioration phenomena. Finally, some microbial metabolites may neutralize the effect of other products of microbial dissimilation. This also leads to a lack of correlation between community structure and spoilage pattern.

Another instance of absence of correlation between population and deterioration activity is observed when, as indicated in Section B, biological structures in foods allow adhesion of micro-organisms to particular receptors contained therein (Butler et al. 1979; McMeekin et al. 1979; Thomas & McMeekin 1980; Firstenberg-Eden 1981; Thomas & McMeekin 1981a). As

TABLE 3

Classification of the major foods in order of increasing microbiological keeping quality

Class	'Processing' including heat treatment, compositional modification and packing	Stability characteristic		Examples	Predominant microbial community structure at consumer outlet
		Temp. (°C)	Time of spoilage-free storage		
1	None of functional nature	<10	10–40 h	Fresh meat, milk, fish, poultry, eggs, vegetables	Psychrotrophic, non-fermentative Gram negative rods
2	Pasteurization, followed by hermetic packing	<10	3 days to 2 weeks	Dairy products, sliced cured meat products	Sporing rods and Lancefield group D streptococci
3	Reduction of water activity to ca. 0·95, pH reduction and addition of preservatives, in combination with hermetic packing	<10	a few weeks	'Gaffelbitter' and similar semipreserved fish products	Lactobacilli, streptococci, yeasts and moulds
4	Reduction of water activity to ca. 0·85, pH/a_w/lactic acid combinations of equivalent microbistatic effect, pasteurization	25	many weeks	Condensed milk, mayonnaise, margarine, smoked sausage	Yeasts, moulds
5	Reduction of water activity to ca. 0·80, sometimes in combination with pH reduction	25	unlimited, i.e. until chemical reactions interfere	Shelf-stable products such as salami, stockfish, sauces	Moulds
6	Reduction of water activity, to <0·60	35	unlimited	Dehydrated foods	Bacilli, group D streptococci, mould spores
7	Appertization	35	unlimited	Canned cured meat products and fruits	An occasional spore, i.e. c.f.u. count ≪10^2/g
8	Sterilization	any	unlimited	Canned milk, soups, meat, vegetables, and fish	None

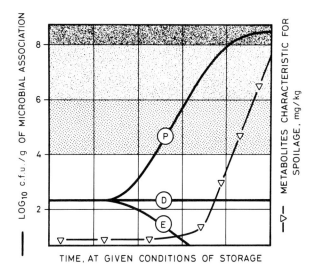

Fig. 2. The fate of the initial community structure of a fresh food as affected by different intrinsic and extrinsic parameters (drawn curves) and the ensuing metabolic activity (triangles) in one particular instance, *viz.* P.

a result of such attachment phenomena sessile microbial communities (Ridgway & Olson 1981) can create micro-environments wherein susceptibility to prevailing antimicrobial effects is considerably modified (Notermans & Kampelmacher 1975; Butler *et al.* 1980; Notermans *et al.* 1980; Cheng *et al.* 1981; Gwynn *et al.* 1981). Once such stratified colonization is attained, *in situ* concentrations of the metabolites formed may greatly exceed those encountered secondary to homogeneous colonization under the same extrinsic conditions (Odlaug & Pflug 1979; Veilleux & Rowland 1981). This may noticeably influence the intensity of dissimilative processes.

   The character of the spoilage tastes and odours ultimately produced in foods, and the rate at which they reach organoleptically detectable levels, may thus vary greatly with circumstances, and important consequences for strictly practical situations follow from this. The most important result is the length of time to onset of spoilage ('shelf life'). An additional important practical effect is that dangerous colonization of foods by pathogens is, as a rule, not expressed by an early organoleptic warning (Greenberg *et al.* 1959; Pivnick & Bird 1965; Lerke & Farber 1971; Collins-Thomson *et al.* 1974; Roberts *et al.* 1976; Ali & Vanduyne 1981; Kautter *et al.* 1981; Hauschild *et al.* 1982; Pelroy *et al.* 1982).

D. The analytical approach
recommended for further studies

Recent studies indicate that much remains to be assessed in the area of microbial colonization of, and metabolic processes occurring in, foods. Investigations aiming at identifying organoleptically relevant compositional changes in foods will have to rely on validated quantitative differential, mostly instrumental, chemical-analytical methods (Niven *et al*. 1949; Fields *et al*. 1968; Rosen *et al*. 1968; Nicol *et al*. 1970; Miller *et al*. 1973; Cox *et al*. 1975; Dainty *et al*. 1975; Freeman *et al*. 1976; Lee *et al*. 1979; Andersson 1980; Ackland *et al*. 1981; Hsu *et al*. 1981; Juven *et al*. 1981; Juven & Weisslowicz 1981; Stanley *et al*. 1981; Thomas & McMeekin 1981; Venkataramaiah & Kempton 1981; Izaguirre *et al*. 1982; Pittard *et al*. 1982). Furthermore, as already indicated, numbers of colony forming units of various groups of organisms have to be related to their metabolic activities.

Quantifying microbial populations, and particularly their shifts, in foods, however, remain important tasks. In this context it is essential to acknowledge that most, if not all, organisms occurring in foods are to a greater or lesser extent sublethally stressed. This is obvious for those organisms surviving processing for safety. However, stress phenomena are no less important in cells exposed during storage and distribution to suboptimal values of intrinsic and extrinsic factors (Mossel & Corry 1977; Ray 1979; Hurst 1980). A summary of the main influences leading to injury is presented in Table 4.

The effects of sublethal damage in microbial populations include: (i) increased lag times; (ii) incompetence to develop quantitatively on selective media that do not exert any inhibitory effect on fully vital populations of the same taxon. Consequently the assessment of numbers of colony forming units (c.f.u.) of injured cells using selective media requires previous repair of the lesions incurred as a result of the stress to which the organisms were exposed. This is often referred to as 'resuscitation'. It can be achieved by allowing restoration in a suitable liquid environment ('liquid medium repair': LMR) or, when possible, preferably by spreading food macerates on an appropriate solid recovery medium ('solid medium repair': SMR; Speck *et al*. 1975; Hackney *et al*. 1979; Mossel *et al*. 1980*b*) (see Fig. 3).

Under no condition should fortuitous repair, i.e. occurring while preparing decimal dilutions, be relied upon. This might lead to insufficient recovery of the more severely stressed part of a population (Nadir & Gilbert 1982), yet allow proliferation of cells that are repaired, or injured slightly or not at all. The most reliable and versatile resuscitation procedure is doubtless SMR. When facultatively anaerobic organisms are to be enumerated, plates with resuscitation medium can simply be overlayered with the usual selective

TABLE 4

*Types of sublethal damage caused by physical and chemical agents**

	Cell wall or cell wall synthesis	Membrane leakage	Protein or protein synthesis	RNA (ribosomes) or RNA synthesis	DNA
Heating		+	+	+	+
Freezing		+		+	+
Drying	(+)	+		+	+
Freeze-drying	+	+		+	+
Gamma-irradiation	+	(+)	(+)	?	+
Change of pO_2		?+			
Osmotic shock	+	+			
Senescence (starvation)			?	+	
Phenols	+	+			
Cationic detergents		+			
Chlorhexidine		+			
Chlorine and iodine			+		
Nitrite		+			+
p-Hydroxybenzoic acid esters		+			
Acetic acid				+	
Penicillins, cephalosporins, phosphoromycin, D-cycloserine	+				
Bacitracin, vancomycin	+	+			
Polyenes		+			
Chloramphenicol			+	+	
Tetracyclines		(+)	(+)	+	
Actinomycin-D			+	+	+
Rifampicin				+	+
Streptomycin, neomycin, kanamycin		(+)		+	
Erythromycin				+	
Tyrocidin, gramicidin, polymyxin		+			
Relative frequency		26/33	7/33	15/33	8/33

+, major effect; (+), minor effect; ?, possible effect.
* Higher doses or exposure to increased concentrations may cause different lesions.
After Mossel & Corry (1977).

medium, held at *ca.* 47°C. Strictly aerobic organisms should be replicated on the usual selective enumeration medium (Mossel *et al.* 1965*a*; Brewer & Turner 1973). A wide variety of convenient procedures is available for this purpose (Beech *et al.* 1955; Lighthart 1968; Bowie *et al.* 1969; Burman & Oestensson 1978; Exner 1980).

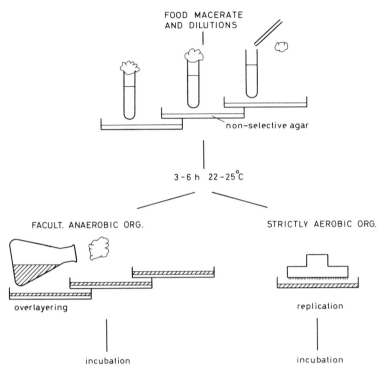

Fig. 3. The solid medium repair technique for the resuscitation of sublethally stressed facultatively, resp. strictly aerobic organisms. Longer resuscitation times may be required in
some particular instances; cf. Mossel & van Netten (1983).

2. The Main Concern of Food Ecologists in the Next Decades: Longitudinally Integrated Processing for Safety and Quality

A. *Identification of the most pressing problem and its solution*

Many classes of fresh foods contain pathogenic microbial associations.
These include, first and foremost, meat and poultry as these are the most
common sources of salmonellosis, campylobacteriosis and related febrile
gastro-enteritis (Bryan 1980, 1981). Indeed salmonellas are frequently
isolated from raw pork (Weissman & Carpenter 1969; Edel *et al.* 1976), veal
(Hobbs 1964) and chicken (van Schothorst *et al.* 1976; Thomas & McMeekin
1981*a,b*; Kraft *et al.* 1982). Pork and poultry are also the main sources of
Campylobacter (Simmons & Gibbs 1979; Grant *et al.* 1980; Stern 1981*b*;
Svedhem & Kaijser 1981) and *Yersinia* (Christensen 1980; Schiemann 1980;
Doyle *et al.* 1981; Stern 1981*a*). Furthermore there is some reason for

concern about prepared vegetables destined to be eaten without exposure to heat treatment of marked microbial lethality (Kominos *et al.* 1972; Ercolani 1976; Wright *et al.* 1976; Tamminga *et al.* 1978).

Efforts to control completely the wholesomeness of such commodities by the hygienic measures taken by producers and consumers alike have so far been notably unsuccessful. As far as fresh meats and poultry are concerned the prospects for essentially improved hygiene at the slaughter line are certainly promising. None the less, such measures will only bring about a marked reduction in the percentage of contaminated fresh meat reaching the consumer as long as the heavy, so-called epidemiological pressure (Silliker 1973; Werner *et al.* 1974; Lachapelle 1979; Jarl & Arnold 1982), particularly on slaughter animals, continues to prevail (Guinée *et al.* 1976; Dockerty *et al.* 1970; Foster 1972; Edel *et al.* 1973, 1976; Peel & Simmons 1978; Roberts 1980; Smeltzer *et al.* 1980; Gerats *et al.* 1981; Craven & Hurst 1982).

Hence a third line of defence has to be introduced: processing for safety as is currently practised in the dairy (Wilson 1933; Davis 1950; Westhoff 1978) and egg (Lee 1974) industries. Procedures of this sort should be 'longitudinally integrated', which means that measures should be taken all along the processing line to prevent occurrences that would nullify the beneficial effect of processing, such as recontamination of the commodities (Roberts 1972; Maxcy 1976; Summer *et al.* 1982) and proliferation of the very low numbers of pathogenic organisms that invariably survive processing (Mossel & Drion 1979; Summer *et al.* 1982).

*B. Evaluation of procedures available for
terminal processing for safety*

Various terminal processing procedures for attaining safety were introduced decades ago and have since been validated by long experience. These procedures include all the heat treatments. Examples are the pasteurization of milk referred to earlier and the so-called appertization (Goresline *et al.* 1964) of cured meat products packed in hermetically sealed containers. Processing for safety by heat has, more recently, also been found effective for the decontamination of fresh meat and poultry surfaces (Klose *et al.* 1971; Cox *et al.* 1974; Snijders & Gerats 1977; Smith & Graham 1978; Eustace 1981) though it often leads to surface denaturation. The chlorination of drinking water, generally accepted, may be useful in the cleaning–decontamination of vegetables, especially lettuce, cauliflower and similar produce (cf. 2.A) that are frequently fertilized by night soil. This is, clearly, an unacceptable practice, yet it seems as hard to eliminate as the production of safe raw milk is difficult to attain (Sharpe *et al.* 1980). More recently, additional modes of processing for safety have been developed. The decontamination treatment of packaged fresh meats and poultry by gamma irradiation is one of the modern methods of processing and it is, in principle, very attractive indeed. At a dose of about 3–5 kGy it eliminates all non-sporing bacterial enteric pathogens (Mossel 1960; Mossel *et al.* 1965a; Billon & de la Sierra Serrano 1968; Ley *et al.* 1970; Mossel *et al.* 1972; Maxcy & Tiwari 1973; Mossel 1977; Mulder 1982). These include not only the almost ubiquitous *Salmonella* spp. but also enteropathogenic types of *E. coli* (Scotland *et al.* 1981), *Yersinia* spp. (Schiemann *et al.* 1981) and *Campylobacter* spp. (Williams & Deacon 1980; Blaser & Reller 1981; Mosenthal *et al.* 1981; Kendall & Tanner 1982) that have plagued the consumer increasingly since 1950. No induced radioactivity or toxicity resulting from radiolytic reactions (Merritt *et al.* 1978) threatens the process (Barna 1979; Elias 1980; Anon. 1981). In addition, the keeping quality of the products is increased considerably as a result of the marked reduction of the psychrotrophic Gram negative, rod-shaped bacteria that are the main components of the spoilage association (Niemand *et al.* 1981). The low virucidal effect of food irradiation (Sullivan *et al.* 1971; Di Girolamo *et al.* 1972) is of some concern. However, emotionally determined rejection by the consumer remains the main problem to be solved. This can probably only be achieved by research in the field of human motivation and subconscious fears (Seligman 1971; Wilson 1973; Douglas 1978).

Another method of processing for safety and quality which has been developed is the decontamination of carcass meats by surface treatment with lactic acid (Mossel & de Bruin 1960; Juven *et al.* 1974; Ockerman *et al.* 1974; Patterson & Gibbs 1979; Snijders *et al.* 1979; Dezeure-Wallays & van Hoof 1980; van Netten & Mossel 1980). When properly applied (Mossel & van Netten 1982) such a process can reduce populations of pathogens by 3 to 4 logarithmic units (base 10). This treatment poses no problems of toxicity, nor does it seem to incite consumers' rejection. Lactic acid decontamination also eliminates, as does gamma irradiation, a considerable part of the spoilage association and thus improves the keeping quality of meat and poultry.

C. Analytical contributions required from
the food microbiologist

The assignments of microbial ecologists in this area are at least four-fold.

(i) *Assessment of process efficacy*

First and foremost the required lethality of the process, defined as $\Lambda =$

log $N_0.N_f$ for the relevant pathogenic agents (Mossel & de Groot 1965; Roberts & Ingram 1965; Yawger 1978; Reyes *et al.* 1981) has to be established. Next, lethalities are to be assessed by the procedure termed risk analysis (Mossel & Drion 1979; Roberts 1980; Roberts *et al.* 1981). This involves estimating the probability (Q) that never in a lifetime will a member of a population be exposed at any one moment to sufficient numbers of infective units of a food-transmitted pathogen to incite disease. Such a number is termed a minimal infectious dose (MID). Obviously, Q depends on N_f, the number of portions of a given food eaten in a given period of time and on the MID of the relevant organism. The latter is, in turn, determined by attributes of the organism itself and also by the vulnerability of the consumer. Vulnerability is affected by age, nutritional status and condition of the gastro-intestinal tract. Recently it has become clear that whether the stomach is empty or not at the time of ingesting the food greatly influences MID-values of a given strain of a specified organism (Mossel & Oei 1975). This may well be the cause of the frequently observed very low MID, i.e. $\leq 10^3$ for the enteric organisms generally considered to have at least 10^3-times higher threshold infectious doses (Armstrong *et al.* 1970; Levine *et al.* 1973; Craven *et al.* 1975; Lipson & Meikle 1977; Fontaine *et al.* 1978, 1980; Robinson 1981).

Table 5 presents a lead to the use of risk analysis in assessing safe processes. The decisive role of the MID has been taken into account in this matrix.

(ii) *Evaluation of the ecological impact of processing for safety*

An important task of the food ecologist is to analyse whether modes of processing for safety, such as irradiation, lactic acid decontamination or compositional modification by the addition of antimicrobial compounds will result in such changes in the microbial association of the commodity that will decrease previously existing resistance against post-process recolonization by pathogens.

For this purpose the general ecological approach outlined in Section 1 of this paper can be applied, because the mechanisms which govern colonization of foods by spoilage organisms are identical with those which determine the fate of pathogenic types in foods. For studying the population changes in foods during storage (Fig. 1), basic procedures have been developed called flora-analysis (Cavett 1963; Corlett *et al.* 1965*a,b*; Splittstoesser & Gadjo 1966; Barnes & Corry 1969; Mossel 1970; Kleeberger & Busse 1975; Wright *et al.* 1976; Hanna *et al.* 1977; van Spreekens 1977; Patterson & Gibbs 1978; Otte *et al.* 1979; Chyr *et al.* 1981; Gillespie 1981; Stiles & Ng 1981*a,b*; Lee *et al.* 1982; Hanna *et al.* 1982; Myers *et al.* 1982) and flora-shift assessment (Blankenagel & Okello-Uma 1969; Harrison & Lee 1969; Mossel 1970; Mourgues *et al.* 1977; van Netten & Mossel 1980; Beckers *et al.* 1981; Tolle

TABLE 5

Matrix for providing guidance with respect to required processing for safety

---

1. A *survey* on the distribution of index organisms (Ix) in a particular commodity provides spread between $_mN_0^{Ix}$ and $_MN_0^{Ix}$, with, e.g. 95th percentile $= \Phi_{Ix}$ (c.f.u./g).
2. A *survey* on ϵ-factors (c.f.u. index/c.f.u. pathogen) in that particular commodity provides spread between $_m\epsilon$ and $_M\epsilon$ with 5th percentile Φ_ϵ.
3. This allows *calculation* of probable N_0^P, i.e. c.f.u./g of pathogen under review, from $N_0^P = \Phi_\epsilon^{-1}\cdot\Phi_{Ix}$.
4. *Assessment* of process lethality, Λ, leads to spread between Λ_m and Λ_M, with 95th percentile at Φ_Λ, equalling n overall decimal reductions.
5. Consequently N_{sf}^P (subfinal level of pathogen, i.e. in processed product immediately after processing) is *calculated* from $N_{sf}^P = \Phi_\Lambda\cdot N_0^P = \Phi_\Lambda\cdot\Phi_\epsilon^{-1}\cdot\Phi_{Ix}$.
6. Ecological *line studies* on the fate of the pathogen during distribution results in 95% probability of change equalling Δ_Φ, i.e. either increase or decline, in c.f.u./g.
7. The pathogen level ultimately reaching the consumer is *calculated* from $N_f^P = \Delta_\Phi\cdot N_{sf}^P = \Delta_\Phi\cdot\Phi_\Lambda\cdot\Phi_\epsilon^{-1}\cdot\Phi_{Ix}$.
8. Values of N_f^P have to be *evaluated* against the data below.

Acceptable levels of contamination, expressed in N_f c.f.u. *per portion eaten*, as determined by the amount of food consumed and the MID of the pathogen concerned, at a level of consumer protection equalling exposure of 72% of the population to the pathogen not exceeding once in about 100 years, when one portion is eaten daily during the entire time span of 100 years

	Population ($\times 10^6$)	
MID	15	250
1	$3\cdot4 \times 10^{-12}$	$2\cdot1 \times 10^{-13}$
10	$3\cdot3 \times 10^{-1}$	$2\cdot5 \times 10^{-1}$
100	$4\cdot6 \times 10$	$4\cdot4 \times 10$
1000	$7\cdot9 \times 10^2$	$7\cdot8 \times 10^2$

After Drion & Mossel (1977) and Mossel & Drion (1979).

et al. 1981; Mossel & van Netten 1982). Figure 4 illustrates our primary plating procedure on media selective for Gram negative and Gram positive organisms (Mossel *et al.* 1977a) and the first stage grouping of isolates relying on Gram stain and catalase reaction. Tables 6 and 7 summarize our methods of identification of isolates at genus level (Mossel *et al.* 1977b).

This procedure was, for example, applied to meat surfaces decontaminated by lactic acid treatment. It was found that the immediate flora-shift and the population change occurring during chilled storage were in the direction of Gram positive, catalase negative bacteria. This was most reassuring evidence from the point of view of consumer protection, because it increases the colonization resistance of meats decontaminated by lactic acid treatment against pathogenic organisms that may recontaminate the product after processing (Table 2).

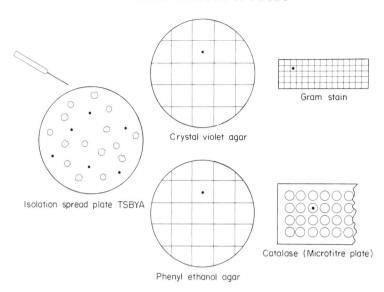

Gram stain

Crystal violet agar

Isolation spread plate TSBYA

Catalase (Microtitre plate)

Phenyl ethanol agar

Fig. 4. The first step in flora-analysis of foods, relying on (i) plating on to a general purpose medium (TSBYA, tryptone soya peptone yeast extract agar) to be incubated aerobically and when required also anaerobically; (ii) replicating on to media, more or less selective for Gram negative (crystal violet agar) and Gram positive (phenyl ethanol agar) organisms; (iii) tentative taxonomic grouping based on morphology and catalase reaction.

Methods used by others for the assessment of the initial composition of, and shifts as a result of customary storage in, the microbial community structure of the main classes of foods are collected in Table 8.

(iii) *Studies on the significance of marker organisms for food safety and quality*

The microbial ecologist should also continue to pay attention to the ultimate resolution of the much-discussed problem of the significance of marker organisms for monitoring the safety and quality of foods.

Following Ingram's superb approach, unfortunately never published due to his untimely death (Mossel 1981), marker organisms are now divided into two groups. These include: (i) index organisms, pointing to the possible occurrence of ecologically related *pathogenic* types; and (ii) indicator organisms, conveying information on the microbiological *condition* of a food *in general*. There is no doubt of the merit of indicator organisms in assessing the safety of manufactured foods, when used with the intention of verifying that good manufacturing practices have been followed. The utility of the search for index organisms in foods manufactured under conditions which are not, or not exactly, known, e.g. when monitoring imported foods, often requires more study.

TABLE 6

Grouping at genus level of Gram negative bacteria of significance for foods

Morphology	Biochemistry[1]	Type genera
Rods, no spores	Oxidase θ	
	glucose type 1–2[2]	
	motile	*Xanthomonas*
	non-motile	*Acinetobacter*
	glucose type 3–4[2]	
	lactose +	
	citrate θ	*Escherichia*[3]
	citrate +	
	MR +	*Citrobacter/Kluyvera*[4]
	MR θ	
	motile	*Enterobacter/Erwinia*[5]
	non-motile	*Klebsiella*
	lactose θ	
	βGal. θ	
	motile	
	urea θ	*Salmonella*
	urea +	*Proteus*
	non-motile	*Shigella*
	βGal. +	
	H_2S +	*Citrobacter*[6]
	H_2S θ	
	LDC θ	*Yersinia*
	LDC +	
	arabinose +	*Hafnia*
	arabinose θ	*Serratia*[7]
Rods, no spores	Oxidase +	
	glucose type 1	
	motile	*Alcaligenes/Comamonas*[8]
	non-motile	*Moraxella/Flavobacter*
	glucose type 2	*Pseudomonas*[9]
	glucose type 3	*Beneckea/Vibrio*[8]
	glucose type 4	*Aeromonas*[10]
Rods, no spores	Ethanol oxidized to acetic acid at pH <4·5	*Acetobacter/Gluconobacter*[8]

θ = negative.

1. Reactions for *ca.* 90% of the most frequently isolated strains.
2. Glucose type: 1, no attack; 2, oxidative dissimilation; 3, fermentative anaerogenic dissimilation; 4, aerogenic fermentative dissimilation.
3. *E. coli*: Aerogenic fermentation of lactose and formation of indole at 44 ± 0·1°C in over 90% of strains.
4. H_2S negative.
5. *Enterobacter* invariably aerogenic; *Erwinia* predominantly anaerogenic.
6. Formerly: the 'Bethesda–Ballerup' group.
7. DNAse positive.
8. Respectively peritrichous and polar flagella.
9. Glucose type 1 types of *Pseudomonas* also exist; *Ps. aeruginosa*: acetamide deaminated at 42°C.
10. Anaerogenic *Aeromonas* types are encountered occasionally; hence *Aeromonas* is ultimately to be distinguished from *Vibrio* based on its insensitivity to pteridine compound 0/129.

TABLE 7

Grouping at genus level of Gram positive bacteria of significance for foods

Morphology	Biochemistry[1]	Type genera
Rods, no spores	Catalase +	*Corynebacterium/Brevibacterium/Arthrobacter*
		Cellulomonas/Curtobacterium
		Microbacterium[2]
		Brochothrix[3]
Rods, no spores	Catalase θ	*Lactobacillus*
Rods, with spores	Catalase +	*Bacillus*[4]
Rods, with spores	Catalase θ	
	Glucose type[5]: 5–1	*Clostridium*[6]
Coccus	Catalase +	
	glucose type 1–2	*Micrococcus*
	glucose type 3	*Staphylococcus*[7]
Coccus	Catalase θ	
	glucose type 3	*Streptococcus*[8]
	glucose type 4	*Leuconostoc*

θ = negative.

1. Reactions for *ca.* 90% of the most frequently isolated strains.
2. Mesophilic: $D_{65} \geqslant 2$ min.
3. Rods changing to cocci in old cultures; psychrotrophic; D_{50} approx. 2 min.
4. *B. cereus*: no attack on mannitol, but dissimilation of egg-yolk.
5. Glucose type: 1, no attack; 2, oxidative dissimilation; 3, fermentative anaerogenic dissimilation; 4, aerogenic fermentative dissimilation; 5, strict anaerobic dissimilation.
6. *Cl. perfringens*: non-motile, indole negative.
7. *Staph. aureus*: coagulase +; mode of attack on mannitol type 3, i.e. fermentative, but anaerogenic in over 90% of strains.
8. Group D: growth with black halos on kanamycin–aesculin–azide agar at 42°C; group N: growth on sorbic acid tomato juice agar of pH = 4·5; group A: β-haemolytic on sulfamethoxazole–trimethoprim–blood agar.

For instance, it has been reported many times that salmonellas were found in a food sample with 'no detectable' *E. coli*. Such an observation calls for ecological studies on the community structure, particularly the Enterobacteriaceae part, of such commodities. If the ratio between c.f.u.'s of *E. coli* and salmonellas, the so called ϵ-value (Drion & Mossel 1977), may appear to be much smaller than usual, i.e. well below 10^5, increasing the sensitivity of the test by examining a larger portion of the food may resolve the problem. If an ϵ-value is so low that such an approach is due to fail, a direct test for the presence of *Salmonella* spp. is the only available method. This has, clearly, always to be done when raw foods, and particularly commodities of unknown manufacturing history, are to be monitored. However, it can, as a rule, be missed and replaced by a well-designed test for Enterobacteriaceae as indicator organisms when examining foods processed for safety (Mossel 1982).

TABLE 8

A summary of methods of assessment of (i) the composition of the microbial community structure of foods; and (ii) flora shifts resulting from processing, storage and distribution

Foods studied	References
Beef	Haines (1933); Patterson & Gibbs (1978); Hanna *et al.* (1977, 1981, 1982*a*); Johnston *et al.* (1982); Kennedy *et al.* (1982); Vanderzant *et al.* (1982)
Dairy products, various	Otte *et al.* (1979); Elliott *et al.* (1981*a*)
Dried foods	Holtzapffel & Coutinho (1969)
Edible offals	Bijker (1981); Hanna *et al.* (1982*b*); Gill & De Lacy (1982); Oblinger *et al.* (1982)
Egg albumen	Barnes & Corry (1969)
Fish	Adams *et al.* (1964); Corlett *et al.* (1965*a,b*); Chai *et al.* (1968); Gillespie & Macrea (1975); Van Spreekens (1977); Lannelongue *et al.* (1982); McMeekin *et al.* (1982)
Foods, general	Mossel *et al.* (1977*a,b*); Van Netten & Mossel (1981)
Meat products, cured	Niven *et al.* (1949); Cavett (1962, 1963); Mol *et al.* (1971); Chyr; *et al.* (1981); Draughon *et al.* (1981*b*)
Minced meat	Kleeberger & Busse (1975); Mossel *et al.* (1979); Harrison *et al.* (1981)
Pork	Buttiaux & Catsaras (1964); Vanderzant & Nichelson (1969); Stiles & Ng (1981*a,b*); Myers *et al.* (1982)
Poultry	Mulder (1982)
Salads and sauces	Holtzapffel & Mossel (1968); Draughon *et al.* (1981*a*)
Shrimp	Harrison & Lee (1969)
Vegetables, mostly frozen	Cavett *et al.* (1965); Splittstoesser *et al.* (1966, 1967); Splittstoesser (1970)
Water	Halls & Board (1973); van der Kooij (1979); Lechevallier *et al.* (1980); Schwaller & Schmidt-Lorenz (1980)

(iv) *Drafting reference values for safety and quality of finished products*

It is also a cardinal responsibility of the microbial food ecologist to provide justified reference values ('microbial limits') for the final products as an element of the monitoring of the various processes. Accurate, ecologically justified procedures for the assessment of microbiological reference values are available. They rely on: (i) surveys on an adequate number of specimens processed and handled according to previously established good manufacturing practices ('GMPs'); (ii) deducing, by methods of frequency distribution analysis, as in Fig. 5, attainable low levels of residual organisms of ecological significance.

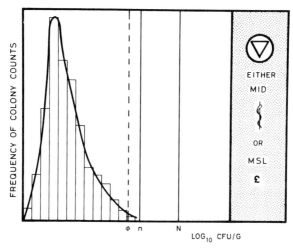

Fig. 5. Distribution plot of the results of microbiological surveys on a given type of food, allowing the assessment of reference values: ø = 95th percentile; n = Reference value, low level; N = Reference value, maximum to be expected under conditions of GMP; MID = minimal infectious dose; MSL = minimal spoilage level.

The *determination* of such distribution curves marks the end of the approach of GMP. An entirely different matter is *decision-making* in cases where the microbial load of samples substantially exceeds even the maximal degree of colonization to be expected under conditions of GMP (*N* in Fig. 5). In these instances the decision to accept or reject consignments of food products should take into account factors of an ecological nature, different from those determining GMPs. These include: (i) the fate of the organisms under consideration during further transport and distribution of the commodity, as affected by the intrinsic and extrinsic parameters discussed in Section 1; (ii) the microbicidal effect of the customary mode of culinary preparation; (iii) the possibility of nullifying this beneficial effect by recontamination and microbial growth after culinary preparation; (iv) the susceptibility of the most vulnerable group of consumers likely to eat the commodity (Mossel 1982).

3. Retrospect and Prospect Linked

When reviewing the past 50 years and the future needs in research into the microbial ecology of foods, it is obvious that Sir Graham Wilson's theorem, formulated for the first time in 1933, is still completely valid. It can be summarized in stressing that the main concern in microbiological quality assurance should be intervention, i.e. the designing of *measures* of control. *Measurements* are of secondary significance only, as they serve solely to monitor if the processing procedures have been carried out correctly. The ingrained habit of attempting to *attain* good microbiological quality by microbiological examination techniques which, in reality, can do no more than to assess a microbiological condition, should definitely be abandoned. Bacteriological examinations are an act of inspection, similar to taking someone's temperature or blood pressure or examining urine for endocrine metabolites characteristic of pregnancy, for example, such analyses will have no effect whatsoever in reducing a fever, controlling hypertension or avoiding pregnancy—control will always require intervention.

The best illustration of Sir Graham's theorem is found in the failure to control salmonellosis. For over 35 years a deluge of papers on the quantification of salmonellas in foods, feeds and water has appeared and increasingly sophisticated, although still far from perfect methods of examination (Roberts *et al.* 1975; Thomason & Dodd 1976; Rigby & Pettit 1980; van Schothorst *et al.* 1980; Andrews *et al.* 1981; Harvey & Price 1981; Sveum & Kraft 1981; Vassiliadis *et al.* 1981; D'Aoust *et al.* 1982; Entis *et al.* 1982; Harvey & Price 1982; Minnich *et al.* 1982; Mulder 1982; Xirouchaki *et al.* 1982) have been developed to that end. All these efforts were notably unsuccessful in reducing salmonellosis morbidity (Hobbs 1964; Foster 1972; Edel *et al.* 1976; van Oye & Ghysels 1978; McKinley *et al.* 1980; Rowe *et al.* 1980; Silliker 1980; Bonland *et al.* 1981; Le Minor & Le Minor 1981; Oosterom *et al.* 1981; Craven & Hurst 1982; Jones *et al.* 1982; Narucka 1982). Again, such exertions could have been much better spent on intervention, in this particular instance, designing processing for safety along the lines indicated in the previous section (Mossel & Kampelmacher 1981).

Consequently it seems to be time for qualified microbiologists and others who have to concern themselves with food microbiology to look beyond their previous parochial concern with 'plate' counts and the demonstration of 'coliform' bacteria (Sharpe 1980). Rather should a valid ecological and taxonomical approach to the basic safety and quality problems in food microbiology be undertaken.

This should be strongly supported by a greater emphasis on microbial ecology during the academic *training* of food microbiologists. It should centre on familiarization with the ecological essentials needed for the assurance and assessment of microbiological safety, quality and acceptability of foods.

Similarly academic *research* should concentrate on these areas. For example, it seems futile to continue carrying out extensive, and expensive, surveys on the contamination rate of pork and poultry with salmonellas and campylobacters; rather, more attention should be paid to preventing infection of the citizen by these organisms through his food supply (Mossel & van

Netten 1982). Likewise, the protection of world crops against colonization by moulds (Christensen & Kaufmann 1969; Milton & Jarrett 1969, 1970; Abramson *et al.* 1980), which may result in the exposure of the consumer to the risk of chronic mycotoxicoses (Lillehoj *et al.* 1976; Wyllie & Morehouse 1978; Moreau & Moss 1979; Hayes 1980; Cole & Cox 1981; Osborne 1982; Thiel *et al.* 1982), should be studied extensively. When effective reduction of the prevailing a_w (Mossel 1982) fails studies on the use of innocuous fungistatic additives (May & Bullerman 1982) seem justified. Neither should effective control of obvious deterioration phenomena, including the perennial problem of microbial spoilage of fish (Adams *et al.* 1964; Chai *et al.* 1968; Gillespie & Macrae 1975; van Spreekens 1977; Shewan & Murray 1979; Banks *et al.* 1980), which is often the only available source of protein for entire populations, be neglected.

Thorough analytical development work should be encouraged. The aim of such research should be the elaboration of simple and rapid techniques, particularly those that allow very early diagnosis of failure of GMP. Hence processing-line sampling, rather than end-product monitoring, should be the first priority. Accepted methods should obviously be reasonably accurate and reproducible, but obsessive pressing for sophisticated, extremely precise methods should be avoided. It should be understood during academic and in-service training that there is neither need nor hope for microbiological monitoring ever to attain the accuracy of chemical analysis of foods. The reasons for this, which should be explained amply and demonstrated in practical classes, are: extreme heterogeneity of colonization, dynamic state of virtually all microbial community structures in foods, and the intrinsic difficulty in enumerating colony forming units, resulting from the definition of microbial vitality, which is method-dependent.

However, procedures for the prevention of diseases transmitted by foods (Bauman 1974; Bobeng & David 1977) and of food spoilage should be described very accurately. Furthermore, it is essential that catering staff and those employed by the food manufacturing industry are expertly motivated to follow these required procedures. This was the strategy elaborated from sound ecological principles between 1933 and 1955 by Sir Graham Wilson. If it continues to be neglected, food microbiologists may well find themselves unable to play the important role in society that they should in the long overdue control of food-transmitted disease (Bryan 1980, 1981) and of loss of precious food as a result of microbial deterioration.

4. References

ABRAMSON, D., SINHA, R. N. & MILLS, J. T. 1980 Mycotoxin and odor formation in moist cereal grain during granary storage. *Cereal Chemistry* **57**, 346–351.

ACKLAND, M. R., TREWHELLA, E. R., REEDER, J. & BEAN, P. G. 1981 The detection of microbial spoilage in canned foods using thin-layer chromatography. *Journal of Applied Bacteriology* **51**, 277–281.

ADAMS, R., FARBER, L. & LERKE, P. 1964 Bacteriology of spoilage of fish muscle. II. Incidence of spoilers during spoilage. *Applied Microbiology* **12**, 277–279.

ALEXANDER, M. 1971 Colonization. In *Microbial Ecology* Ch. 3. New York: Wiley.

ALI, F. S. & VANDUYNE, F. O. 1981 Microbial quality of ground beef after simulated freezer failure. *Journal of Food Protection* **44**, 62–65.

ANDERSSON, R. E. 1980 Lipase production, lipolysis and formation of volatile compounds by *Pseudomonas fluorescens* in fat containing media. *Journal of Food Science* **45**, 1694–1701.

ANDREWS, W. H., POELMA, P. L. & WILSON, C. R. 1981 Comparative efficiency of brilliant green, bismuth sulfite, Salmonella-Shigella, Hektoen enteric and xylose lysine desoxycholate agars for the recovery of Salmonella from foods: collaborative study. *Journal of the Association of Official Analytical Chemists* **64**, 899–928.

ANON. 1981 *Wholesomeness of Irradiated Food*. World Health Organization Technical Report Series No. 659.

ARIES, V., CHENLY, P. A. & MOSSEL, D. A. A. 1982 Ecological studies on the occurrence of bacteria utilizing lactic acid at pH values below 4·5. *Journal of Applied Bacteriology* **52**, 345–351.

ARMSTRONG, R. W., FODGE, T., CURLIN, G. T., COHEN, A. B., MORRIS, G. K., MARTIN, W. T. & FELDMAN, J. 1970 Epidemic Salmonella gastroenteritis due to contaminated imitation ice cream. *American Journal of Epidemiology* **91**, 300–307.

BAAS BECKING, L. G. M. 1934 *Geobiologie* (English version: *Geology and Microbiology*, New Zealand Oceanographic Institute Memoir No. 3 (1959) pp. 48–64).

BANKS, H., NICKELSON, R. & FINNE, G. 1980 Shelf-life studies on carbon dioxide packaged finfish from the Gulf of Mexico. *Journal of Food Science* **45**, 157–162.

BARNA, J. 1979 Compilation of bioassay data on the wholesomeness of irradiated food items. *Acta Alimentaria* **8**, 205–315.

BARNES, E. M. & CORRY, J. E. L. 1969 Microbial flora of raw and pasteurized egg albumen. *Journal of Applied Bacteriology* **32**, 193–205.

BARNES, E. M. & MELTON, W. 1971 Extracellular enzymic activity of poultry spoilage bacteria. *Journal of Applied Bacteriology* **34**, 599–609.

BARNES, E. M., MEAD, G. C. IMPEY, C. S. & ADAMS, B. W. 1979 Spoilage organisms of refrigerated poultry meat. In *Cold Tolerant Microbes in Spoilage and the Environment* ed. Russell, A. D. & Fuller, R. pp. 101–106. London & New York: Academic Press.

BAUMAN, H. E. 1974 The HACCP concept and microbiological hazard categories. *Food Technology* **28**, 30–34, 74.

BEACHY, E. H. 1981 Bacterial adherence: adhesin–receptor interactions mediating the attachment of bacteria to mucosal surfaces. *Journal of Infectious Diseases* **143**, 325–345.

BECKERS, H. J., VAN SCHOTHORST, M., VAN SPREEKENS, K. J. A. & OOSTERHUIS, J. J. 1981 Microbiological quality of frozen precooked and peeled shrimp from South-East Asia and from the North Sea. *Zentralblatt für Bakteriologie und Hygiene. Abt. I. Orig. B* **172**, 401–410.

BEECH, F. W., CARR, J. G. & CODNER, R. C. 1955 A multipoint inoculator for plating bacteria or yeasts. *Journal of General Microbiology* **13**, 408–410.

BEIJERINCK, M. W. 1908 Fermentation lactique dans le lait. *Archives néerlandaises des Sciences exactes et naturelles, Séries II* **13**, 356–378.

BEIJERINCK, M. W. 1913 De infusies en de ontdekking der bakteriën. *Jaarboek Koninklijke Academie van Wetenschappen*, 1–28.

BELL, R. G. & GILL, C. O. 1982 Microbial spoilage of luncheon meat prepared in an impermeable plastic casing. *Journal of Applied Bacteriology* **53**, 97–102.

BENNETT, R. W. & AMOS, W. T. 1982 *Staphylococcus aureus* growth and toxin production in nitrogen-packed sandwiches. *Journal of Food Protection* **45**, 157–161.

BERKELEY, R. C. W., LYNCH, J. M., MELLING, J., RUTTER, P. R. & VINCENT, B. (eds) 1980 *Microbial Adhesion to Surfaces*. Chichester: Horwood.

BERRY, J. A. 1933 Destruction and survival of micro-organisms in frozen pack foods. *Journal of Bacteriology* **26**, 459–470.

BEUCHAT, L. R. 1981a Combined effects of solutes and food preservatives on rates of inactivation of and colony formation by heated spores and vegetative cells of molds. *Applied and Environmental Microbiology* **41**, 472–477.

BEUCHAT, L. R. 1981b Synergistic effects of potassium sorbate and sodium benzoate on thermal inactivation of yeasts. *Journal of Food Science* **46**, 771–777.

BIJKER, P. G. H. 1981 *Hygienic aspects of edible offals*. Ph.D. Thesis, Veterinary Medicine, University of Utrecht.

BILLON, J. & DE LA SIERRA SERRANO, D. 1968 Experimental radurization of minced beef. Influence on preservation and on the radicidation of different serotypes of Salmonella. *Food Irradiation* **8**, 22–27.

BITTON, G. & MARSHALL, K. C. (eds) 1980 *Adsorption of Micro-organisms to Surfaces*. New York: Wiley.

BLACKMAN, F. F. 1905 Optima and limiting factors. *Annals of Botany* **19**, 281–295.

BLANKENAGEL, G. & OKELLO-UMA, I. 1969 Gram-negative bacteria in raw milk. *Canadian Institute of Food Science and Technology Journal* **2**, 69–71.

BLASER, M. J. & RELLER, L. B. 1981 Campylobacter enteritis. *New England Journal of Medicine* **305**, 1444–1452.

BOARD, R. G., LOSEBY, S. & MILES, V. R. 1979 A note on microbial growth on hen egg-shells. *British Poultry Science* **20**, 413–420.

BOBENG, B. J. & DAVID, B. D. 1977 HACCP models for quality control of entree production in foodservice systems. *Journal of Food Protection* **40**, 632–638.

BOCHEM, H. P., SCHOBERTH, S. M., SPREY, B. & WENGLER, P. 1982 Thermophilic biomethanation of acetic acid: morphology and ultrastructure of a granular consortium. *Canadian Journal of Microbiology* **28**, 500–510.

BONLAND, J., CENTOFANTI, J., HUNT, B., LEWIS, G., SALZMAN, M., DE MELFI, T., MCCARTHY, M. A., WITTE, E. J., CRAVETZ, M., SMITH, K., JIVOK, S., LYONS, J., GUZEWICH, J., HAMEL, H., MORSE, D., SQUIRE, A., TOLY, M., ROTHENBERG, R., COONS, C., POMAR, D., STONE, K. & VOGT, R. L. 1981 Multiple outbreaks of salmonellosis associated with precooked roast beef—Pennsylvania, New York, Vermont. *Morbidity and Mortality Weekly Reports* **30**, 569–570.

BOURS, J. & MOSSEL, D. A. A. 1973 A comparison of methods for the determination of lipolytic properties of yeasts, mainly isolated from margarine, moulds and bacteria. *Archiv. für Lebensmittelhygiene* **24**, 197–203.

BOWIE, I. S., LOUTIT, M. W. & LOUTIT, J. S. 1969 Identification of aerobic heterotrophic soil bacteria to generic level by using multipoint inoculation techniques. *Canadian Journal of Microbiology* **15**, 297–302.

BOYLAN, S. L., ACOTT, K. A. & LABUZA, 1976 *Staphylococcus aureus* challenge study in an intermediate moisture food. *Journal of Food Science* **41**, 918–921.

BREWER, J. A. & TURNER, A. G. 1973 Replicating Rodac plates for identifying and enumerating bacterial contamination. *Health Laboratory Science* **10**, 195–202.

BROCKLEHURST, T. F. & LUND, B. M. 1981 Properties of pseudomonads causing spoilage of vegetables stored at low temperature. *Journal of Applied Bacteriology* **50**, 259–266.

BRYAN, F. L. 1980 Foodborne diseases in the United States associated with meat and poultry. *Journal of Food Protection* **43**, 140–150.

BRYAN, F. L. 1981 Current trends in foodborne salmonellosis in the United States and Canada. *Journal of Food Protection* **44**, 394–401.

BUCK, J. D., MEYERS, S. P. & KAMP, K. M. 1962 Marine bacteria with antiyeast activity. *Science, New York* **138**, 1339–1340.

BUNCH, W. L., MATTHEWS, M. E. & MARTH, E. H. 1977 Fate of *Staphylococcus aureus* in beef-soy leaves subjected to procedures used in hospital chill foodservice systems. *Journal of Food Science* **42**, 565–566.

BURMAN, L. G. & OESTENSSON, R. 1978 Time- and media-saving testing and identification of micro-organisms by multipoint inoculation on undivided agar plates. *Journal of Clinical Microbiology* **8**, 219–227.

BUTLER, J. L., STEWART, J. C., VANDERZANT, C., CARPENTER, Z. L. & SMITH, G. C. 1979 Attachment of micro-organisms to pork skin and surfaces of beef and lamb carcasses. *Journal of Food Protection* **42**, 401–406.

BUTLER, J. L., VANDERZANT, C., CARPENTER, Z. L., SMITH, G. C., LEWIS, R. E. & DUTSON, T. R. 1980 Influence of certain processing steps on attachment of micro-organisms to pork skin. *Journal of Food Protection* **43**, 699–705.

BUTTIAUX, R. & CATSARAS, M. 1964 Les bactéries psychrotrophes des carcasses de porc entreposées en chambre froide. *Annales de l'Institut Pasteur, Lille* **15**, 165–173.

CAVETT, J. J. 1962 The microbiology of vacuum packed sliced bacon. *Journal of Applied Bacteriology* **25**, 282–289.

CAVETT, J. J. 1963 A diagnostic key for identifying the lactic acid bacteria of vacuum packed bacon. *Journal of Applied Bacteriology* **26**, 453–470.

CAVETT, J. J., DRING, G. J. & KNIGHT, A. W. 1965 Bacterial spoilage of thawed frozen peas. *Journal of Applied Bacteriology* **28**, 241–251.

CHAI, T., CHEN, C., ROSEN, A. & LEVIN, R. E. 1968 Detection and incidence of specific species of spoilage bacteria on fish. II. Relative incidence of *Pseudomonas putrefaciens* and fluorescent pseudomonads on haddock fillets. *Applied Microbiology* **16**, 1738–1741.

CHENG, K. J., IRVIN, R. T. & COSTERTON, J. W. 1981 Autochthonous and pathogenic colonization of animal tissues by bacteria. *Canadian Journal of Microbiology* **27**, 461–490.

CHOPIN, A., MOCQUOT, G. & GRAET, Y. LE 1977 Destruction de *Microbacterium lacticum*, *Escherichia coli* et *Staphylococcus aureus* au cours du séchage de lait par atomisation. II. Influence des conditions de séchage. *Canadian Journal of Microbiology* **23**, 755–762.

CHOPIN, A., TESONE, S., VILA, J. P., LE GRAET, Y. & MOCQUOT, G. 1978 Survie de *Staphylococcus aureus* au cours de la préparation et de la conservation de lait écrémé en poudre. Problèmes posés par le dénombrement des survivants. *Canadian Journal of Microbiology* **24**, 1371–1380.

CHRISTENSEN, S. G. 1980 *Yersinia enterocolitica* in Danish pigs. *Journal of Applied Bacteriology* **48**, 377–382.

CHRISTENSEN, C. M. & KAUFMANN, H. H. 1969 *Grain storage. The Role of Fungi in Quality Loss*, Minneapolis: University of Minnesota Press.

CHRISTIAN, J. H. B. & STEWART, B. J. 1973 Survival of *Staphylococcus aureus* and *Salmonella newport* in dried foods, as influenced by water activity and oxygen. In *The Microbiological Safety of Food* ed. Hobbs, B. C. & Christian, J. H. B. pp. 107–119. London & New York: Academic Press.

CHYR, C. Y., WALKER, H. W. & SEBRANEK, J. G. 1981 Bacteria associated with spoilage of Braunschweiger. *Journal of Food Science* **46**, 468–470, 474.

CLARKE, P. H. 1981 Adaptation. *Journal of General Microbiology* **126**, 5–20.

CLAYSON, D. H. F. 1955 Food perishability in terms of micro-ecology. *Journal of the Science of Food and Agriculture* **6**, 565–575.

CLAYSON, D. H. F. & BLOOD, R. M. 1957 Food perishability: the determination of the vulnerability of food surfaces to bacterial infection. *Journal of the Science of Food and Agriculture* **8**, 404–414.

COLE, R. J. & COX, R. H. 1981 *Handbook of Toxic Fungal Metabolites*. New York & London: Academic Press.

COLLINS-THOMPSON, D. L., CHANG, P. C., DAVIDSON, C. M., LARMOND, E. & PIVNICK, H. 1974 Effect of nitrite and storage temperature on the organoleptic quality and toxinogenesis by *Clostridium botulinum* in vacuum-packaged side bacon. *Journal of Food Science* **39**, 607–609.

COLLINS-THOMPSON, D. L. & RODRIGUEZ LOPEZ, G. 1980 Influence of sodium nitrite, temperature and lactic acid bacteria on the growth of *Brochothrix thermosphacta* under anaerobic conditions. *Canadian Journal of Microbiology* **26**, 1416–1421.

CORLETT, D. A., LEE, J. S. & SINNHUBER, R. O. 1965a Application of replica plating and computer analysis for rapid identification of bacteria in some foods. I. Identification scheme. *Applied Microbiology* **13**, 808–817.

CORLETT, D. A., LEE, J. S. & SINNHUBER, R. O. 1965b Application of replica plating and computer analysis for rapid identification of bacteria in some foods. II. Analysis of microbial flora in irradiated Dover sole (*Microstomus pacificus*). *Applied Microbiology* **13**, 818–822.

COX, N. A., MERCURI, A. J., THOMSON, J. E. & GREGORY, D. W. 1974 Quality of broiler carcasses as affected by hot water treatments. *Poultry Science* **53**, 1566–1571.

Cox, N. A., Juven, B. J., Thomson, J. E., Mercuri, A. J. & Chew, V. 1975 Spoilage odours in poultry meat produced by pigmented and nonpigmented *Pseudomonas*. *Poultry Science* **54**, 2001–2006.

Craven, J. A. & Hurst, D. B. 1982 The effect of time in lairage on the frequency of *Salmonella* infection in slaughtered pigs. *Journal of Hygiene* **88**, 107–111.

Craven, P. C., Mackel, D. C., Baine, W. B., Barker, W. H., Gangarosa, E. J., Goldfield, M., Rosenfeld, H., Altman, R., Lachapelle, G., Davies J. W. & Swanson, R. C. 1975 International outbreak of *Salmonella eastbourne* infection traced to contaminated chocolate. *Lancet* **i**, 788–793.

Cravioto, A., Scotland, S. M. & Rowe, B. 1982 Hemagglutination activity and colonization factor antigens I and II in enterotoxigenic and non-enterotoxigenic strains of *Escherichia coli* isolated from humans. *Infection and Immunity* **36**, 189–197.

Dack, G. M. & Lippitz, G. 1962 Fate of staphylococci and enteric micro-organisms introduced into slurry of frozen pot pies. *Applied Microbiology* **10**, 472–479.

Dahl, C. A., Matthews, M. E. & Marth, E. H. 1980 Fate of *Staphylococcus aureus* in beef loaf, potatoes and frozen and canned green beans after microwave-heating in a simulated cook/chill hospital food service system. *Journal of Food Protection* **43**, 916–923, 932.

Dainty, R. H. & Hibbard, C. M. 1980 Aerobic metabolism of *Brochothrix thermosphacta* growing on meat surfaces and in laboratory media. *Journal of Applied Bacteriology* **48**, 387–396.

Dainty, R. H., Shaw, B. G., de Boer, K. A. & Scheps, E. S. J. 1975 Protein changes caused by bacterial growth on beef. *Journal of Applied Bacteriology* **39**, 73–81.

D'Aoust, J. Y., Maishment, C., Stotland, P. & Boville, A. 1982 Surfactants for the effective recovery of *Salmonella* in fatty foods. *Journal of Food Protection* **45**, 249–252, 259.

Daud, H. B., McMeekin, T. A. & Olley, J. 1978 Temperature function integration and the development and metabolism of poultry spoilage bacteria. *Applied and Environmental Microbiology* **36**, 650–654.

Daud, H. B., McMeekin, T. A. & Thomas, C. J. 1979 Spoilage association of chicken skin. *Applied and Environmental Microbiology* **37**, 399–401.

Davidson, C. M. & Witty, J. K. 1977 Fate of salmonellae in vacuum-packaged ground meat during storage and subsequent cooking. *Canadian Institute of Food Science and Technology Journal* **10**, 223–225.

Davis, J. G. 1950 Developments in the bacteriological testing of milk. *Journal of the Royal Sanitary Institute London* **70**, 227–256.

Delarras, C. 1982 Etude de l'activité lipolytique des Micrococcaceae. *Revue française des corps gras* **29**, 165–167.

Dennis, C. & Buhagiar, R. W. M. 1980 Yeast spoilage of fresh and processed fruits and vegetables. In *Biology and Activities of Yeasts* ed. Skinner, F. A., Passmore, S. M. & Davenport, R. R. pp. 123–133. London & New York: Academic Press.

Dennis, C. & Davies, R. P. 1977 Susceptibility of strawberry varieties to post-harvest fungal spoilage. *Journal of Applied Bacteriology* **42**, 197–206.

Dezeure-Wallays, B. & van Hoof, J. 1980 Effect of lactic acid sprays on beef carcass contamination. *Proceedings of the 26th European Meeting of Meat Research Workers* Vol. 2, pp. 316–319.

Dickinson, R. J., Varian, S. A., Axon, A. T. R. & Cooke, E. M. 1980 Increased incidence of faecal coliforms with *in vitro* adhesive and invasive properties in patients with ulcerative colitis. *Gut* **21**, 787–792.

di Girolamo, R., Liston, J. & Matches, J. 1972 Effects of irradiation on the survival of virus in West Coast oysters. *Applied Microbiology* **24**, 1005–1006.

Dockerty, T. R., Ockerman, H. W., Cahill, V. R., Kunkle, L. E. & Weiser, H. H. 1970 Microbial level of pork skin as affected by the dressing process. *Journal of Animal Science* **30**, 884–890.

Douglas, M. T. 1978 *Purity and Danger. An Analysis of Concepts of Pollution and Taboo*. London: Routledge & Kegan Paul.

Dowdell, M. J. & Board, R. G. 1971 The microbial associations in British fresh sausages. *Journal of Applied Bacteriology* **34**, 317–337.

DOYLE, M. P., HUGDAHL, M. B. & TAYLOR, S. L. 1981 Isolation of virulent *Yersinia enterocolitica* from porcine tongues. *Applied and Environmental Microbiology* **42**, 661–666.
DOYLE, M. P., BAINS, N. J., SCHOENI, J. L. & FOSTER, E. M. 1982 Fate of *Salmonella typhimurium* and *Staphylococcus aureus* in meat salads prepared with mayonnaise. *Journal of Food Protection* **45**, 152–156, 168.
DRAUGHON, F. A., ELAHI, M. & MCCARTY, I. E. 1981*a* Microbial spoilage of Mexican-style sauces. *Journal of Food Protection* **44**, 284–287.
DRAUGHON, F. A., MELTON, C. C. & MAXEDON, D. 1981*b* Microbial profiles of country-cured hams aged in stockinettes, barrier bags and paraffin wax. *Applied and Environmental Microbiology* **41**, 1078–1080.
DRION, E. F. & MOSSEL, D. A. A. 1977 The reliability of the examination of foods, processed for safety, for enteric pathogens and Enterobacteriaceae: a mathematical and ecological study. *Journal of Hygiene, Cambridge* **78**, 301–324.
DUBOIS, G., BEAUMIER, H. & CLARBONNEAU, B. 1979 Inhibition of bacteria isolated from ground meat by streptococcaceae and lactobacillaceae. *Journal of Food Science* **44**, 1649–1652.
DUCHA SARDANA, J. 1976 Aportaciones al estudio de la microflora en leche condensada. *Revista española de Lecheria* No. 99, pp. 3–33; No. 100, pp. 101–109.
DUITSCHAEVER, C. L. & IRVINE, D. M. 1971 A case study: effect of mold on growth of coagulase-positive staphylococci in Cheddar cheese. *Journal of Milk and Food Technology* **34**, 583.
EDEL, W., GUINÉE, P. A. M., VAN SCHOTHORST, M. & KAMPELMACHER, E. H. 1973 Salmonella cycles in foods with special reference to the effects of environmental factors, including feeds. *Canadian Institute of Food Science and Technology Journal* **6**, 64–67.
EDEL, W., VAN SCHOTHORST, M. & KAMPELMACHER, E. H. 1976 Epidemiological studies on *Salmonella* in a certain area ("Walcheren Project") I. The presence of *Salmonella* in man, pigs, insects, seagulls and in foods and effluents. *Zentralblatt für Bakteriologie und Parasitenkunde, Abt. 1, Orig. A* **235**, 476–484.
EGAN, A. F. & SHAY, B. J. 1982 Significance of lactobacilli and film permeability in the spoilage of vacuum-packaged beef. *Journal of Food Science* **47**, 1119–1122, 1126.
EGAN, A. F., FORD, A. L. & SKAY, B. J. 1980 A comparison of *Microbacterium thermosphactum* and lactobacilli as spoilage organisms of vacuum-packaged sliced luncheon meats. *Journal of Food Science* **45**, 1745–1748.
EIJKMAN, C. 1904. Ueber thermolabile Stoffwechselprodukte als Ursache der natürlichen Wachstumshemmung der Mikroorganismen. *Centralblatt für Bakteriologie und Parasitenkunde Abt. I*, **37**, 436–449.
ELIAS, P. S. 1980 The wholesomeness of irradiated food. *Ecotoxicology and Environmental Safety* **4**, 172–183.
ELLIOTT, J. A., MILLARD, G. E. & HOLLEY, R. A. 1981*a* Late gas defect in Cheddar cheese caused by an unusual bacterium. *Journal of Dairy Science* **64**, 2278–2283.
ELLIOTT, K., O'CONNOR, M. & WHELAN, J. (eds) 1981*b* *Adhesion and Micro-organism Pathogenicity*. London: Pitman Medical Ltd.
ENTIS, P., BRODSKY, M. H., SHARPE, A. N. & JARVIS, G. A. 1982 Rapid detection of *Salmonella* spp. in food by use of the ISO-GRID hydrophobic grid membrane filter. *Applied and Environmental Microbiology* **43**, 261–268.
ERCOLANI, G. L. 1976 Bacteriological quality assessment of fresh marketed lettuce and fennel. *Applied and Environmental Microbiology* **31**, 847–852.
EUSTACE, I. J. 1981 Control of bacterial contamination of meat during processing. *Food Technology in Australia* **33**, 28–32.
EXNER, M. 1980 Zur kulturellen Differenzierung von Abklatschkolonien mit einer Stempelmethode. *Zentralblatt für Bakteriologie, Abt. I, Orig. B* **170**, 88–92.
FARMILOE, F. J., CORNFORD, S. J., COPPOCK, J. B. M. & INGRAM, M. 1954 The survival of *Bacillus subtilis* spores in the baking of bread. *Journal of the Science of Food and Agriculture* **5**, 292–304.
FIELDS, M. L., RICHMOND, B. S. & BALDWIN, R. E. 1968 Food quality as determined by metabolic by-products of micro-organisms. *Advances in Food Research* **16**, 161–229.

FIRSTENBERG-EDEN, R. 1981 Attachment of bacteria to meat surfaces: a review. *Journal of Food Protection* **44**, 602–607.

FONTAINE, R. E., ARNON, S., MARTIN, W. T., VERNON, T. M., GANGAROSA, E. J., FARMER, J. J., MORAN, A. B., SILLIKER, J. H. & DECKER, D. L. 1978 Raw hamburger: an interstate common source of human salmonellosis. *American Journal of Epidemiology* **107**, 36–45.

FONTAINE, R. E., COHEN, M. L., MARTIN, W. T. & VERNON, T. M. 1980 Epidemic salmonellosis from cheddar cheese: surveillance and prevention. *American Journal of Epidemiology* **111**, 247–253.

FOSTER, E. M. 1972 The need for science in food safety. *Food Technology* **26**, 81–87.

FRANK, J. F. & MARTIN, E. H. 1977 Inhibition of enteropathogenic *Escherichia coli* by homofermentative lactic acid bacteria in skim milk. *Journal of Food Protection* **40**, 749–753, 754–759.

FREEMAN, L. R., SILVERMAN, G. J., ANGELINI, P., MERRITT, C. & ESSELEN, W. B. 1976 Volatiles produced by micro-organisms isolated from refrigerated chicken at spoilage. *Applied and Environmental Microbiology* **32**, 222–231.

GALESLOOT, TH. E. & STADHOUDERS, J. 1968 The microbiology of spray-dried milk products with special reference to *Staphylococcus aureus* and Salmonellae. *Netherlands Milk and Dairy Journal* **22**, 158–172.

GEESON, J. D. 1979 The fungal and bacterial flora of stored white cabbage. *Journal of Applied Bacteriology* **46**, 189–193.

GENIGEORGIS, C., RIEMANN, H. & SADLER, W. W. 1969 Production of enterotoxin-B in cured meats. *Journal of Food Science* **34**, 62–68.

GENIGEORGIS, C., SAVOUKIDIS, M. & MARTIN, S. 1971 Initiation of staphylococcal growth in processed meat environments. *Applied Microbiology* **21**, 940–942.

GEORGALA, D. L. & HURST, A. 1963 The survival of food poisoning bacteria in frozen foods. *Journal of Applied Bacteriology* **26**, 346–358.

GERATS, G. E., SNIJDERS, J. M. A. & VAN LOGTESTIJN, J. G. 1981 Slaughter techniques and bacterial contamination of pig carcasses. *Proceedings of the 27th European Meeting of Meat Research Workers*, pp. 198–200.

GIBBS, P. A., PATTERSON, J. T. & HARVEY, J. 1980 Interactive growth of *Staphylococcus aureus* strains with a poultry skin microflora in a diffusion apparatus. *Journal of Applied Bacteriology* **48**, 191–205.

GILL, C. O. 1976 Substrate limitation of bacterial growth at meat surfaces. *Journal of Applied Bacteriology* **41**, 401–410.

GILL, C. O. 1979 Intrinsic bacteria in meat. *Journal of Applied Bacteriology* **47**, 367–378.

GILL, C. O. & DE LACY, K. M. 1982 Microbial spoilage of whole sheep livers. *Applied and Environmental Microbiology* **43**, 1262–1266.

GILL, C. O. & NEWTON, K. G. 1977 The development of aerobic spoilage flora on meat stored at chill temperatures. *Journal of Applied Bacteriology* **43**, 189–195.

GILL, C. O. & NEWTON, K. G. 1978 The ecology of bacterial spoilage of fresh meat at chill temperatures. *Meat Science* **2**, 207–217.

GILL, C. O., PENNEY, N. & NOTTINGHAM, P. M. 1978 Tissue sterility in uneviscerated carcasses. *Applied and Environmental Microbiology* **36**, 356–359.

GILL, C. O., PENNEY, N. & WAUTERS, A. M. 1981 Survival of clostridial spores in animal tissues. *Applied and Environmental Microbiology* **41**, 90–92.

GILLESPIE, N. C. 1981 A numerical taxonomic study of *Pseudomonas*-like bacteria isolated from fish in Southeastern Queensland and their association with spoilage. *Journal of Applied Bacteriology* **50**, 29–44.

GILLESPIE, N. C. & MACRAE, I. C. 1975 The bacterial flora of some Queensland fish and its ability to cause spoilage. *Journal of Applied Bacteriology* **39**, 91–100.

GILLILAND, S. E. & SPECK, M. L. 1977 Antagonistic action of *Lactobacillus acidophilus* toward intestinal and food-borne pathogens in associative cultures. *Journal of Food Protection* **40**, 820–823.

GOEPFERT, J. M. & KIM, H. U. 1975 Behavior of selected food-borne pathogens in raw ground beef. *Journal of Milk and Food Technology* **38**, 449–452.

GOODFELLOW, S. J. & BROWN, W. L. 1978 Fate of *Salmonella* inoculated into beef for cooking. *Journal of Food Protection* **41**, 598–605.

GORESLINE, H. E., INGRAM, M., MACUCH, P., MOCQUOT, G., MOSSEL, D. A. A., NIVEN, C. F. & THATCHER, F. S. 1964 Tentative classification of food irradiation processes with microbiological objectives. *Nature, London* **204**, 237–238.

GRAHAM, J. M. 1978 Inhibition of *Clostridium botulinum* type C by bacteria isolated from mud. *Journal of Applied Bacteriology* **45**, 205–211.

GRANT, I. H., RICHARDSON, N. J. & BOKKENHEUSER, V. D. 1980 Broiler chickens as potential source of *Campylobacter* infections in humans. *Journal of Clinical Microbiology* **11**, 508–510.

GRAU, F. 1979 Fresh meats: bacterial association. *Archiv für Lebensmittelhygiene* **30**, 87–92.

GRAU, F. H. 1980 Inhibition of the anaerobic growth of *Brochothrix thermosphacta* by lactic acid. *Applied and Environmental Microbiology* **40**, 433–436.

GRAU, F. H. 1981 Microbial ecology and interactions in chilled meat. *Food Research Quarterly* **41**, 12–18.

GREENBERG, R. A., SILLIKER, J. H. & FATTA, L. D. 1959 The influence of sodium chloride on toxin production and organoleptic breakdown in perishable cured meat inoculated with *Clostridium botulinum*. *Food Technology* **13**, 509–511.

GREER, G. G. & JEREMIAH, L. E. 1981 Proper control of retail case temperature improves beef shelf life. *Journal of Food Protection* **44**, 297–299.

GUINÉE, P. A. M., EDEL, W. & KAMPELMACHER, E. H. 1967 Studies on *Salmonella* infections in fattening calves. *Netherlands Journal of Veterinary Medicine* **92**, 158–167.

GWYNN, M. N., WEBB, L. T. & ROLINSON, G. N. 1981 Regrowth of *Pseudomonas aeruginosa* and other bacteria after the bactericidal action of carbenicillin and other β-lactam antibiotics. *Journal of Infectious Diseases* **144**, 263–269.

HACKNEY, C. R., RAY, B. & SPECK, M. L. 1979 Repair detection procedure for enumeration of fecal coliforms and enterococci from seafoods and marine environments. *Applied and Environmental Microbiology* **37**, 947–953.

HAINES, R. B. 1933 The bacterial flora developing on stored lean meat, especially with regard to "slimy" meat. *Journal of Hygiene* **33**, 175–182.

HAINES, W. C. & HARMON, L. G. 1973 Effect of selected lactic acid bacteria on growth of *Staphylococcus aureus* and production of enterotoxin. *Applied Microbiology* **25**, 436–441.

HALLS, N. A. & BOARD, R. G. 1973 The microbial associations developing on experimental trickling filters irrigated with domestic sewage. *Journal of Applied Bacteriology* **36**, 465–474.

HANDFORD, P. M. & GIBBS, B. M. 1964 Antibacterial effects of smoke constituents on bacteria isolated from bacon. In *Microbial Inhibitors in Food* ed. Molin, N. & Erichsen, A. pp. 333–346. Stockholm: Almqvist & Wiksell.

HANNA, M. O., STEWART, J. C., ZINK, D. L., CARPENTER, Z. L. & VANDERZANT, C. 1977*a* Development of *Yersinia enterocolitica* on raw and cooked beef and pork at different temperatures. *Journal of Food Science* **42**, 1180–1184.

HANNA, M. O., VANDERZANT, C., CARPENTER, Z. L. & SMITH, G. C. 1977*b* Characteristics of psychrotrophic, Gram-positive, catalase-positive, pleomorphic coccoid rods from vacuum-packaged wholesale cuts of beef. *Journal of Food Protection* **40**, 94–97.

HANNA, M. O., VANDERZANT, C., SMITH, G. C. & SAVELL, J. W. 1981 Packaging of beef loin steaks in 75% O_2 plus 25% CO_2. II. Microbiological properties. *Journal of Food Protection* **44**, 928–933.

HANNA, M. O., SMITH, G. C., SAVELL, J. W., MCKEITH, F. K. & VANDERZANT, C. 1982*a* Microbial flora of livers, kidneys and hearts from beef, pork and lamb: effects of refrigeration, freezing and thawing. *Journal of Food Protection* **45**, 63–73.

HANNA, M. O., SMITH, G. C., SAVELL, J. W., MCKEITH, F. K. & VANDERZANT, C. 1982*b* Effects of packaging methods on the microbial flora of livers and kidneys from beef or pork. *Journal of Food Protection* **45**, 74–81.

HARRISON, J. M. & LEE, J. S. 1969 Microbiological evaluation of Pacific shrimp processing. *Applied Microbiology* **18**, 188–192.

HARRISON, M. A., MELTON, C. C. & DRAUGHON, F. A. 1981 Bacterial flora of ground beef and soy extended ground beef during storage. *Journal of Food Science* **46**, 1088–1090.

HARVEY, R. W. S. & PRICE, T. H. 1981 Comparison of selenite F, Muller–Kauffmann tetrathionate and Rappaport's medium for *Salmonella* isolation from chicken giblets after preenrichment in buffered peptone water. *Journal of Hygiene* **87**, 219–224.

HARVEY, R. W. S. & PRICE, T. H. 1982 Influence of multiple plating from fluid media on *Salmonella* isolation from animal feeding stuffs. *Journal of Hygiene* **88**, 113–119.

HAUSCHILD, A. H. W., HILSHEIMER, R., JARVIS, G. & RAYMOND, D. P. 1982 Contribution of nitrite to the control of *Clostridium botulinum* in liver sausage. *Journal of Food Protection* **45**, 500–506.

HAYES, A. W. 1980 Mycotoxins: a review of biological effects and their role in human diseases. *Clinical Toxicology* **17**, 45–83.

HEATH, O. V. S. 1969 Interaction of factors. In *The Physiological Aspects of Photosynthesis* pp. 113–130. London: Heinemann.

HERBERT, R. A., HENDRIE, M. S., GIBSON, D. M. & SHEWAN, J. M. 1971 Bacteria active in the spoilage of certain sea foods. *Journal of Applied Bacteriology* **34**, 41–50.

HOBBS, B. C. 1964 *Salmonella* in foods. *Proceedings of National Conference on Salmonellosis*, United States Department of Health, Education and Welfare, pp. 84–93.

HOLDING, A. J. 1973 The isolation and identification of certain soil Gram negative bacteria. In *Sampling—Microbiological Monitoring of Environments* ed. Board, R. G. & Lovelock, D. W. pp. 137–141. London & New York: Academic Press.

HOLTZAPFFEL, D. & COUTINHO, H. E. 1969 Investigations on the microbiological condition of dehydrated vegetables with particular reference to the fate of organisms during storage in air and under nitrogen. In *The Microbiology of Dried Foods* ed. Kampelmacher, E. H., Ingram, M. & Mossel, D. A. A. pp. 387–406. Haarlem: Grafische Industrie.

HOLTZAPFFEL, D. & MOSSEL, D. A. A. 1968 The survival of pathogenic bacteria in, and the microbial spoilage of salads containing meat, fish and vegetables. *Journal of Food Technology* **3**, 223–239.

HONE, J. D., OCKERMAN, H. W., CAHILL, V. R., BORTON, R. J. & PROCTOR, G. O. 1975 A rapid method for the aseptic collection of tissue. *Journal of Milk and Food Technology* **38**, 664–666.

HSU, E. J., GODSEY, J. H., CHANG, E. K. & LANDUYT, S. L. 1981 Differentiation of pseudomonads by amplification of metabolic profiles. *International Journal of Systematic Bacteriology* **31**, 43–55.

HURST, A. 1980 Injury and its effect on survival and recovery. In *Microbial Ecology of Foods* Vol. 1, pp. 205–214. New York & London: Academic Press.

HURST, A. & COLLINS-THOMPSON, D. L. 1979 Food as a bacterial habitat. *Advances in Microbial Ecology* **3**, 79–134.

INGRAM, M. 1934 The microbiology of pork and bacon. *Reports of the Food Investigation Board*, London, p. 77.

INGRAM, M. 1948 Fatigue musculaire, pH et proliferation bactérienne dans la viande. *Annales de l'Institut Pasteur, Paris* **75**, 139–146.

INGRAM, M. 1949 Science in the imported meat industry. Hygiene and storage. *Journal of the Royal Sanitary Institute, London* **69**, 39–47.

INGRAM, M. 1955 Microbial association of semi-preserved meats. *Annales de l'Institut Pasteur, Lille* **7**, 32–45.

INGRAM, M. & BROOKS, J. 1952 Bacteriological standards for perishable foods. Eggs and egg products. *Journal of the Royal Sanitary Institute, London* **72**, 411–421.

INGRAM, M. & DAINTY, R. H. 1971 Changes caused by microbes in spoilage of meats. *Journal of Applied Bacteriology* **34**, 21–39.

IZAGUIRRE, G., HWANG, C. J., KRASNER, S. W. & McGUIRE, M. J. 1982 Geosmin and 2-methylisoborneol from cyanobacteria in three water supply systems. *Applied and Environmental Microbiology* **43**, 708–714.

JARL, D. L. & ARNOLD, E. A. 1982 Influence of drying plant environment on salmonellae contamination of dry milk products. *Journal of Food Protection* **45**, 16–18, 22.

JAYNE-WILLIAMS, D. J. 1973 A medium for overcoming the *in vitro* inhibition of *Clostridium perfringens* by *Streptococcus faecalis* var. *zymogenes* and a note on the *in vivo* interaction of the two organisms. *Journal of Applied Bacteriology* **36**, 575–583.

JOHNSTON, R. W., HARRIS, M. E., MORAN, A. B., KRUMM, G. W. & LEE, W. H. 1982. A comparative study of the microbiology of commercial vacuum-packaged and hanging beef. *Journal of Food Protection* **45**, 223–228.

JONES, G. W., RICHARDSON, L. A. & UHLMAN, D. 1981 The invasion of HeLa cells by *Salmonella typhimurium*: reversible and irreversible bacterial attachment and the role of bacterial motility. *Journal of General Microbiology* **127**, 351–360.

JONES, P. W., COLLINS, P., BROWN, G. T. H. & AITKEN, M. 1982 Transmission of *Salmonella mbandaka* to cattle from contaminated feed. *Journal of Hygiene* **88**, 255–263.

JUVEN, B. J. 1979 Fate of spoilage micro-organisms in frozen and chilled concentrated tomato juices. *Journal of Food Protection* **42**, 19–22.

JUVEN, B. J. & WEISSLOWICZ, H. 1981 Chemical changes in tomato juices caused by lactic acid bacteria. *Journal of Food Science* **46**, 1543–1545.

JUVEN, B. J., COX, N. A., MERCURI, A. J. & THOMSON, J. E. 1974 A hot acid treatment for eliminating *Salmonella* from chicken meat. *Journal of Milk and Food Technology* **37**, 237–239.

JUVEN, B. J., GORDIN, S., ROSENTHAL, I. & LAUFER, A. 1981 Changes in refrigerated milk caused by Enterobacteriaceae. *Journal of Dairy Science* **64**, 1781–1784.

KAFEL, S. & AYRES, J. C. 1969 The antagonism of enterococci on other bacteria in canned hams. *Journal of Applied Bacteriology* **32**, 217–232.

KAUTTER, D. A., LYNT, R. K., LILLY, T. & SOLOMON, H. M. 1981*a* Evaluation of the botulism hazard from nitrogen-packed sandwiches. *Journal of Food Protection* **44**, 59–61.

KAUTTER, D. A., LYNT, R. K., LILLY, T. & SOLOMON, H. M. 1981*b* Evaluation of the botulism hazard from imitation cheeses. *Journal of Food Science* **46**, 749–750, 764.

KENDALL, E. J. C. & TANNER, E. I. 1982 *Campylobacter* enteritis in general practice. *Journal of Hygiene* **88**, 155–163.

KENNEDY, J. E., OBLINGER, J. L. & WEST, R. L. 1980 Fate of *Salmonella infantis*, *Staphylococcus aureus* and *Hafnia alvei* in vacuum packaged beef plate pieces during refrigerated storage. *Journal of Food Science* **45**, 1273–1277, 1300.

KENNEDY, J. E., OBLINGER, J. L. & WEST, R. L. 1982 Microbiological comparison of hot-boned and conventionally processed beef plate cuts during extended storage. *Journal of Food Protection* **45**, 607–614.

KING, A. D., MICHENER, H. D., BAYNE, H. G. & MIHARA, K. L. 1976 Microbial studies on shelf life of cabbage and coleslaw. *Applied and Environmental Microbiology* **31**, 404–407.

KITCHELL, A. G. & INGRAM, M. 1956 A comparison of bacterial growth on fresh meat and on frozen meat after thawing. *Annales de l'Institut Pasteur, Lille* **8**, 121–131.

KLEEBERGER, A. & BUSSE, M. 1975 Keimzahl und Florazusammensetzung bei Hackfleisch unter besonderer Berücksichtigung von Enterobakterien und Pseudomonaden. *Zeitschrift für Lebensmittel Untersuchung und Forschung* **158**, 321–331.

KLOSE, A. A., KAUFMAN, V. F., BAYNE, H. G. & POOL, M. F. 1971 Pasteurization of poultry meat by steam under reduced pressure. *Poultry Science* **50**, 1156–1160.

KOMINOS, S. D., COPELAND, C. E., GROSIAK, B. & POSTIE, B. 1972 Introduction of *Pseudomonas aeruginosa* into a hospital via vegetables. *Applied Microbiology* **24**, 567–570.

KORNACKI, J. L. & MARTH, E. H. 1982 Fate of nonpathogenic and enteropathogenic *Escherichia coli* during the manufacture of Colby-like cheese. *Journal of Food Protection* **45**, 310–316.

KRAFT, A. A., OBLINGER, J. L., WALKER, H. W., KAWAL, M. C., MOON, N. J. & REINBOLD, G. W. 1976 Microbial interactions in foods: meats, poultry and dairy products. In *Microbiology in Agriculture, Fisheries and Food* ed. Skinner, F. A. & Carr, J. G. pp. 141–150. London & New York: Academic Press.

KRAFT, A. A., REDDY, K. V., HASIAK, R. J., LIND, K. D. & GALLOWAY, D. E. 1982 Microbiological quality of vacuum packaged poultry with or without chlorine treatment. *Journal of Food Science* **47**, 380–385.

KWAN, P. L. & LEE, J. S. 1974 Compound inhibitory to *Clostridium botulinum* type E produced by a *Moraxella* species. *Applied Microbiology* **27**, 329–332.

LABADIE, J., GONET, PH. & FOURNAND, J. 1977 Blood poisonings at slaughter and their consequences. *Zentralblatt für Bakteriologie und Parasitenkunde, Abt. I, Orig. B* **164**, 390–396.

LACHAPELLE, G. 1979 *Salmonella* dans les usines de lait en poudre au Québec. *Canadian Institute of Food Science and Technology Journal* **12**, 177–179.

LANNELONGUE, M., HANNA, M. O., FINNE, G., NICKELSON, R. & VANDERZANT, C. 1982 Storage characteristics of finfish fillets (*Archosargus probatocephalus*) packaged in modified gas atmospheres containing carbon dioxide. *Journal of Food Protection* **45**, 440–444, 449.

LA ROCCO, K. A. & MARTIN, S. E. 1981 Effects of potassium sorbate alone and in combination with sodium chloride on the growth of *Salmonella typhimurium* 7136. *Journal of Food Science* **46**, 568–570.

LECHEVALLIER, M. W., SEIDLER, R. J. & EVANS, T. M. 1980 Enumeration and characterization of standard plate count bacteria in chlorinated and raw water supplies. *Applied and Environmental Microbiology* **40**, 922–930.

LEE, C. Y., FUNG, D. Y. C. & KASTNER, C. L. 1982 Computer-assisted identification of bacteria on hot-boned and conventionally processed beef. *Journal of Food Science* **47**, 363–367, 373.

LEE, J. A. 1974 Recent trends in human salmonellosis in England and Wales: the epidemiology of prevalent serotypes other than *Salmonella typhimurium*. *Journal of Hygiene* **72**, 185–195.

LEE, J. S. & HARRISON, J. M. 1968 Microbial flora of pacific hake (*Merluccius productus*). *Applied Microbiology* **16**, 1937–1938.

LEE, J. S. & PFEIFER, D. K. 1973 Aerobic microbial flora of smoked salmon. *Journal of Milk and Food Technology* **36**, 143–145.

LEE, M. L., SMITH, D. L. & FREEMAN, L. R. 1979 High-resolution gas chromatographic profiles of volatile organic compounds produced by micro-organisms at refrigerated temperatures. *Applied and Environmental Microbiology* **37**, 85–90.

LE MINOR, L. & LE MINOR, S. 1981 Origine et répartition en sérotypes des souches isolées en France et reçues au Centre National des *Salmonella* de 1977 à 1979. *Revue d'Epidémiologie et de Santé Publique* **29**, 45–55.

LEPOVETSKY, B. C., WEISER, H. H. & DEATHERAGE, F. E. 1953 A microbiological study of lymph nodes, bone marrow and muscle tissue obtained from slaughtered cattle. *Applied Microbiology* **1**, 57–59.

LERKE, P. & FARBER, L. 1971 Heat pasteurization of crab and shrimp from the Pacific coast of the United States: public health aspects. *Journal of Food Science* **36**, 277–279.

LERKE, P., ADAMS, R. & FARBER, L. 1965 Bacteriology of spoilage of fish muscle. III. Characterization of spoilers. *Applied Microbiology* **13**, 625–630.

LEVETT, P. N. & DANIEL, R. R. 1981 Adhesion of vibrios and aeromonads to isolated rabbit brush borders. *Journal of General Microbiology* **125**, 167–172.

LEVIN, R. E. 1972 Correlation of DNA base composition and metabolism of *Pseudomonas putrefaciens* isolates from food, human clinical specimens, and other sources. *Antonie van Leeuwenhoek* **38**, 121–127.

LEVINE, M. M., DU PONT, H. L., FORMAL, S. B., HORNICK, R. B., TAKEUCHI, A., GANGAROSA, E. J., SNIJDER, M. J. & LIBONATI, J. P. 1973 Pathogenesis of *Shigella dysenteriae* 1 (Shiga) dysentery. *Journal of Infectious Diseases* **127**, 261–270.

LEY, F. J., KENNEDY, T. S., KAWASHIMA, K., ROBERTS, D. & HOBBS, B. C. 1970 The use of gamma radiation for the elimination of *Salmonella* from frozen meat. *Journal of Hygiene* **68**, 293–311.

LI CARI, J. J. & POTTER, N. N. 1970 *Salmonella* survival during spray drying and subsequent handling of skim milk powder. III. Effects of storage temperature on *Salmonella* and dried milk properties. *Journal of Dairy Science* **53**, 877–882.

LIGHTHART, B. 1968 Multipoint inoculator system. *Applied Microbiology* **16**, 1797–1798.

LILLEHOJ, E. B., FENNELL, D. I. & HESSELTINE, C. W. 1976 *Aspergillus flavus* infection and aflatoxin production in mixtures of high-moisture and dry maize. *Journal of Stored Products Research* **12**, 11–18.

LIPSON, A. & MEIKLE, H. 1977 Porcine pancreatin as a source of *Salmonella* infection in children with cystic fibrosis. *Archives of Diseases of Childhood* **52**, 569–572.

LUND, B. M. 1971 Bacterial spoilage of vegetables and certain fruits. *Journal of Applied Bacteriology* **34**, 9–20.

LUND, B. M. 1982 The effect of bacteria on post-harvest quality of vegetables and fruits, with particular reference to spoilage. In *Bacteria and Plants* ed. Rhodes-Roberts, M. E. & Skinner, F. A. pp. 133–153. London & New York: Academic Press.

MABBITT, L. A. 1981 Metabolic activity of bacteria in raw milk. *Kieler milchwirtschaftliche Forschungsberichte* **33**, 273–280.

MCINERNEY, M. J. & BRYANT, M. P. 1981 Anaerobic degradation of lactate by syntrophic associations of *Methanosarcina barkeri* and *Desulfovibrio* species and effect of H₂ on acetate degradation. *Applied and Environmental Microbiology* **41**, 346–354.

MCINERNEY, M. J., MACKIE, R. I. & BRYANT, M. P. 1981 Syntrophic association of a butyrate-degrading bacterium and *Methanosarcina* enriched from bovine rumen fluid. *Applied and Environmental Microbiology* **41**, 826–828.

MCKINLEY, G. A., FAGERBERG, D. J., QUARLES, C. L., GEORGE, B. A., WAGNER, D. E. & ROLLINS, L. D. 1980 Incidence of salmonellae in fecal samples of production swine and swine at slaughter plants in the United States in 1978. *Applied and Environmental Microbiology* **40**, 562–566.

MCMEEKIN, T. A. 1975 Spoilage association of chicken breast muscle. *Applied Microbiology* **29**, 44–47.

MCMEEKIN, T. A. 1977 Spoilage association of chicken leg muscle. *Applied and Environmental Microbiology* **33**, 1244–1246.

MCMEEKIN, T. A. & THOMAS, C. J. 1979 Aspects of the microbial ecology of poultry processing and storage: a review. *Food Technology in Australia* **31**, 35–43.

MCMEEKIN, T. A., THOMAS, C. J. & MCCALL, D. 1979 Scanning electron microscopy of micro-organisms on chicken skin. *Journal of Applied Bacteriology* **46**, 195–200.

MCMEEKIN, T. A., HULSE, L. & BREMNER, H. A. 1982 Spoilage association of vacuum packed sand flathead (*Platycephalus bassensis*) fillets. *Food Technology of Australia* **34**, 278–282.

MATCHES, J. R. 1982 Effects of temperature on the decomposition of Pacific coast shrimp (*Pandalus jordani*). *Journal of Food Science* **47**, 1044–1047, 1069.

MAXCY, R. B. 1976 Fate of post-cooking microbial contaminants of some major menu items. *Journal of Food Science* **41**, 375–378.

MAXCY, R. B. & TIWARI, N. P. 1973 Irradiation of meats for Public Health Protection. In *Radiation Preservation of Food*. Publication STI/317, pp. 491–503. Vienna: International Atomic Energy Agency.

MEAD, G. C., ADAMS, B. W. & HAQUE, Z. 1982 Vorkommen, Ursprung und Verderbspotential psychrotropher Enterobacteriaceae auf verarbeitetem Geflügel. *Fleischwirtschaft* **62**, 1173–1177.

MEANS, E. G. & OLSON, B. H. 1981 Coliform inhibition by bacteriocin-like substances in drinking water distribution systems. *Applied and Environmental Microbiology* **42**, 506–512.

MEGEE, R. D., DRAKE, J. F., FREDERICKSON, A. G. & TSUCHIYA, H. M. 1972 Studies in intermicrobial symbiosis, *Saccharomyces cerevisiae* and *Lactobacillus casei*. *Canadian Journal of Microbiology* **18**, 1733–1742.

MERRITT, C., ANGELINI, P. & GRAHAM, R. A. 1978 Effect of radiation parameters on the formation of radiolysis products in meat and meat substances. *Journal of Agricultural and Food Chemistry* **26**, 29–35.

MICHENER, H. D. 1979 Micro-organisms in frozen and refrigerated food. In *Food Microbiology and Technology* ed. Jarvis, B., Christian, J. H. B. & Michener, H. D. pp. 175–193. Parma: Medicina Viva.

MILDENHALL, J. P., PRIOR, B. A. & TROLLOPE, L. A. 1981 Water relations of *Erwinia chrysanthemi*: growth and extracellular pectic acid lyase production. *Journal of General Microbiology* **127**, 27–34.

MILLER, A., SCANLAN, R. A., LEE, J. S. & LIBBEY, L. M. 1973 Volatile compounds produced in sterile fish muscle (*Sebastes melanops*) by *Pseudomonas putrefaciens*, *Pseudomonas fluorescens* and an *Achromobacter* species. *Applied Microbiology* **26**, 18–21.

MILLER, J. S. & LEDFORD, R. A. 1977 Potential for growth of enterotoxigenic staphylococci in Cheddar cheese whey. *Journal of Dairy Science* **60**, 1689–1692.

MILTON, R. F. & JARRETT, K. J. 1969 Storage and Transport of maize—I. Temperature, humidity and microbiological spoilage. *World Crops* **21**, 356–357.

MILTON, R. F. & JARRETT, K. J. 1970 Storage and transport of maize. *World Crops* **22**, 48–49, 96, 98–99.

MINNICH, S. A., HARTMAN, P. A. & HEIMSCH, R. C. 1982 Enzyme immunoassay for detection of salmonellae in foods. *Applied and Environmental Microbiology* **43**, 877–883.

MOL, J. H. H. & TIMMERS, C. A. 1970 Assessment of the stability of pasteurized comminuted meat products. *Journal of Applied Bacteriology* **33**, 233–247.

MOL, J. H. H., HIETBRINK, J. E. A., MOLLEN, H. W. M. & VAN TINTEREN, J. 1971 Observations on the microflora of vacuum packed sliced cooked meat products. *Journal of Applied Bacteriology* **34**, 377–397.

MONTVILLE, T. J. 1982 Metabiotic effect of *Bacillus lichenformis* on *Clostridium botulinum*: implications for home-canned tomatoes. *Applied Environmental Microbiology* **44**, 334–338.

MOON, N. J. & REINBOLD, G. W. 1976 Commensalism and competition in mixed cultures of *Lactobacillus bulgaricus* and *Streptococcus thermophilus*. *Journal of Milk and Food Technology* **39**, 337–341.

MOREAU, C. & MOSS, M. 1979 *Moulds, Toxins and Food*. Chichester: Wiley.

MOSENTHAL, A. C., MONES, R. L. & BOKKENHEUSER, V. D. 1981 *Campylobacter fetus jejuni* enteritis in New York City. *New York State Journal of Medicine* **81**, 321–323.

MOSSEL, D. A. A.1960 The destruction of *Salmonella* bacteria in refrigerated liquid whole egg with gamma radiation. *International Journal of Applied Radiation and Isotopes* **9**, 109–112.

MOSSEL, D. A. A. 1969 Nahrungsmittel als Umwelt für Mikroorganismen, die Lebensmittel gesundheitsschädlich machen. *Alimenta* **8**, 8–19.

MOSSEL, D. A. A. 1970 The role of microbiology and hygiene in the manufacture of margarine. In *Margarine Today* ed. Coenen, J. W. E., Feron, R. & Mossel, D. A. A. pp. 104–124. Leiden: Brill.

MOSSEL, D. A. A. 1971 Physiological and metabolic attributes of microbial groups associated with foods. *Journal of Applied Bacteriology* **34**, 95–118.

MOSSEL, D. A. A. 1977 The elimination of enteric bacterial pathogens from food and feed of animal origin by gamma irradiation with particular reference to *Salmonella* radicidation. *Journal of Food Quality* **1**, 85–104.

MOSSEL, D. A. A. 1981 Coliform test for cheese and other foods. *Lancet* **ii**, 1425.

MOSSEL, D. A. A. 1982 *Microbiology of Foods. The Ecological Essentials of Assurance and Assessment of Safety and Quality* 3rd edn. Utrecht University Press.

MOSSEL, D. A. A. & CORRY, J. E. L. 1977 Detection and enumeration of sublethally injured pathogenic and index bacteria in foods and water processed for safety. *Alimenta* **16**, Special Issue on Microbiology, 19–34.

MOSSEL, D. A. A. & DE BRUIN, A. S. 1960 The survival of Enterobacteriaceae in acid liquid foods stored at different temperatures. *Annales de l'Institut Pasteur, Lille* **11**, 65–72.

MOSSEL, D. A. A. & DE GROOT, A. P. 1965 The use of pasteurizing doses of gamma radiation for the destruction of salmonellae and other Enterobacteriaceae in some foods of low water activity. In *Radiation Preservation of Foods*. Proceedings of the International Conference, Boston, Mass. 1964. Publ. 1273, pp. 233–264. Washington D.C.: National Academy of Science.

MOSSEL, D. A. A. & DRION, E. F. 1979 Risk analysis. Its application to the protection of the consumer against food-transmitted diseases of microbial aetiology. *Antonie van Leeuwenhoek* **45**, 321–323.

MOSSEL, D. A. A. & INGRAM, M. 1955 The physiology of the microbial spoilage of foods. *Journal of Applied Bacteriology* **18**, 232–268.

MOSSEL, D. A. A. & KAMPELMACHER, E. H. 1981 Prevention of salmonellosis. *Lancet* **i**, 208.

MOSSEL, D. A. A. & OEI, H. Y. 1975 Person-to-person transmission of enteric bacterial infection. *Lancet* **i**, 751.

MOSSEL, D. A. A. & SCHOLTS, H. H. 1964 Diagnostic, pronostic et prévention des altérations microbiennes des boissons hygiéniques. *Annales de l'Institut Pasteur, Lille* **15**, 11–30.

MOSSEL, D. A. A. & VAN NETTEN, P. 1982 Whither protection of the consumer against enteropathogenic bacteria on fresh meats and poultry by processing for safety. In *Food Irradiation Now* pp. 2–19. The Hague: Nijhoff.

MOSSEL, D. A. A. & VAN NETTEN, P. 1983 Harmful effects of selective media on stressed micro-organisms—nature and remedies. In *Revival of Injured Microbes*, eds Russell, A. D. & Andrew, M. H. E. London & New York, Academic Press, in press.

MOSSEL, D. A. A. & WESTERDIJK, J. 1949 The physiology of microbial spoilage in foods. *Antonie van Leeuwenhoek* **15**, 190–202.

MOSSEL, D. A. A., BUECHLI, K. & DE WAART, J. 1965a Dose-range-finding tests for the elimination of salmonellae from proteinaceous foods with a low a_w by irradiation with Co^{60} gamma rays. *Antonie van Leeuwenhoek* **31**, 220.

MOSSEL, D. A. A., JONGERIUS, E. & KOOPMAN, M. J. 1965b Sur la nécessité d'une revivification préalable pour le dénombrement des Enterobacteriaceae dans les aliments deshydratés irradiés ou non. *Annales de l'Institut Pasteur, Lille* **16**, 119–125.

MOSSEL, D. A. A., KROL, B. & MOERMAN, P. C. 1972 Bacteriological and quality perspectives of *Salmonella* radicidation of frozen boneless meats. *Alimenta* **11**, 51–59.

MOSSEL, D. A. A., VAN DOORNE, H., EELDERINK, I. & DE VOR, H. 1977a The selective enumeration of Gram-positive and Gram-negative bacteria in foods, water and medicinal and cosmetic preparations. *Pharmaceutisch Weekblad* **112**, 41–48.

MOSSEL, D. A. A., EELDERINK, I. & SUTHERLAND, J. P. 1977b Development and use of single, 'polytropic' diagnostic tubes for the approximate taxonomic grouping of bacteria isolated from foods, water and medicinal preparations. *Zentralblatt für Bakteriologie und Parasiten-kunde, Abt. I, Orig. A* **238**, 66–79.

MOSSEL, D. A. A., EELDERINK, I., KOOPMANS, M. T. A. G. F. & VAN ROSSEM, F. 1979 Influence of carbon source, bile salts and incubation temperature on recovery of Enterobac-teriaceae from foods, using MacConkey-type agars. *Journal of Food Protection* **42**, 470–475.

MOSSEL, D. A. A., VELDMAN, A. & EELDERINK, I. 1980a Comparison of the effects of liquid medium repair and the incorporation of catalase in McConkey-type media on the recovery of Enterobacteriaceae sublethally stressed by freezing. *Journal of Applied Bacteriology* **49**, 405–419.

MOSSEL, D. A. A., WIJERS, B. & BOUWER-HERTZBERGER, S. 1980b The enumeration and identification of stressed *Escherichia coli* in foods using solid medium repair and Eijkman's thermotrophic criteria. *Journal of Applied Bacteriology* **49**, xvii–xviii.

MOURGUES, R., VASSAL, L., AUCLAIR, J., MOCQUOT, G. & VANDENWEGHE, J. 1977 Origine et développement des bactéries coliformes dans les fromages à pâte molle. *Lait* **57**, 131–149.

MULDER, R. W. A. W. 1982 *Salmonella radicidation of poultry carcasses*. Ph.D. Thesis, Agriculture, Wageningen Agricultural University.

MULDER, S. J. & KROL, B. 1975 Der Einfluss der Milchsäure auf die Keimflora und die Farbe frischen Fleisches. *Fleischwirtschaft* **55**, 1255–1258.

MYERS, B. R., MARSHALL, R. T., EDMONDSON, J. E. & STRINGER, W. C. 1982 Isolation of pectinolytic *Aeromonas hydrophila* and *Yersinia enterocolitica* from vacuum-packaged pork. *Journal of Food Protection* **45**, 33–37.

NADIR, M. T. & GILBERT, P. 1982 Injury and recovery of *Bacillus megaterium* from mild chlorhexidine treatment. *Journal of Applied Bacteriology* **52**, 111–115.

NARUCKA, U. 1982 Studies on the incidence of *Salmonella* in a number of lymph nodes and tonsillar tissues of slaughtered pigs. *Netherlands Journal of Veterinary Medicine* **107**, 220–224.

NEWTON, K. G. & GILL, C. O. 1978 The development of the anaerobic spoilage flora of meat stored at chill temperatures. *Journal of Applied Bacteriology* **44**, 91–95.

NEWTON, K. G. & GILL, C. O. 1981 The microbiology of DFD fresh meats: a review. *Meat Science* **5**, 223–232.

NEWTON, K. G., HARRISON, J. C. L. & SMITH, K. M. 1977 The effect of storage in various gaseous atmospheres on the microflora of lamb chops held at −1°C. *Journal of Applied Bacteriology* **43**, 53–59.

NICOL, D. J., SHAW, M. K. & LEDWARD, D. A. 1970 Hydrogen sulfide production by bacteria and sulfmyoglobin formation in prepacked chilled beef. *Applied Microbiology* **19**, 937–939.

NIEMAND, J. G., VAN DER LINDE, H. J. & HOLZAPFEL, W. H. 1981 Radurization of prime beef cuts. *Journal of Food Protection* **44**, 677–681.

NIEVES, B. M., GIL, F. & CASTILLO, F. J. 1981 Growth inhibition activity and bacteriophage and bacteriocinlike particles associated with different species of *Clostridium*. *Canadian Journal of Microbiology* **27**, 216–225.

NIVEN, C. F., CASTELLANI, A. G. & ALLANSON, V. 1949 A study of the lactic acid bacteria that cause surface discolourations of sausages. *Journal of Bacteriology* **58**, 633–641.

NORTHOLT, M. D., VAN EGMOND, H. P. & PAULSCH, W. E. 1979 Ochratoxin A production by some fungal species in relation to water activity and temperature. *Journal of Food Protection* **42**, 485–490.

NOTERMANS, S. & KAMPELMACHER, E. H. 1975 Heat destruction of some bacterial strains attached to broiler skin. *British Poultry Science* **16**, 351–361.

NOTERMANS, S., TERBIJHE, R. J. & VAN SCHOTHORST, M. 1980 Removing faecal contamination of broiler chickens by spray-cleaning during evisceration. *British Poultry Science* **21**, 115–121.

NOTTINGHAM, P. M. 1960 Bone-taint in beef—II. Bacteria in ischiatic lymph nodes. *Journal of the Science of Food and Agriculture* **11**, 436–441.

NURMIKKO, V. 1956 Biochemical factors affecting symbiosis among bacteria. *Experientia* **12**, 245–249.

OBLINGER, J. L. & KRAFT, A. A. 1970 Inhibitory effects of *Pseudomonas* on selected *Salmonella* and bacteria isolated from poultry. *Journal of Food Science* **35**, 30–32.

OBLINGER, J. L., KENNEDY, J. E., ROTHENBERG, C. A., BERRY, B. W. & STERN, N. J. 1982 Identification of bacteria isolated from fresh and temperature abused variety meats. *Journal of Food Protection* **45**, 650–654.

OCKERMAN, H. W., BORTON, R. J., CAHILL, V. R., PARRETT, N. A. & HOFFMAN, H. D. 1974 Use of acetic and lactic acid to control the quantity of micro-organisms on lamb carcasses. *Journal of Milk and Food Technology* **37**, 203–204.

ODLAUG, T. E. & PFLUG, I. J. 1979 *Clostridium botulinum* growth and toxin production in tomato juice containing *Aspergillus gracilis*. *Applied and Environmental Microbiology* **37**, 496–504.

ODUM, E. P. 1969 The strategy of ecosystem development. *Science, New York* **164**, 262–270.

OOSTEROM, J., VAN ERNE, E. H. W. & VAN SCHOTHORST, M. 1981 Studies on the possibility of fattening pigs free from *Salmonella*. *Netherlands Journal of Veterinary Medicine* **106**, 599–612.

OSBORNE, B. G. 1982 Mycotoxins and the cereals industry—a review. *Journal of Food Technology* **17**, 1–9.

OTTE, I., TOLLE, A. & HAHN, G. 1979 Zur Analyse der Mikroflora von Milch und Milchprodukten. 2. Miniaturisierte Primärtests zur Bestimmung der Gattung. *Milchwissenschaft* **34**, 152–156.

OYE, E. VAN & GHYSELS, G. 1978 Incidence of *Salmonella* infection in man in Belgium 1962–1976. *International Journal of Epidemiology* **7**, 189–193.

PASTEUR, L. 1858 Mémoire sur la fermentation appelée lactique. *Annales de Chimie et de Physiologie*, 3e Sér. **52**, 404–418.

PATTERSON, J. T. & GIBBS, P. A. 1978 Sources and properties of some organisms isolated in two abattoirs. *Meat Science* **2**, 263–273.

PATTERSON, J. T. & GIBBS, P. A. 1979 Vacuum-packaging of bovine edible offal. *Meat Science* **3**, 209–222.

PEEL, B. & SIMMONS, G. C. 1978 Factors in the spread of *Salmonella* in meatworks with special reference to contamination of knives. *Australian Veterinary Journal* **54**, 106–110.

PELROY, G. A., EKLUND, M. W., PARANJPYE, R. N., SUZUKI, E. M. & PETERSON, M. E. 1982 Inhibition of *Clostridium botulinum* types A and E toxin formation by sodium nitrite and sodium chloride in hot-process (smoked) salmon. *Journal of Food Protection* **45**, 833–841.

PIERSON, M. D., COLLINS-THOMPSON, D. L. & ORDAL, Z. J. 1970 Microbiological, sensory and pigment changes of aerobically and anaerobically packaged beef. *Food Technology* **24**, 1171–1175.

PITTARD, B. T., FREEMAN, L. R., LATER, D. W. & LEE, M. L. 1982 Identification of volatile organic compounds produced by fluorescent pseudomonads on chicken breast muscle. *Applied Environmental Microbiology* **43**, 1504–1506.

PIVNICK, H. & BIRD, H. 1965 Toxinogenesis by *Clostridium botulinum* types A and E in perishable cooked meats vacuum-packed in plastic pouches. *Food Technology* **19**, 1156–1164.

POST, F. J., BLISS, A. H. & O'KEEFE, W. B. 1961 Studies on the ecology of selected food poisoning organisms in foods. I. Growth of *Staphylococcus aureus* in cream and a cream product. *Journal of Food Science* **26**, 436–441.

PRICE, R. J. & LEE, J. S. 1970 Inhibition of *Pseudomonas* species by hydrogen peroxide producing lactobacilli. *Journal of Milk and Food Technology* **33**, 13–18.

QADRI, R. B., BUCKLE, K. A. & EDWARDS, R. A. 1976 Bacteriological changes during storage of live and chucked oysters. *Food Technology in Australia* **28**, 283–287.

QURESHI, J. V. & GIBBONS, R. J. 1981 Differences in the adsorptive behavior of human strains of *Actinomyces viscosus* and *Actinomyces naeslundii* to saliva treated hydroxyapatite surfaces. *Infection and Immunity* **31**, 261–266.

RACCACH, M., BAKER, R. C., REGENSTEIN, J. M. & MULNIX, E. J. 1979 Potential application of microbial antagonism to extended storage stability of a flesh type food. *Journal of Food Science* **44**, 43–46.

RAY, B. 1979 Methods to detect stressed micro-organisms. *Journal of Food Protection* **42**, 346–355.

RAY, B., JEZESKI, J. J. & BUSTA, F. F. 1971 Isolation of Salmonellae from naturally contaminated dried milk products. II. Influence of storage time on the isolation of Salmonellae. *Journal of Milk and Food Technology* **34**, 423–427.

RAY, L. L. & BULLERMAN, L. B. 1982 Preventing growth of potentially toxic molds using antifungal agents. *Journal of Food Protection* **45**, 953–963.

RAYMAN, M. K., D'AOUST, J. Y., ARIS, B., MAISHMENT, C. & WASIK, R. 1979 Survival of micro-organisms in stored pasta. *Journal of Food Protection* **42**, 330–334.

REED, W. P., SELINGER, D. S., ALBRIGHT, E. L., ABDIN, Z. H. & WILLIAMS, R. C. 1980 Streptococcal adherence to pharyngeal cells of children with acute rheumatic fever. *Journal of Infectious Diseases* **142**, 803–810.

REEVES, P. 1979 The concept of bacteriocins. *Zentralblatt für Bakteriologie und Hygiene, Abt. I, Orig. A* **244**, 78–89.

REYES, A. L., CRAWFORD, R. G., WEHBY, A. J., PEELER, J. T., WIMSATT, J. C., CAMPBELL, J. E. & TWEDT, R. M. 1981 Heat resistance of *Bacillus* spores at various relative humidities. *Applied and Environmental Microbiology* **42**, 692–697.

RIDGWAY, H. F. & OLSON, B. H. 1981 Scanning electron microscope evidence for bacterial colonization of a drinking-water distribution system. *Applied and Environmental Microbiology* **41**, 274–287.

RIGBY, C. E. & PETTIT, J. R. 1980 Delayed secondary enrichment for the isolation of salmonellae from broiler chickens and their environment. *Applied and Environmental Microbiology* **40**, 783–786.

ROBERTS, B. F. 1980 Food hygiene-quantifying the risks. *Environmental Health* **88**, 243–246.

ROBERTS, D. 1972 Observations on procedures for thawing and spit-roasting frozen dressed chickens and post-cooking care and storage: with particular reference to food-poisoning bacteria. *Journal of Hygiene* **70**, 565–588.

ROBERTS, D., BOAG, K., HALL, M. L. M. & SHIPP, C. R. 1975 The isolation of salmonellas from British pork sausages and sausage meat. *Journal of Hygiene* **75**, 173–184.

ROBERTS, D., WATSON, G. N. & GILBERT, R. J. 1982 Contamination of food plants and plant products with bacteria of public health significance. In *Bacteria and Plants* ed. Rhodes-Roberts, M. E. & Skinner, F. A. pp. 169–195. London & New York: Academic Press.

ROBERTS, T. A. 1980 Contamination of meat. The effects of slaughter practices on the bacteriology of the red meat carcass. *Royal Society of Health Journal* **100**, 3–9.

ROBERTS T. A. & INGRAM, M. 1965 Radiation resistance of spores of *Clostridium* species in aqueous suspension. *Journal of Food Science* **30**, 879–885.

ROBERTS, T. A. & SMART, J. L. 1976 The occurrence and growth of *Clostridium* spp. in vacuum-packed bacon with particular reference to *Cl. perfringens* (*welchii*) and *Cl. botulinum*. *Journal of Food Technology* **11**, 229–244.

ROBERTS, T. A., JARVIS, B. J. & RHODES, A. C. 1976 Inhibition of *Clostridium botulinum* by curing salts in pasteurized pork slurry. *Journal of Food Technology* **11**, 25–40.

ROBERTS, T. A., GIBSON, A. M. & ROBINSON, A. 1981 Factors controlling the growth of *Clostridium botulinum* types A and B in pasteurized, cured meats. *Journal of Food Technology* **16**, 239–281.

ROBINSON, D. A. 1981 Infective dose of *Campylobacter jejuni* in milk. *British Medical Journal* **282**, 1584.

ROSEN, A. A., SAFFERMAN, R. S., MASHNI, C. I. & ROMANO, A. H. 1968 Identity of odorous substance produced by *Streptomyces griseoluteus*. *Applied Microbiology* **16**, 178–179.

ROTH, L. A. & CLARK, D. S. 1975 Effect of lactobacilli and carbon dioxide on the growth of *Microbacterium thermosphactum* on fresh beef. *Canadian Journal of Microbiology* **21**, 629–632.

ROWE, B., HALL, M. L. M., WARD, L. R. & DE SA, J. D. H. 1980 Epidemic spread of *Salmonella hadar* in England and Wales. *British Medical Journal* **280**, 1065–1066.

SAHL, H. G. & BRANDIS, H. 1981 Production, purification and chemical properties of an antistaphylococcal agent produced by *Staphylococcus epidermidis*. *Journal of General Microbiology* **127**, 377–384.

SALIT, I. E. & MORTON, G. 1981 Adherence of *Neisseria meningitidis* to human epithelial cells. *Infection and Immunity* **31**, 430–435.

SAMISH, Z., ETINGER-TULCZYNSKA, R. & BICK, M. 1963 The microflora within the tissue of fruits and vegetables. *Journal of Food Science* **28**, 259–266.

SAND, F. E. M. J. & VAN GRINSVEN, A. M. 1976 Comparison between the yeast flora of Middle Eastern and Western European soft drinks. *Antonie van Leeuwenhoek* **442**, 523–532.

SCHAEFFER, A. J., JONES, J. M. & DUNN, J. K. 1981 Association of *in vitro Escherichia coli* adherence to vaginal and buccal epithelial cells with susceptibility of women to recurrent urinary-tract infections. *New England Journal of Medicine* **304**, 1062–1066.

SCHELHORN, M. VON 1953 Hemmende und abtötende Wirkung von Konservierungsmitteln. *Archiv für Mikrobiologie* **19**, 30–44.

SCHIEMANN, D. A. 1980 Isolation of toxigenic *Yersinia enterocolitica* from retail pork products. *Journal of Food Protection* **43**, 360–365, 369.

SCHIEMANN, D. A., DEVENISH, J. A. & TOMA, S. 1981 Characteristics of virulence in human isolates of *Yersinia enterocolitica*. *Infection and Immunity* **32**, 400–403.

SCHWALLER, P. & SCHMIDT-LORENZ, W. 1980 Flore microbienne de quatre eaux minérales non gazéifiées et mises en bouteilles. I. Dénombrement de colonies, composition grossière de la flore et caractères du groupe des bactéries Gram négatif pigmentées en jaune. *Zentralblatt für Bakteriologie und Hygiene, I, Abteilung C* **1**, 330–347.

SCHOTHORST, M. VAN, NORTHOLT, M. D., KAMPELMACHER, E. H. & NOTERMANS, S. 1976 Studies on the estimation of the hygienic condition of frozen broiler chickens. *Journal of Hygiene* **76**, 57–63.

SCHOTHORST, M. VAN, VAN LEUSDEN, F. M., GHOSH, A. C., HOFSTEE, M. P. M., PRICE, T. H., SIMON, I., GILBERT, R. J., HARVEY, R. W. S., PIETZSCH, O. & KAMPELMACHER, E. H. 1980 Laboratory induced variations in a standardized *Salmonella* isolation method. *Zentralblatt für Bakteriologie und Hygiene, Abt. I, Orig, B* **171**, 224–230.

SCOTLAND, S. M., DAY, N. P., CRAVIOTO, A., THOMAS, L. V. & ROWE, B. 1981 Production of heat-labile or heat-stable enterotoxins by strains of *Escherichia coli* belonging to serogroups 044, 0114 and 0128. *Infection and Immunity* **31**, 500–503.

SEILER, D. A. L. 1980 Yeast spoilage of bakery products. In *Biology and Activities of Yeasts* ed. Skinner, F. A., Passmore, S. M. & Davenport, R. P. pp. 135–152. London & New York: Academic Press.

SELIGMAN, M. E. P. 1971 Phobias and preparedness. *Behaviour Therapy* **2**, 307–320.

SHARP, J. C. M., PATERSON, G. M. & FORBES, G. I. 1980 Milk-borne salmonellosis in Scotland. *Journal of Infection* **2**, 333–340.

SHARPE, A. N. 1980 *Food Microbiology. A Framework for the Future.* Springfield, Ill.: Thomas.

SHEWAN, J. M. & MURRAY, C. K. 1979 The microbial spoilage of fish with special reference to the role of psychrophiles. In *Cold Tolerant Microbes in Spoilage and the Environment* ed. Russell, A. D. & Fuller, R. pp. 117–136. London & New York: Academic Press.

SILLIKER, J. H. 1973 Environmental testing best way to control *Salmonella*. *Food Processing* **34**, 38.

SILLIKER, J. H. 1980 Status of *Salmonella*—ten years later. *Journal of Food Protection* **43**, 307–313.

SIMMONS, N. A. & GIBBS, F. J. 1979 *Campylobacter* spp. in oven-ready poultry. *Journal of Infection* **1**, 159–162.

SINELL, H. J. & LEVETZOW, R. 1967 Sulfitreduzierende Clostridien in vorverpackten Fleischwaren. *Fleischwirtschaft* **47**, 392–396.

SKOVGAARD, N. 1979 Bacterial association of and metabolic activity in fish in North-Western Europe. *Archiv für Lebensmittelhygiene* **30**, 106–109.

SLOTS, J. & GIBBONS, R. J. 1978 Attachment of *Bacteroides melaninogenicus*, subsp. *asaccharolyticus* to oral surfaces and its possible role in colonization of the mouth and of periodontal pockets. *Infection and Immunity* **19**, 254–264.

SMELTZER, T., THOMAS, R. & COLLINS, G. 1980 The role of equipment having accidental or indirect contact with the carcass in the spread of *Salmonella* in an abattoir. *Australian Veterinary Journal* **56**, 14–17.

SMITH, L. D. 1975 Inhibition of *Clostridium botulinum* by strains of *Clostridium perfringens* isolated from soil. *Applied Microbiology* **30**, 319–323.

SMITH, M. G. & GRAHAM, A. 1978 Destruction of *Escherichia coli* and salmonellae on mutton carcasses by treatment with hot water. *Meat Science* **2**, 119–128.

SMITHER, R. 1978 Bacterial inhibitors formed during the adventitious growth of microorganisms in chicken liver and pig kidney. *Journal of Applied Bacteriology* **45**, 267–277.

SNIJDERS, J. M. A. & GERATS, G. E. 1977 Hygiene bei der Schlachtung von Schweinen. VI. Die Verwendung eines Infrarottunnels in der Schlachtstrasse. *Fleischwirtschaft* **57**, 2216–2219.

SNIJDERS, J. M. A., SCHOENMAKERS, M. J. G., GERATS, G. E. & DE PIJPER, F. W. 1979 Dekontamination schlachtwarmer Rinderkörper mit organischen Säuren. *Fleischwirtschaft* **59**, 656–663.

SOBEL, J. D., MYERS, P. G., KAYE, D. & LEVISON, M. E. 1981 Adherence of *Candida albicans* to human vaginal and buccal epithelial cells. *Journal of Infectious Diseases* **143**, 76–82.

SON, N. T. & FLEET, G. H. 1980 Behavior of pathogenic bacteria in the oyster, *Crassostrea commercialis*, during depuration, re-laying and storage. *Applied and Environmental Microbiology* **40**, 994–1002.

SORRELLS, K. M. & SPECK, M. L. 1970 Inhibition of *Salmonella gallinarum* by culture filtrates of *Leuconostoc citrovorum*. *Journal of Dairy Science* **53**, 239–241.

SPECK, M. L., RAY, B. & READ, R. B. 1975 Repair and enumeration of injured coliforms by a plating procedure. *Applied Microbiology* **29**, 549–550.

SPICHER, G. 1980 Die Faktoren des Wachstums der Schimmelpilze als Ansatzpunkte für Massnahmen zur Unterbindung der Schimmelbildung bei Backwaren. *Getreide, Mehl und Brot* **34**, 128–137.

SPLITTSTOESSER, D. F. 1970 Predominant micro-organisms on raw plant foods. *Journal of Milk and Food Technology* **33**, 500–505.

SPLITTSTOESSER, D. F. 1973 The microbiology of frozen vegetables. *Food Technology* **27**, 54–56, 60.

SPLITTSTOESSER, D. F. & GADJO, I. 1966 The groups of micro-organisms composing the "total"-count population in frozen vegetables. *Journal of Food Science* **31**, 234–249.

SPLITTSTOESSER, D. F. & SEGEN, B. 1970 Examination of frozen vegetables for salmonellae. *Journal of Milk and Food Technology* **33**, 111–113.

SPLITTSTOESSER, D. F., WEXLER, M., WHITE, J. & COLWELL, R. R. 1967 Numerical Taxonomy of Gram-positive and catalase-positive rods isolated from frozen vegetables. *Applied Microbiology* **15**, 158–162.

SPREEKENS, K. J. A. VAN 1977 Characterization of some fish and shrimp spoiling bacteria. *Antonie van Leeuwenhoek* **43**, 283–303.

STADHOUDERS, J. 1975 Microbes in milk and dairy products. An ecological approach. *Netherlands Milk and Dairy Journal* **29**, 104–126.

STADHOUDERS, J., CORDES, M. M. & VAN SCHOUWENBURG-VAN FOEKEN, A. W. J. 1978 The effect of manufacturing conditions on the development of staphylococci in cheese. Their inhibition by starter bacteria. *Netherlands Milk and Dairy Journal* **32**, 192–203.

STANLEY, G., SHAW, K. J. & EGAN, A. F. 1981 Volatile compounds associated with spoilage of vacuum-packaged sliced luncheon meat by *Brochothrix thermosphacta*. *Applied and Environmental Microbiology* **41**, 816–818.

STEELE, J. E. & STILES, M. E. 1981 Food poisoning potential of artificially contaminated vacuum-packaged sliced ham in sandwiches. *Journal of Food Protection* **44**, 430–434, 444.

STERN, N. J. 1981a Isolation of potentially virulent *Yersinia enterocolitica* from a variety of meats. *Journal of Food Science* **46**, 41–42, 51.

STERN, N. J. 1981*b* Recovery rate of *Campylobacter fetus* ssp. *jejuni* on eviscerated pork, lamb and beef carcasses. *Journal of Food Science* **46**, 1291, 1293.

STERN, N. J., PIERSON, M. D. & KOTULA, A. W. 1980 Growth and competitive nature of *Yersinia enterocolitica* in whole milk. *Journal of Food Science* **45**, 972–974.

STILES, M. E. & NG, L. K. 1979 Fate of enteropathogens inoculated into chopped ham. *Journal of Food Protection* **42**, 624–630.

STILES, M. E. & NG, L. 1981*a* Biochemical characteristics and identification of Enterobacteriaceae isolated from meats. *Applied and Environmental Microbiology* **41**, 639–645.

STILES, M. E. & NG, L. 1981*b* Enterobacteriaceae associated with meats and meat handling. *Applied and Environmental Microbiology* **41**, 867–872.

STILES, M. E., NG, L. K. & PARADIS, D. C. 1979 Survival of enteropathogenic bacteria on artificially contaminated bologna. *Canadian Institute of Food Science and Technology Journal* **12**, 128–130.

SUGIYAMA, H., WOODBURN, M., YANG, K. H. & MOVROYDIS, C. 1981 Production of botulinum toxin in inoculated pack studies of foil-wrapped baked potatoes. *Journal of Food Protection* **44**, 896–898, 902.

SULLIVAN, R., FASSOLITIS, A. C., LARKIN, E. P., READ, R. B. & PEELER, J. T. 1971 Inactivation of thirty viruses by gamma radiation. *Applied Microbiology* **22**, 61–65.

SUMNER, J. L., SAMARAWEERA, I., JAYAWEERA, V. & FONSEKA, G. 1982 A survey of process hygiene in the Sri Lanka prawn industry. *Journal of the Science of Food and Agriculture* **33**, 802–808.

SURESH, E. R., ONKARAYYA, H. & ETHIRAJ, S. 1982 A note on the yeast flora associated with fermentation of mango. *Journal of Applied Bacteriology* **52**, 1–4.

SVEDHEM, A. & KAIJSER, B. 1981 Isolation of *Campylobacter jejuni* from domestic animals and pets: probable origin of human infection. *Journal of Infection* **3**, 37–40.

SVEUM, W. H. & KRAFT, A. A. 1981 Recovery of salmonellae from foods using a combined enrichment technique. *Journal of Food Science* **46**, 94–99.

SWAMINATHAN, B., HOWE, J. M. & ESSLING, C. M. 1981 Mayonnaise, sandwiches and *Salmonella*. *Journal of Food Protection* **44**, 115–117, 127.

TABATABAI, L. B. & WALKER, H. W. 1974 Inhibition of *Clostridium perfringens* by *Streptococcus faecalis* in a medium containing curing salts. *Journal of Milk and Food Technology* **37**, 387–391.

TAMMINGA, S. K., BEUMER, R. R., KAMPELMACHER, E. H. & VAN LEUSDEN, F. M. 1976 Survival of *Salmonella eastbourne* and *Salmonella typhimurium* in chocolate. *Journal of Hygiene, Cambridge* **76**, 41–47.

TAMMINGA, S. K., BEUMER, R. R., KAMPELMACHER, E. H. & VAN LEUSDEN, F. M. 1977 Survival of *Salmonella eastbourne* and *Salmonella typhimurium* in milk chocolate prepared with artificially contaminated milk powder. *Journal of Hygiene, Cambridge* **79**, 333–337.

TAMMINGA, S. K., BEUMER, R. R. & KAMPELMACHER, E. H. 1978 The hygienic quality of vegetables grown in or imported into the Netherlands: a quantitative survey. *Journal of Hygiene, Cambridge* **80**, 143–154.

TAMMINGA, S. K., BEUMER, R. R. & KAMPELMACHER, E. H. 1982 Microbiological studies on hamburgers. *Journal of Hygiene, Cambridge* **88**, 125–142.

TANAKA, N., GOEPFERT, J. M., TRAISMAN, E. & HOFFBECK, W. M. 1979 A challenge of pasteurized process cheese spread with *Clostridium botulinum* spores. *Journal of Food Protection* **42**, 787–789.

TERVET, I. W. & HOLLIS, J. P. 1948 Bacteria in the storage organs of healthy plants. *Phytopathology* **38**, 960–967.

TESONE, S., HUGHES, A. & HURST, A. 1981 Salt extends the upper temperature limit for growth of food-poisoning bacteria. *Canadian Journal of Microbiology* **27**, 970–972.

THIEL, P. G., MEYER, C. J. & MARASAS, W. F. O. 1982 Natural occurrence of moniliformin together with deoxynivalenol and zearalenone in Transkeian corn. *Journal of Agricultural and Food Chemistry* **30**, 308–312.

THOMAS, C. J. & MCMEEKIN, T. A. 1980 Contamination of broiler carcass skin during commercial processing procedures: an electron microscopic study. *Applied and Environmental Microbiology* **40**, 133–144.

THOMAS, C. J. & MCMEEKIN, T. A. 1981a Spoilage of chicken skin at 2°C: electron microscopic study. *Applied and Environmental Microbiology* **41**, 492–503.

THOMAS, C. J. & MCMEEKIN, T. A. 1981b Attachment of *Salmonella* spp. to chicken muscle surfaces. *Applied and Environmental Microbiology* **42**, 130–134.

THOMAS, C. J. & MCMEEKIN, T. A. 1981c Production of off odours by isolates from poultry skin with particular reference to volatile sulphides. *Journal of Applied Bacteriology* **51**, 529–534.

THOMAS, W. D. & GRAHAM, R. W. 1952 Bacteria in apparently healthy pinto beans. *Phytopathology* **42**, 214.

THOMASON, B. M. & DODD, D. J. 1976 Comparison of enrichment procedures for fluorescent antibody and cultural detection of salmonellae in raw meat and poultry. *Applied and Environmental Microbiology* **31**, 787–788.

THOMPSON, D. R., WILLARDSEN, R. R., BUSTA, F. F. & ALLEN, C. E. 1979 *Clostridium perfringens* population dynamics during constant and rising temperatures in beef. *Journal of Food Science* **44**, 646–651.

THOMPSON, S. S., HARMON, L. G. & STINE, C. M. 1978 Survival of selected organisms during spray drying of skim milk and storage of non-fat dry milk. *Journal of Food Protection* **41**, 16–19.

THORNE, S. N. 1972 studies of the behaviour of stored carrots with respect of their invasion of *Rhizopus stolonifer* Lind. *Journal of Food Technology* **7**, 139–151.

TOLLE, A. 1979 Bacterial association and metabolic activity in milk and dairy products. *Archiv für Lebensmittelhygiene* **30**, 84–87.

TOLLE, A., OTTE, I. & SUHREN, G. 1981 Zur Dynamik der produktspezifischen Keimflora und von coliformen Keimen/E. coli während des Herstellungsprozesses von Camembertkäse. *Milchwissenschaft* **36**, 5–9.

TRELEAVEN, B. E., DIALLO, A. A. & RENSHAW, E. C. 1980 Spurious hydrogen sulphide production by *Providencia* and *Escherichia coli* species. *Journal of Clinical Microbiology* **11**, 750–752.

TROLLER, J. A. & STINSON, J. V. 1978 Influence of water activity on the production of extracellular enzymes by *Staphylococcus aureus*. *Applied and Environmental Microbiology* **35**, 521–526.

TUYNENBURG MUYS, G. 1971 Microbial safety in emulsions. *Process Biochemistry* **6** (no. 6), 25–28.

VAN DER KOOIJ, D. 1979 Characterization and classification of fluorescent pseudomonads isolated from tap water and surface water. *Antonie van Leeuwenhoek* **45**, 225–240.

VANDERZANT, C. & NICKELSON, R. 1969 A microbiological examination of muscle tissue of beef, pork and lamb. *Journal of Milk and Food Technology* **32**, 357–361.

VANDERZANT, C., HANNA, M. O., EHLERS, J. G., SAVELL, J. W., SMITH, G. C., GRIFFIN, D. B., TERRELL, R. N., LIND, K. D. & GALLOWAY, D. E. 1982. Centralized packaging of beef loin steaks with different oxygen-barrier films: microbiological characteristics. *Journal of Food Science* **47**, 1070–1079.

VAN NETTEN, P. & MOSSEL, D. A. A. 1980 The ecological consequences of decontaminating raw meat surfaces with lactic acid. *Archiv für Lebensmittelhygiene* **31**, 190–191.

VAN NETTEN, P. & MOSSEL, D. A. A. 1981 Ecological studies on lactic acid decontamination of fresh meat surfaces. *Abstracts of the Annual Meeting of the American Society of Microbiology*, p. 200.

VASSILIADIS, P., KALAPOTHAKI, V., TRICHOPOULOS, D., MAVROMMATTI, C. & SERIE, C. 1981a Improved isolation of salmonellae from naturally contaminated meat products by using Rappaport–Vassiliadis enrichment broth. *Applied and Environmental Microbiology* **42**, 615–618.

VASSILIADIS, P., TRICHOPOULOS, D., KALAPOTHAKI, V. & SERIE, C. 1981b Isolation of *Salmonella* with the use of 100 ml of the R10 modification of Rappaport's enrichment medium. *Journal of Hygiene, Cambridge* **87**, 35–41.

VEILLEUX, B. G. & ROWLAND, I. 1981 Simulation of the rat intestinal ecosystem using a two-stage continuous culture system. *Journal of General Microbiology* **123**, 103–115.

VENKATARAMAIAH, G. N. & KEMPTON, A. G. 1981 Gas chromatographic detection of pre-processing spoilage of bacterial origin in pet foods. *Journal of Food Protection* **44**, 750–755.

VERRIPS, C. T. & KWAST, R. H. 1982 Recovery of heat-injured *Citrobacter freundii* cells. *Journal of Applied Bacteriology* **52**, 15–20.

VESIKARI, T., NURMI, T., MAEKI, M., SKURNIK, M., SUNDQVIST, C., GRANFORS, K. & GROENROOS, P. 1981 Plasmids in *Yersinia enterocolitica* serotypes 0 : 3 and 0 : 9: correlation with epithelial cell adherence *in vitro*. *Infection and Immunity* **33**, 870–876.

WALLACE, G. O. & PARK, S. E. 1933 Microbiology of frozen foods. IV. Longevity of certain pathogenic bacteria in frozen cherries and frozen cherry juice. *Journal of Infectious Diseases* **52**, 146–149.

WEISSMAN, M. A. & CARPENTER, J. A. 1969 Incidence of salmonellae in meat and meat products. *Applied Microbiology* **17**, 899–902.

WERNER, S. B. *et al.* 1974 National outbreak of *Salmonella eastbourne* infection related to contaminated chocolate. *Salmonella Surveillance*, Rep. No. 120, 5–7.

WESTERDIJK, J. 1949 The concept "association" in mycology. *Antonie van Leeuwenhoek* **53**, 187–189.

WESTHOFF, D. C. 1978 Heating milk for microbial destruction: a historical outline and update. *Journal of Food Protection* **41**, 122–130.

WILLIAMS, G. V. & DEACON, G. J. 1980 *Campylobacter*: common cause of enteritis in an infectious diseases hospital. *Medical Journal of Australia* **2**, 268–271.

WILSON, G. D. (ed.) 1973 *The Psychology of Conservatism*. London & New York: Academic Press.

WILSON, G. S. 1933 The necessity for a safe milk-supply. *Lancet* ii, 829–832.

WILSON, G. S. 1955 Symposium on food microbiology and public health: general conclusion. *Journal of Applied Bacteriology* **18**, 629–630.

WINTER, A. R., WEISER, H. H. & LEWIS, M. 1953 The control of bacteria in chicken salad. II. *Salmonella*. *Applied Microbiology* **1**, 278–281.

WRIGHT, C., KOMINOS, S. D. & YEE, R. B. 1976 Enterobacteriaceae and *Pseudomonas aeruginosa* recovered from vegetable salads. *Applied and Environmental Microbiology* **31**, 453–454.

WYATT, C. J. & GUY, V. H. 1981*a* Growth of *Salmonella typhimurium* and *Staphylococcus aureus* in retail pumpkin pies. *Journal of Food Protection* **44**, 418–421.

WYATT, C. J. & GUY, V. H. 1981*b* Incidence and growth of *Bacillus cereus* in retail pumpkin pies. *Journal of Food Protection* **44**, 422–424, 429.

WYLLIE, T. D. & MOREHOUSE, L. G. (eds) 1978 *Mycotoxic Fungi, Mycotoxins, Mycotoxicoses. An Encyclopedic Handbook* Vols I–III. New York: Marcel Dekker.

XIROUCHAKI, E., VASSILIADIS, P., TRICHOPOULOS, D. & MAVROMMATI, C. 1982 A note on the performance of Rappaport's medium, compared with Rappaport–Vassiliadis broth, in the isolation of salmonellas from meat products, after preenrichment. *Journal of Applied Bacteriology* **52**, 125–127.

YAMAZAKI, Y., EBISU, S. & OKADA, H. 1981 *Eikenella corrodens* adherence to human buccal epithelial cells. *Infection and Immunity* **31**, 21–27.

YAWGER, E. S. 1978 Bacteriological evaluation for thermal process design. *Food Technology* **32**, no. 6, 59–62.

YEOH, H. T., BUNGAY, H. R. & KRIEG, N. R. 1968 A microbial interaction involving combined mutualism and inhibition. *Canadian Journal of Microbiology* **14**, 491–492.

YILDIZ, F. & WESTHOFF, D. 1981 Associative growth of lactic acid bacteria in cabbage juice. *Journal of Food Science* **46**, 962–963.

Food-borne Infections and Intoxications—Recent Trends and Prospects for the Future

R. J. GILBERT

Food Hygiene Laboratory, Central Public Health Laboratory, London, UK

Contents

1. Introduction

UNTIL THE mid nineteenth century, the food laws of the United Kingdom were largely concerned with protecting the purchaser against short weight in bread and various other commodities, and the fraudulent adulteration or dilution of expensive foodstuffs and beverages with cheaper substitutes. Interestingly, it was an outbreak of poisoning which brought the debate on the 1855 Report of the Select Committee on Food Adulteration to an abrupt halt and precipitated the first Adulteration of Food and Drink Act in 1860. Twenty-one people died and several hundred were taken ill after eating peppermint lozenges purchased at a market stall (Turner 1980). The maker of the lozenges had been supplied with white arsenic in mistake for the plaster of Paris with which he normally adulterated the product! Between 1880 and 1900 the link between bacteria such as *Salmonella* and *Clostridium botulinum* and food poisoning became apparent. Today, more

FOOD MICROBIOLOGY
ISBN 0 12 589670 0

than 80 years later, new agents of food-borne illness are still being described. In the last decade both the rise in the incidence of food poisoning and accounts of individual outbreaks attracted wide publicity in an increasingly consumer conscious nation. This led to increased demand for a general improvement in standards of food hygiene to reduce the toll of preventable illness.

Much has been written about food poisoning in this country and a few selected references are given in Table 1. The purpose of this review is to discuss the recent trends and present day problems and to consider the prospects for the future.

TABLE 1

Selected references since 1974 on bacterial food poisoning and salmonella infection in the UK

Agent	Reference
Bacillus cereus	Mortimer & McCann (1974);
	Kramer *et al.* (1982)
B. licheniformis and *B. subtilis*	Gilbert *et al.* (1981)
Clostridium botulinum	Ball *et al.* (1979); Gilbert & Willis (1980)
Cl. perfringens	Hobbs (1979); Stringer *et al.* (1980, 1982)
Salmonella spp.	McCoy (1975); Harbour *et al.* (1977);
	Turnbull (1979); Rowe *et al.* (1980);
	Sharp *et al.* (1980)
Staphylococcus aureus	Gilbert (1974); de Saxe *et al.* (1982)
Vibrio parahaemolyticus	Barrow (1974); Hooper *et al.* (1974)
General information	
Statistics	
England and Wales	Vernon (1977); Hepner (1980);
	Anon. (1980, 1981*a*)
Scotland	Sharp & Collier (1981)
Investigation of outbreaks	Elias-Jones *et al.* (1980);
	Pinegar & Suffield (1982)

2. Food Poisoning and Salmonella Infection, 1980

The three sources of information on food poisoning and salmonellosis in England and Wales are (a) statutory notifications of clinical food poisoning to medical officers for environmental health, published by the Office of Population Censuses and Surveys; (b) annual reports of outbreaks and sporadic cases to the Department of Health and Social Security; and (c) voluntary reporting to the Communicable Disease Surveillance Centre

(CDSC) of the Public Health Laboratory Service of outbreaks and isolations of food poisoning organisms by public health (PHLS) and hospital laboratories. These sources do not provide information on all the food poisoning which occurs. Some cases will be reported by all three systems but at present these cannot be identified and the degree of overlap between the reporting sources cannot be estimated.

All sources of data show similar trends in the 1970s with an annual increase since the low point recorded in 1972 (Anon. 1980). Reports from public health and hospital laboratories show that the increase in numbers was due to an increase in sporadic cases. Indeed, while the number of general outbreaks has remained fairly stable, family outbreaks have decreased by about 30% since 1973. Almost all of the sporadic cases and most of the family outbreaks were caused by salmonellas.

In 1980 laboratories reported 10,856 cases of bacterial food poisoning and salmonella infection, 9% fewer than in 1979 but a level similar to that for 1978 (Anon. 1981*a*). There were 221 general outbreaks, 297 family outbreaks and 6971 sporadic cases (Table 2). Reports of cases of infection with *Salm. typhimurium* increased in 1980 but those due to other serotypes fell, so that for all salmonellas there was a 4% fall from the record level in 1979. The number of reported cases of *Cl. perfringens* and *Staphylococcus aureus* food poisoning also fell in 1980.

TABLE 2

*Bacterial food poisoning and salmonella infection in England and Wales, 1980; laboratory reports**

Organism	General outbreaks	Family outbreaks	Sporadic cases	All cases†‡
Salmonella typhimurium	45	102	2454	3161
Other *Salmonella* spp.	104	188	4502	6379
Clostridium perfringens	53	2	2	1056
Staphylococcus aureus	7	4	6	189
Bacillus cereus	12	1	—	64
Other *Bacillus* spp.	—	—	7	7
TOTAL	221	297	6971	10,856

* Anon. (1981*a*).
† Excluding symptomless excreters of salmonellas.
‡ In some outbreaks the total number was not known.

Salmonella outbreaks outside the home occurred most commonly in restaurants and hotels and at receptions. Twenty outbreaks occurred in hospitals but in only three was there good evidence that food poisoning was the initial mode of infection; the remainder were probably incidents of cross-infection. As in previous years the most common locations of *Cl. perfringens* outbreaks were hospitals, followed by restaurants and hotels.

In January 1980 a new food poisoning and salmonellosis reporting system was introduced by which medical officers for environmental health and environmental health officers were encouraged to report outbreaks and cases to the CDSC as part of a general rationalization of food poisoning statistics. The reporting forms were specifically designed to fit the requirements of the WHO European Region food-borne surveillance programme and will be used for the contribution from England and Wales to this programme. The new report forms have been particularly helpful in describing outbreaks of gastro-enteritis thought to be food-borne but where no bacterial cause was found and in documenting factors contributing to outbreaks in more detail and more consistently than has been possible in laboratory reports (Anon. 1981*a*).

In 1980 the outbreaks reported by the 'new' system (357) and by PHLS and hospital laboratories (570) were, in general, different incidents and therefore the two reporting systems appear to be complementary. Only 64 outbreaks were reported by both systems. Thus 863 separate outbreaks were ascertained in 1980 which is 51% more than laboratories alone reported (Anon. 1981*b*).

Four aspects of food poisoning and salmonellosis in recent years warrant special attention: (a) the high contamination rate of raw meat and poultry with salmonellas (Table 3) and the increase in the number of outbreaks incriminating these foods, particularly cooked chicken and turkey; (b) the large number of outbreaks following the consumption of raw milk (Table 4); (c) the epidemic spread of *Salm. hadar* to become the second most common salmonella serotype isolated from patients (Rowe *et al.* 1980); and (d) the occurrence of several new and some less common types of bacterial and non-bacterial food-borne infections and intoxications (see Section 3).

TABLE 3

Incidence of Salmonella *spp. in meat and poultry examined in the Food Hygiene Laboratory*

Food	Period	Number of samples	Positive No.	%	Number of serotypes
Minced beef or pork	1977–78	279	21	7·5	13
Frozen chicken	1979–80	100	79	79·0	18
Pork sausages	1981	208	25	12·0	14

TABLE 4

Infections from unpasteurized milk in the UK

Organism	Period	Number of outbreaks	Number of cases
Salmonella spp.	1970–80	>100	>3500
Campylobacter spp.	1977–80	>20	>4000

3. New or Less Common Food-borne Infections and Intoxications

A. Bacillus cereus *food poisoning*

In recent years *Bacillus cereus* has become well established as an agent of food poisoning. The organism causes two distinct forms of illness characterized either by nausea and vomiting or by diarrhoea and abdominal pain. The first type occurs 1–5 h and the second 8–16 h after ingestion of contaminated food. Incidents reported in this country have almost all been of the vomiting type and associated with the consumption of fried or boiled rice from Chinese restaurants or 'take-away' shops. Since 1971 more than 120 such episodes have been reported and similar incidents have been described in several countries throughout the world (Gilbert 1979).

The spores of *B. cereus*, which is a natural contaminant of raw rice and other cereals, can survive normal cooking procedures (Gilbert *et al.* 1974). Vegetative cell growth is rapid in cooked rice stored at room temperature and is enhanced by the addition of beef, chicken or egg (Morita & Woodburn 1977). Heat resistance studies on spores of *B. cereus* from food poisoning outbreaks and from raw rice have shown that spores from serotype 1 strains are markedly more heat resistant than other serotypes (Parry & Gilbert 1980). This would account for the preponderance of serotype 1 in outbreaks in this and other countries.

Results from various studies in the last decade indicate that the activities of at least two enterotoxins are responsible for the distinct clinical forms of *B. cereus* food poisoning. Evidence that a toxin was responsible for the 'emetic-syndrome' form of illness was first presented by Melling *et al.* (1976) after monkey feeding tests. Preliminary characterization of this toxin has shown it to be a highly stable, small molecular weight (<10,000) compound able to withstand exposure to 126°C for 90 min without appreciable loss of activity (Melling & Capel 1978). This heat stability correlates well with the epidemiological evidence that contaminated cooked rice which has been stored and reheated produces food poisoning in man. The various toxins produced by *B. cereus* have been reviewed extensively by Turnbull (1981). A number of reports, reviewed recently by Gilbert *et al.* (1981) and Kramer *et al.* (1982), have implicated *B. subtilis*, *B. licheniformis* and *B. brevis* as potential food poisoning agents. On many occasions the epidemiological evidence for incriminating these organisms as 'causative' was inadequate. However, when taken together, the pattern of their repeated occurrence and their isolation in large numbers, sometimes in almost pure culture, in the implicated foods suggests a significant involvement.

B. Vibrio parahaemolyticus *food poisoning*

Illness associated with *Vibrio parahaemolyticus* was first recorded in the UK in 1972 following an outbreak affecting air travellers who ate *hors d'oeuvres* containing dressed crab prepared in Bangkok; they became ill on arrival in London (Peffers *et al.* 1973). Between 1972 and 1980 some 200 cases were reported including three outbreaks affecting 19, 24 and 40 persons who became ill after eating prawn cocktails made with imported cooked prawns thawed and served without further cooking. In another outbreak at least 12 holidaymakers suffered acute gastro-enteritis 16–18 h after eating dressed crab from the same fishmonger at a seaside resort; the crab was of British origin. Faecal specimens from six patients yielded Kanagawa-positive strains of *V. parahaemolyticus* (Hooper *et al.* 1974). In most of the sporadic cases reported the patients were infected abroad, usually in the Far East or Africa, and in some there was a definite history of consumption of seafood.

Vibrio parahaemolyticus is a marine bacterium recognized as a cause of food poisoning throughout the world, but especially in Japan and Thailand. The organism has been isolated on numerous occasions from Malaysian frozen cooked prawns imported into the UK (Turnbull & Gilbert 1982) and from raw shellfish and the coastal waters around the UK (Barrow & Miller 1972; Ayres & Barrow 1978).

C. Botulism

Botulism in man is rare in Britain and prior to the 1978 canned salmon incident (Ball *et al.* 1979) there had been no reports of this illness for more than 20 years. During the summer of 1978 four persons were admitted to hospital in Birmingham suffering from the severe effects of botulism; two of the patients died. Neurological symptoms developed about 12 h after eating salmon from a freshly opened can originating from Alaska. Laboratory tests confirmed the presence of *Cl. botulinum* type E toxin in the sera of all four patients and in remnants of the fish. The organism itself was isolated from the remnants and, rather remarkably, from the wall can-opener in the kitchen of the household affected. The available evidence indicated that contamination of the salmon occurred during processing, most probably by entry of *Cl. botulinum* type E spores through a defect in the can during the cooling period after heat treatment. The defect was a small hole with a surrounding area of damage, but the precise cause of the defect was never ascertained. The strain of *Cl. botulinum* incriminated was non-proteolytic and, therefore, the salmon was not rendered unpalatable.

It is of interest to note that in both the outbreak described above and in the previous outbreak in 1955 (MacKay-Scollay 1958) the foods involved were imported preserved or canned fish.

D. *Campylobacter enteritis*

During the last five years the number of isolations from man of the 'thermophilic' campylobacters (*Campylobacter jejuni* and *C. coli*) reported in England and Wales has risen steadily from 1349 in 1977 to 9506 in 1980. Indeed, the figure of 12,496 isolates for 1981 indicates that campylobacters are now the most common faecal pathogens to be isolated from patients with acute diarrhoea. Infections are more prevalent in the warmer months and this peak has occurred consistently each year.

The sources of the organism in cases of sporadic infection are still unknown and they are likely to remain so until methods of strain identification are better developed (Robinson *et al.* 1980). However, outbreaks afford unique opportunities for tracing sources and, as a result of several particularly thorough investigations, it is clear that unpasteurized milk is one vehicle of infection. Milk-borne incidents have been reviewed by Robinson & Jones (1981), and detailed accounts of two large outbreaks affecting 616 persons in a rural area of Scotland (Wallace 1980) and about 2500 school children in two towns in England (Jones *et al.* 1981) illustrate the importance of the illness. In 1978 the proportion of unpasteurized milk sold in England and Wales was 3·3% (217 million litres) while in Scotland the proportion was 10% or 64 million litres (Sharp *et al.* 1980). The continuing sale of unpasteurized milk is unacceptable on health grounds and outbreaks of salmonellosis and campylobacter enteritis (Table 4) will continue until all milk sold for human consumption is heat-treated.

Although campylobacters can survive in milk and other foods there is no evidence that they can multiply to large numbers. Indeed, apart from milk, food has not definitely been implicated as a source of campylobacter enteritis; to date, only two presumptive food-borne outbreaks have been described in this country (Hayek & Cruickshank 1977; Skirrow *et al.* 1981), both following the consumption of chicken.

Although human challenge experiments with *C. jejuni* indicate that the infective dose can be as little as 500 organisms (Robinson 1981), no evidence is available from source material in outbreak situations. However, some circumstantial evidence has been obtained, including the occurrence of water-borne outbreaks and the occasional person-to-person transmission, which implies that the infective dose is small. The apparent absence of multiplication of campylobacters in milk and food would also support this hypothesis.

The organisms associated with campylobacter enteritis are not new, but their sudden and rapid rise from obscurity to notoriety is none the less remarkable. Useful reviews have been given by Skirrow (1977), Butzler & Skirrow (1979) and in the Proceedings of an International Workshop on Campylobacter infections (Newell 1982).

E. Scombrotoxic fish poisoning

Scombrotoxic fish poisoning generally results from eating fish of the families *Scombridae* and *Scomberesocidae*, which include tuna, bonito and mackerel, the flesh of which has become toxic as result of bacterial contamination and growth. The symptoms are essentially those of histamine toxicity. Except for anecdotal cases and an outbreak among mariners at sea (Henderson 1830) the first detailed accounts of scombrotoxic poisoning in the UK were given by Cruickshank & Williams (1978) and Gilbert *et al.* (1980).

Of the 79 reported incidents affecting 276 persons between 1976 and 1980, some 52 (66%) were associated with smoked mackerel and a further six incidents were from canned or soused mackerel or pâté (Turnbull & Gilbert 1982). The increase in incidents coincides with a dramatic increase in the annual landings of mackerel in Britain from 8800 tonnes in 1972 to 320,900 tonnes in 1978—about 90% of the 1978 catch was exported. This increase is in turn directly related to a decrease in the landings of herring from 145,700 tonnes in 1972 to only 14,700 tonnes in 1978, the result of a depletion of herring stock by over-fishing.

Freshly opened canned tuna and bonito imported from a number of countries have also been implicated in outbreaks, but the most interesting development has been the increasing involvement of canned 'non-scombroid' fish such as sardines, pilchards and anchovies (Murray *et al.* 1982).

The illness is the only known form of fish-flesh poisoning that is caused by bacterial spoilage and should therefore be totally preventable. The 'toxin' is remarkably stable and once formed is not destroyed by curing or by the heat used for processing canned fish.

F. Ciguatera poisoning

Episodes of ciguatera poisoning have long been a well-recognized and serious problem in the tropical and subtropical regions of the world. Various fish, particularly the large and older specimens, acquire ciguatera toxin through a food chain in which they eat smaller herbivorous fish which have fed on dinoflagellates which produce the toxin. Symptoms in man usually begin within a few hours of ingesting the fish but can be almost immediate. Nausea, vomiting and diarrhoea may be severe and various neurosensory disturbances, in particular paraesthesia and dyaesthesia (often reported as sensation of burning on contact with cold), are commonly reported. Severe cases terminating in shock, convulsions, muscular paralysis and death have been described.

The illness is seldom encountered in temperate countries but three cases in Britain have been reported recently. The first occurred in a man who purchased two fish, a speckled moray eel and a greater amberjack, in Antigua; after drying and salting, the fish were brought back to London and consumed (Tatnall et al. 1980). The other cases involved a couple who had eaten barracuda fish also in Antigua just prior to flying home, symptoms developing during the flight (Moon 1981).

The importation of potentially ciguatoxic fish is strongly discouraged in this country. However, with ever increasing international air travel and the popularity of specialized restaurants, more cases may be reported in the future.

G. Gastro-enteritis of viral or unknown aetiology

Most outbreaks of gastro-enteritis associated with the consumption of food are investigated initially as outbreaks of bacterial food poisoning. However, on many occasions neither the traditional nor the less common food poisoning bacteria are isolated from clinical specimens or remnants of food. These outbreaks are being reported with increased frequency (Anon. 1979; Appleton et al. 1981), but probably represent only a small proportion of such episodes since laboratories tend to report only positive findings. Many of the outbreaks follow a pattern of a long incubation period of 24–48 h, a high attack rate and symptoms of vomiting, diarrhoea and abdominal pain.

The available evidence now indicates that certain viruses are a cause of at least some of these incidents and between 1976 and 1980 they were incriminated in nine outbreaks in England and Wales; the foods involved were either cockles or oysters (Appleton & Pereira 1977; Appleton *et al*. 1981). The laboratory procedure used depends on the separation of virus particles from faecal specimens by differential centrifugation and their concentration on caesium-chloride gradients. Selected fractions are then negatively stained and examined by electron microscopy. Using this method, small, round, featureless, virus-like particles were observed in 78 of 90 (87%) faecal specimens from these nine outbreaks. Examination of serial specimens from three patients indicated that maximum excretion of virus particles occurred from 4 to 6 days after onset of symptoms and could continue for at least a month after illness. Virus particles similar to those seen in faecal specimens from patients were detected in oysters from one outbreak, but attempts to detect the virus in samples of cockles from other outbreaks were unsuccessful.

Bivalve molluscs are known to concentrate viruses from polluted water and small round viruses have also been detected in oysters in a very large outbreak of non-bacterial gastro-enteritis in Australia (Murphy *et al*. 1979; Eyles *et al*. 1981). It is important to remember that, unlike bacteria, human enteric viruses do not replicate in food and if present they are likely to be there in only small numbers.

H. Hepatitis

Acute viral hepatitis is a common disease with a global distribution but only hepatitis A has importance in relation to food and infection. Outbreaks of food-borne hepatitis are rare although several shellfish-associated incidents have been reported especially in the USA (Bryan 1980; Shear & Gottlieb 1980).

An epidemiological investigation of notifications of hepatitis A infection in Leeds early in 1978 identified 41 cases associated with the consumption of mussels from a common source (Bostock *et al*. 1979). Further investigation revealed that the suspected mussels had been purchased live from the same wholesale fish market and that they were imported. An additional 41 cases were established as a result of a retrospective examination of case histories in other towns and cities receiving mussels from the same suspected source. It was concluded that the amount of heat used in the preparation of the mussels was sufficient to open the shells but was inadequate to cook them thoroughly. Two further outbreaks involving about 20 persons occurred in two areas in 1980, but in these, cooked cockles appeared to be the common epidemiological link.

A notable outbreak following a dinner in London in 1980 and two smaller outbreaks in the London area were thought on epidemiological grounds to be due to contaminated frozen raspberries but this could not be confirmed by laboratory tests.

I. Poisoning from red kidney beans

Since the first episode in 1976 there has been a steady increase in the number of outbreaks of food poisoning reported in the UK following the consumption of raw or incompletely cooked red kidney beans, *Phaseolus vulgaris* (Noah *et al.* 1980). Twenty-six incidents affecting at least 118 persons have been officially recorded although publicity in a television programme with an invitation to the public to report similar experiences resulted in 355 letters—suggesting 355 incidents—involving 870 persons (Professor A. E. Bender, pers. comm.).

Symptoms suffered by most of the patients included nausea and vomiting followed by diarrhoea and sometimes abdominal pain. A short incubation period of 1–3 h was most common and it is interesting to note that several patients have developed symptoms after consuming only four or five raw soaked beans. The toxic factor responsible is a naturally occurring haemagglutinin that can be destroyed by adequate cooking such as boiling for 10 min.

Two recent trends appear to be responsible for these outbreaks—one is the movement towards 'natural foods' which extols the virtues of consuming raw foods, and the other is the introduction and use of low temperature 'slow cooker' vessels in which temperatures do not usually reach levels sufficient to inactivate the haemagglutinin.

4. Factors Contributing to Outbreaks of Food Poisoning

Published reports on the incidence of food poisoning and salmonellosis in England and Wales give information on the number of outbreaks and cases, location, causal agent and type of food as well as details of outbreaks of special interest and the prevalence of particular serotypes and phage types of the organisms implicated. Until recently, however, little information has been available on the malpractices in food preparation that lead to food poisoning. Between 1970 and 1979 some 6457 general and family outbreaks were reported, but in only 1044 (16%) of these was there sufficient data recorded on the factors which had contributed towards the outbreaks (Roberts 1982a,b). In some outbreaks only one factor was recorded whereas in others as many as five or six were mentioned. The 1044 outbreaks were made up as follows: *Salmonella* spp. 396, *Cl. perfringens* 387, *Staph. aureus* 133, *B. cereus* and other *Bacillus* spp. 53, other bacteria 13 (*Vibrio parahaemolyticus* 8, *Escherichia coli* 3, *Cl. botulinum* 1, *Yersinia enterocolitica* 1), other non-bacterial 55 (scombrotoxic fish poisoning 47, red kidney beans 7, virus particles 1) and not known, 7.

The analysis of contributory factors (Table 5) by Roberts (1982b) provides a valuable and important insight into how and why food poisoning occurs and how its prevalence could be reduced. In more than 60% of the outbreaks the food had been prepared in advance of need, i.e. at least 12 h before it was consumed. This alone would not necessarily cause food poisoning but in combination with inadequate cooling and storage it becomes an extremely important factor. The other main factors were storage at ambient temperature (40%), inadequate cooling (32%), inadequate reheating (29%) and the use of contaminated processed food (19%). The latter included foods such as meat and poultry, pies, smoked fish and 'take-away' meals prepared in premises other than those in which the final dish was consumed, but does not include canned foods. The importance of the different factors showed little change over the 10-year period studied.

In most types of food poisoning, preparation in advance and storage at ambient temperature were the two main factors involved but after these certain factors appeared to be associated with particular organisms. Contaminated processed food and undercooking were most commonly associated with outbreaks caused by *Salmonella*. In *Cl. perfringens* outbreaks multiple factors were frequently recorded with inadequate cooling and inadequate reheating figuring prominently. Canned foods were most frequently implicated in staphylococcal food poisoning.

Infected food handlers did not play a significant role except in staphylococcal food poisoning. Although they were recorded in 126 of the 396 *Salmonella* outbreaks, in only nine was there evidence to suggest that they were the original source of the contaminating organism. In most instances food handlers are victims, not sources, and become infected either from frequent contact with contaminated raw food, from tasting during preparation or from eating left-over cooked food which is contaminated.

TABLE 5

Factors contributing to 1044 outbreaks of food poisoning in England and Wales 1970–1979*

Contributing factor	Number of outbreaks in which factor recorded							Total	%†
	Salmonella spp.	*Clostridium perfringens*	*Staphylococcus aureus*	*B. cereus* and other *Bacillus* spp.	Other—bacterial	Other—non-bacterial	Not known		
Preparation too far in advance	173	338	66	49	4	—	3	633	60·6
Storage at ambient temperature	115	208	54	32	1	2	1	413	39·6
Inadequate cooling	71	236	10	16	—	—	—	333	31·9
Inadequate reheating	47	215	4	29	3	—	2	300	28·7
Contaminated processed food (excluding canned)	105	9	25	4	9	46	1	199	19·1
Undercooking	91	62	1	1	1	4	1	161	15·4
Inadequate thawing	42	22	—	—	—	—	—	64	6·1
Cross-contamination	57	3	2	—	—	—	—	62	5·9
Inadequate warm holding	9	42	—	8	1	—	—	60	5·7
Infected food handlers	9	—	44	1	1	—	—	54	5·2
Use of left-overs	21	20	8	—	—	—	—	50	4·8
Raw food consumed	37	—	—	—	1	8	—	46	4·4
Extra large quantities prepared	13	17	2	—	—	—	—	32	3·1
Contaminated canned food									
(a) freshly opened	2	3	16	1	1	3	3	29	2·8
(b) not freshly opened	—	—	8	—	1	—	—	9	0·9
(c) not known	—	—	7	—	—	—	—	7	0·7
TOTAL	792	1175	247	141	23	63	11	2452	factors

* Roberts (1982*b*).   † The overall total is greater than 100% because multiple factors were reported in most of the outbreaks.

TABLE 6

Counts of various organisms in foods incriminated in outbreaks of food poisoning in the UK

Colony count of specific pathogen per g of food	Number of outbreaks of food poisoning involving		
	Bacillus cereus	Clostridium perfringens	Staphylococcus aureus
$<10^4$	2	6	5
10^4–9.9×10^4	5	17	5
10^5–9.9×10^5	8	27	15
10^6–9.9×10^6	13	20	30
10^7–9.9×10^7	20	20	40
10^8–9.9×10^8	17	14	39
10^9–9.9×10^9	8	2	27
$>10^{10}$	2	0	1
TOTAL	75	106	162
Median count/g	2.5×10^7	1.0×10^6	4.0×10^7

Table 6 shows that in foods incriminated in outbreaks due to B. cereus and Staph. aureus counts of the organism responsible usually range between 10^6 and 10^{10}/g and for Cl. perfringens between 10^5 and 10^9/g. Such levels highlight again some of the gross malpractices of food hygiene in this country and they endorse the results from the analysis of the factors which contribute towards outbreaks.

5. Economic Impact

Food-borne illness costs money. Although difficult to assess, several attempts have been made particularly in the USA and Canada to determine the economic impact of specific outbreaks with respect to the costs of (a) hospital and medical care, (b) loss of earnings and productivity and (c) investigation of the outbreak; Bryan (1978) and Turnbull (1979) have reviewed this information. In this country a considerable amount of time and money is spent each year by doctors, microbiologists, environmental health officers and veterinary officers on the treatment of patients, investigations of outbreaks, inspection of food premises, education of food handlers and on general surveillance and research. However, no information is available to date on the cost of food-borne illness in the UK, but the economic impact on the food industry of five well-investigated outbreaks indicate that this expense must be enormous (Table 7).

TABLE 7

Food-borne illness in the UK: examples of the economic impact on the food industry

Year	Illness	Food	Cost (£)
1964	Typhoid	Canned corned beef	25 million
1974	Salmonellosis (*Salm. infantis*)	Cold roast pork	350,000
	Salmonellosis (*Salm. agona* & *Salm. anatum*)	Cooked ham, pork and beef	50,000
1978	Botulism	Canned salmon	2 million
1979	Staphylococcal food poisoning	Canned corned beef	1 million

6. The Next Decade

The absolute safety of food is an impossible goal. We cannot make sterile food available for general consumption, we do not need it and we do not want it. The average adult in this country consumes about 1·5 kg of food per day, i.e. >0·5 tonnes per year, so the task of ensuring that the 30 million tonnes of food that go to make up our annual food supply is wholesome, safe and nutritious is a mammoth one. Everyone agrees on the need for legislation to ensure the safety of foods, that toxic chemicals should be absent or not present in appreciable amounts and that foodstuffs should be clean and wholesome. However, with respect to the prevention of bacterial food poisoning and for a general improvement in standards of food hygiene most interested parties agree that more education rather than legislation is needed. The magnitude of this educational problem is a great challenge as the number of establishments (shops, supermarkets, hotels, restaurants, hospitals, schools, etc.) selling or providing foods and meals is more than 300,000 and they employ over 2·5 million full- or part-time staff.

The number of cases of food-borne infection and intoxication reported in this country is likely to continue to increase in the next few years, particularly with reference to salmonella infection, campylobacter enteritis, viral gastro-enteritis and hepatitis. However, the total morbidity from food-borne illness is greatly in excess of any published figures and in the USA and Canada the ratio of estimated cases to initially reported cases is considered to be about 25 : 1 (Hauschild & Bryan 1980).

Improvements in catering services should result in a reduction in the number of outbreaks in hospitals and a similar reduction should occur in schools with the decline of the school meals service. In contrast, some increase in the number of incidents may occur because of developments in food technology, consumer trends and changes of eating habits. These include (a) more central processing of food and longer distribution chains; (b) inadequate care with cook–chill catering systems; (c) greater use of convenience foods; and (d) greater consumption of food outside the home or from 'take-away' services.

The most important problem will still be how to reduce the incidence of salmonellosis in man. There are a number of lines of approach beginning at the farm and going right through to the kitchen. These include: (a) prevention of the entry of infection into flocks of turkeys and chickens and herds of cattle and pigs via excreting animals or the use of contaminated feedingstuffs; (b) limitation of the spread of infection if introduced; (c) reduction in the spread of intestinal contents to meat during slaughter; (d) refrigeration of prepared foods to prevent growth and multiplication of organisms; (e) heat destruction of organisms in foods by pasteurization or sterilization and (f) proper cooking and the prevention of re-contamination after cooking. The final line of defence in the prevention of salmonellosis and most other types of bacterial food poisoning is good kitchen hygiene. There is an urgent need to improve the standards of basic food hygiene and to eliminate or reduce many of the gross malpractices. Nevertheless, for the most part food poisoning can be avoided by an awareness of some very simple rules in the kitchen (Fig. 1).

Compared with the situation 125 years ago the deliberate adulteration of food in this country is probably non-existent. However, the epidemic of serious illness in Spain in 1981 from adulterated cooking oil, which affected

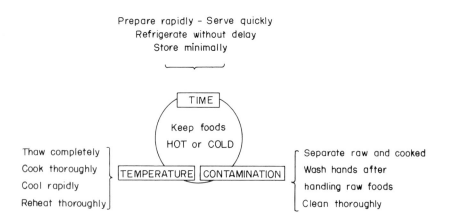

Fig. 1. Prevention of food poisoning.

more than 15,000 persons with 200 deaths (Ross 1981; Trench 1981), is a timely reminder that chemical food poisoning is still a problem elsewhere.

I am grateful to Dr Diane Roberts for various helpful comments and criticisms.

7. References

ANON. 1979 Gastroenteritis of unknown aetiology. *British Medical Journal* **2**, 1008.
ANON. 1980 Surveillance of food poisoning and salmonella infections in England and Wales: 1970–9. *British Medical Journal* **281**, 817–818.
ANON. 1981*a* Food poisoning and salmonellosis surveillance in England and Wales, 1980. *British Medical Journal* **283**, 924–925.
ANON. 1981*b* Outbreaks of food poisoning reported in 1980. *Public Health Laboratory Service Communicable Diseases Report* No. 30, p. 4 (unpublished).
APPLETON, H. & PEREIRA, M. S. 1977 A possible virus aetiology in outbreaks of food poisoning from cockles. *Lancet* **i**, 780–781.
APPLETON, H., PALMER, S. R. & GILBERT, R. J. 1981 Foodborne gastroenteritis of unknown aetiology: a virus infection? *British Medical Journal* **282**, 1801–1802.
AYRES, P. A. & BARROW, G. I. 1978 The distribution of *Vibrio parahaemolyticus* in British coastal waters: report of a collaborative study 1975–6. *Journal of Hygiene, Cambridge* **80**, 281–294.
BALL, A. P., HOPKINSON, R. B., FARRELL, I. D., HUTCHINSON, J. G. P., PAUL, R., WATSON, R. D. S., PAGE, A. J. F., PARKER, R. G. F., EDWARDS, C. W., SNOW, M., SCOTT, D. K., LEONE-GANADO, A., HASTINGS, A., GHOSH, A. C. & GILBERT, R. J. 1979 Human botulism caused by *Clostridium botulinum* type E: the Birmingham outbreak. *Quarterly Journal of Medicine* **48**, 473–491.
BARROW, G. I. 1974 Microbiological and other hazards from seafoods with special reference to *Vibrio parahaemolyticus*. *Postgraduate Medical Journal* **50**, 612–619.
BARROW, G. I. & MILLER, D. C. 1972 *Vibrio parahaemolyticus*: a potential pathogen from marine sources in Britain. *Lancet* **i**, 485–486.
BOSTOCK, A. D., MEPHAM, P., PHILLIPS, S., SKIDMORE, S. & HAMBLING. M. H. 1979 Hepatitis A infection associated with the consumption of mussels. *Journal of Infection* **1**, 171–177.
BRYAN, F. L. 1978 Impact of foodborne diseases and methods of evaluating control programs. *Journal of Environmental Health* **40**, 315–323.
BRYAN, F. L. 1980 Epidemiology of foodborne diseases transmitted by fish, shellfish and marine crustaceans in the United States, 1970–1978. *Journal of Food Protection* **43**, 859–876.
BUTZLER, J. P. & SKIRROW, M. B. 1979 Campylobacter enteritis. *Clinics in Gastroenterology* **8**, 737–765.
CRUICKSHANK, J. G. & WILLIAMS, H. R. 1978 Scombrotoxic fish poisoning. *British Medical Journal* **2**, 739–740.
DE SAXE, M., COE, A. W. & WIENEKE, A. A. 1982 The use of phage typing in the investigation of food poisoning caused by *Staphylococcus aureus* enterotoxins. In *Isolation and Identification Methods for Food Poisoning Organisms* ed. Corry, J. E. L., Roberts, D. & Skinner, F. A. pp. 173–197. Society for Applied Bacteriology Technical Series No. 17. London & New York: Academic Press.
ELIAS-JONES, T. F., ANDERSON, M. E. M., COLLIER, P. W., GILRAY, G., PATTERSON, W. J., SHARP, J. C. M., SUTHERLAND, J. A. & TURNER, A. 1980 *The Investigation and Control of Food Poisoning in Scotland*. Scottish Home and Health Department, pp. 1–45. Edinburgh: Her Majesty's Stationery Office.

EYLES, M. J., DAVEY, G. R. & HUNTLEY, E. J. 1981 Demonstration of viral contamination of oysters responsible for an outbreak of viral gastroenteritis. *Journal of Food Protection* **44**, 294–296.

GILBERT, R. J. 1974 Staphylococcal food poisoning and botulism. *Postgraduate Medical Journal* **50**, 603–611.

GILBERT, R. J. 1979 *Bacillus cereus* gastroenteritis. In *Foodborne Infections and Intoxications* 2nd edn, ed. Riemann, H. & Bryan, F. L. pp. 495–518. New York and London: Academic Press.

GILBERT, R. J. & WILLIS, A. T. 1980 Botulism. *Community Medicine* **2**, 25–27.

GILBERT, R. J., STRINGER, M. F. & PEACE, T. C. 1974 The survival and growth of *Bacillus cereus* in boiled and fried rice in relation to outbreaks of food poisoning. *Journal of Hygiene, Cambridge* **73**, 433–444.

GILBERT, R. J., HOBBS, G., MURRAY, C. K., CRUICKSHANK, J. G. & YOUNG, S. E. J. 1980 Scombrotoxic fish poisoning: features of the first 50 incidents to be reported in Britain (1976–9). *British Medical Journal* **281**, 71–72.

GILBERT, R. J., TURNBULL, P. C. B., PARRY, J. M. & KRAMER, J. M. 1981 *Bacillus cereus* and other *Bacillus* species: their part in food poisoning and other clinical infections. In *The Aerobic Endospore-forming Bacteria: Classification and Identification* ed. Berkeley, R. C. W. & Goodfellow, M. pp. 297–314. Society for General Microbiology Special Publication No. 4. London & New York: Academic Press.

HARBOUR, H. E., ABELL, J. M., CAVANAGH, P., CLEGG, F. G., GOULD, C. M., ELLIS, P., PYKE, M., RILEY, C. T. & LAVER, U. 1977 *Salmonella. The Food Poisoner.* Report by a Study Group of the British Association for the Advancement of Science. pp. 1–51. Berkhamsted: Clunbury Cottrell Press.

HAUSCHILD, A. H. W. & BRYAN, F. L. 1980 Estimate of cases of food and waterborne illness in Canada and the United States. *Journal of Food Protection* **43**, 435–440.

HAYEK, L. J. & CRUICKSHANK, J. G. 1977 Campylobacter enteritis. *British Medical Journal* **2**, 1219.

HENDERSON, P. B. 1830 Case of poisoning from the bonito (*Scomber pelamis*). *Edinburgh Medical Journal* **34**, 317–318.

HEPNER, E. 1980 Food poisoning and salmonella infections in England and Wales, 1976–78. *Public Health, London* **94**, 337–349.

HOBBS, B. C. 1979. *Clostridium perfringens* gastroenteritis. In *Food-borne Infections and Intoxications* 2nd edn, ed. Riemann, H. & Bryan, F. L. pp. 131–171. New York & London: Academic Press.

HOOPER, W. L., BARROW, G. I. & MCNAB, D. J. N. 1974 *Vibrio parahaemolyticus* food poisoning in Britain. *Lancet* **i**, 1100–1102.

JONES, P. H., WILLIS, A. T., ROBINSON, D. A., SKIRROW, M. B. & JOSEPHS, D. S. 1981 Campylobacter enteritis associated with the consumption of free school milk. *Journal of Hygiene, Cambridge* **87**, 155–162.

KRAMER, J. M., TURNBULL, P. C. B., MUNSHI, G. & GILBERT, R. J. 1982 Identification and characterization of *Bacillus cereus* and other *Bacillus* species associated with foods and food poisoning. In *Isolation and Identification Methods for Food Poisoning Organisms* ed. Corry, J. E. L., Roberts, D. & Skinner, F. A. pp. 261–286. Society for Applied Bacteriology Technical Series No. 17. London & New York: Academic Press.

MACKAY-SCOLLAY, E. M. 1958 Two cases of botulism. *Journal of Pathology and Bacteriology* **75**, 482–485.

MCCOY, J. H. 1975 Trends in salmonella food poisoning in England and Wales 1941–72. *Journal of Hygiene, Cambridge* **74**, 271–282.

MELLING, J. & CAPEL, B. J. 1978 Characteristics of *Bacillus cereus* emetic toxin. *Federation of European Microbiological Societies Microbiology Letters* **4**, 133–135.

MELLING, J., CAPEL, B. J., TURNBULL, P. C. B. & GILBERT, R. J. 1976 Identification of a novel enterotoxigenic activity associated with *Bacillus cereus*. *Journal of Clinical Pathology* **29**, 938–940.

MOON, A. J. 1981 Ciguatera poisoning. *Practitioner* **225**, 1176–1178.

MORITA, T. N. & WOODBURN, M. J. 1977 Stimulation of *Bacillus cereus* growth by protein in cooked rice combinations. *Journal of Food Science* **42**, 1232–1235.

MORTIMER, P. R. & McCANN, G. 1974 Food-poisoning episodes associated with *Bacillus cereus* in fried rice. *Lancet* **i**, 1043–1045.

MURPHY, A. M., GROHMANN, G. S., CHRISTOPHER, P. J., LOPEZ, W. A., DAVEY, G. R. & MILLSOM, R. H. 1979 An Australia-wide outbreak of gastroenteritis from oysters caused by Norwalk virus. *Medical Journal of Australia* **2**, 329–333.

MURRAY, C. K., HOBBS, G. & GILBERT, R. J. 1982 Scombrotoxin and scombrotoxin-like poisoning from canned fish. *Journal of Hygiene, Cambridge* **88**, 215–220.

NEWELL, D. G. (ed.) 1982 *Campylobacter: Epidemiology, Pathogenesis and Biochemistry* pp. 1–308. Lancaster: MTP Press.

NOAH, N. D., BENDER, A. E., REAIDI, G. B. & GILBERT, R. J. 1980 Food poisoning from raw red kidney beans. *British Medical Journal* **281**, 236–237.

PARRY, J. M. & GILBERT, R. J. 1980 Studies on the heat resistance of *Bacillus cereus* spores and growth of the organism in boiled rice. *Journal of Hygiene, Cambridge* **84**, 77–82.

PEFFERS, A. S. R., BAILEY, J., BARROW, G. I. & HOBBS, B. C. 1973 *Vibrio parahaemolyticus* gastroenteritis and international air travel. *Lancet* **i**, 143–145.

PINEGAR, J. A. & SUFFIELD, A. 1982 The investigation of food poisoning outbreaks in England and Wales. In *Isolation and Identification Methods for Food Poisoning Organisms* ed. Corry, J. E. L., Roberts, D. & Skinner, F. A. pp. 1–23. Society for Applied Bacteriology Technical Series No. 17. London & New York: Academic Press.

ROBERTS, D. 1982*a* Factors contributing to outbreaks of food poisoning in England and Wales 1970–79. *Public Health Laboratory Service Communicable Disease Report* No. 2, pp. 3–4 (unpublished).

ROBERTS, D. 1982*b* Factors contributing to outbreaks of food poisoning in England and Wales 1970–1979. *Journal of Hygiene, Cambridge* **89**, 491–498.

ROBINSON, D. A. 1981 Infective dose of *Campylobacter jejuni* in milk. *British Medical Journal* **282**, 1584.

ROBINSON, D. A. & JONES, D. M. 1981 Milk-borne campylobacter infection. *British Medical Journal* **282**, 1374–1376.

ROBINSON, D. A., GILBERT, R. J. & SKIRROW, M. B. 1980 Campylobacter enteritis. *Environmental Health* **88**, 140–141.

ROSS, G. 1981 A deadly oil. *British Medical Journal* **283**, 424–425.

ROWE, B., HALL, M. L. M., WARD, L. R. & DE SA, J. D. H. 1980 Epidemic spread of *Salmonella hadar* in England and Wales. *British Medical Journal* **280**, 1065–1066.

SHARP, J. C. M. & COLLIER, P. W. 1981 Food-poisoning in Scotland, 1973–80. *Journal of Infection* **3**, 286–292.

SHARP, J. C. M., PATERSON, G. M. & FORBES, G. I. 1980 Milk-borne salmonellosis in Scotland. *Journal of Infection* **2**, 333–340.

SHEAR, C. L. & GOTTLIEB, M. S. 1980 Shellfishborne disease control in the United States: a commentary. *Medical Hypotheses* **6**, 315–327.

SKIRROW, M. B. 1977 Campylobacter enteritis: a 'new' disease. *British Medical Journal* **2**, 9–11.

SKIRROW, M. B., FIDOE, R. G. & JONES, D. M. 1981 An outbreak of presumptive food-borne campylobacter enteritis. *Journal of Infection* **3**, 234–236.

STRINGER, M. F., TURNBULL, P. C. B. & GILBERT, R. J. 1980 Application of serological typing to the investigation of outbreaks of *Clostridium perfringens* food poisoning 1970–1978. *Journal of Hygiene, Cambridge* **84**, 443–456.

STRINGER, M. F., WATSON, G. N. & GILBERT, R. J. 1982 *Clostridium perfringens* type A: serological typing and methods for the detection of enterotoxin. In *Isolation and Identification Methods for Food Poisoning Organisms* ed. Corry, J. E. L., Roberts, D. & Skinner, F. A. pp. 111–135. Society for Applied Bacteriology Technical Series No. 17. London & New York: Academic Press.

TATNALL, F. M., SMITH, H. G., WELSBY, P. D. & TURNBULL, P. C. B. 1980 Ciguatera poisoning. *British Medical Journal* **281**, 948–949.

TRENCH, B. 1981 Spain's toxic oil: the death-toll mounts. *New Scientist* **92**, 604–606.

TURNBULL, P. C. B. 1979 Food poisoning with special reference to *Salmonella*—its epidemiology, pathogenesis and control. *Clinics in Gastroenterology* **8**, 663–714.

TURNBULL, P. C. B. 1981 *Bacillus cereus* toxins. *Pharmacology and Therapeutics* **13**, 453–505.

TURNBULL, P. C. B. & GILBERT, R. J. 1982 Fish and shellfish poisoning in Britain. In *Adverse Effects of Foods* ed. Jelliffe, E. F. P. & Jelliffe D. B. pp. 297–306. New York: Plenum Press.

TURNER, A. 1980 Food and public safety. *Institute of Food Science and Technology Proceedings* **13**, 235–250.

VERNON, E. 1977 Food poisoning and salmonella infections in England and Wales, 1973–75. *Public Health, London* **91**, 225–235.

WALLACE, J. M. 1980 Milk-associated campylobacter infection. *Health Bulletin* **38**, 57–61.

Mechanisms of Action of Food Preservation Procedures

G. W. GOULD, M. H. BROWN AND B. C. FLETCHER

*Unilever Research Laboratory, Colworth House,
Sharnbrook, Bedford, UK*

Contents

1. Introduction

ALTHOUGH there is a long history behind the majority of procedures employed to preserve foods, most of the techniques that are currently used were developed empirically. Consequently, it is not surprising that their mechanisms at the biophysical, biochemical or molecular level, are usually not known. At the same time, food preservation forms the basis of such a large industry contributing to the well-being of mankind, that it is perhaps surprising to find it is based on such superficial understanding.

One is led to ask: if we knew more about the mechanisms by which food preservation techniques act, could we make more effective use of them? Would improved understanding facilitate the development of new techniques, giving more acceptable flavours with improved efficiency and safety? We do not intend to attempt to provide answers to these questions now, but rather, to summarize some recent advances in understanding that are beginning to shed light on how preservatives work and how the target micro-organisms adapt to minimize their effects on growth and survival. We hope that, in the fullness of time, application of such knowledge will lead to real improvements in practical techniques.

It is useful first to categorize preservation methods currently used (Table 1). Examination of Table 1 reveals the striking fact that, of the major antimicrobial food preservation techniques in use today, only a few operate

FOOD MICROBIOLOGY
ISBN 0 12 589670 0

TABLE 1

Classification of preservation factors

Mode of action	Preservation factor	Mode of achievement
Inactivation of micro-organisms	Heat	Pasteurization Sterilization
	Radiation	Radicidation Radurization Radappertization
Inhibition or slowing of growth of micro-organisms	Cool	Chill Freeze
	Restrict water (reduce water activity)	Dry Add salt Add sugar Add glycerol Add other solutes, or use combinations of the above
	Restrict oxygen	Vacuum pack Nitrogen pack
	Increase carbon dioxide	CO_2 pack
	Acidify	Add acids Lactic fermentation Acetic fermentation
	Alcohol	Fermentation Fortification
	Add preservatives	Inorganic (e.g. sulphite, nitrite) Organic (e.g. sorbate, benzoate, parabens, etc.) Antibiotics (e.g. nisin) Smoke
Restriction of access of micro-organisms to product	Microstructure control	Emulsions (w/o)
	Decontamination	Ingredients Packaging materials, e.g. by chemicals (HCl, H_2O_2), heat, irradiation (ionizing or X; non-ionizing).
	Aseptic or clean handling	Superclean processing Aseptic processing Aseptic or clean packaging
	Packaging	

through the killing of micro-organisms. The majority of usable techniques are based on the inhibition or slowing of growth of spoilage or pathogenic micro-organisms though, if unable to multiply, such cells will of course die at some unspecified rate. For example (Table 1), the only widely used food preservation procedure that acts directly by killing micro-organisms is heat. The use of ionizing radiation is steadily, if very slowly, increasing, but is still unlikely to overtake or even seriously compete with thermal processing as a major preservation technique within the next few decades.

In contrast, the range of preservation techniques in current use that act primarily by slowing or preventing growth, rather than killing micro-organisms, is long and diverse, including essentially physical procedures (e.g. chilling, freezing, reduction of water activity through drying or addition of solutes, packaging in reduced levels of oxygen, use of other gases) as well as essentially chemical or biological ones (e.g. addition of acids, alcohol, chemical preservatives and antibiotics such as nisin and pimaricin, and the use of fermentation). In addition, restriction of access of micro-organisms to a foodstuff by the use of suitable packaging is generally a valuable adjunct to any method of preservation, as may be restriction of the availability of nutrients to the contaminating flora.

The total range of procedures used, particularly to inhibit microbial growth in foods, is therefore quite impressive although, as pointed out above, there has been relatively little research conducted into their detailed mechanisms of action. If food preservation techniques are to be better used in the future, acquisition of sound understanding should arguably be a much higher priority objective.

In this paper, therefore, we review some of the most recent work on mechanisms of action selecting, in particular, illustrative examples from (i) heat, (ii) depression of water activity and (iii) reduction of pH, and also some aspects of the mechanisms of action of some preservatives and of the specially effective combination preservation techniques that are being increasingly used.

2. Water Relations of Cells and the Heat-resistance of Bacterial Endospores

Bacterial endospores represent the most heat-resistant forms of life on earth. The thermal processing industry exists to kill them. What do we know about the mechanisms of their heat-resistance?

Until recently dipicolinic acid (DPA; pyridine 2,6-dicarboxylic acid) was thought to play a central role. It is present, often at concentrations as high as 10% of the dry weight, along with an approximately equivalent amount of calcium with which it will form a stable chelate, in the central protoplast of the spore. When DPA-negative spore mutants of *Bacillus cereus* were isolated by Halvorson & Swanson (1969) and found to be heat-sensitive, this seemed to confirm most workers' expectations. However, Hanson *et al.* (1972) subsequently isolated a number of revertants from the DPA-negative heat-sensitive strain, the spores of which were still DPA-negative and low in calcium and yet had heat-resistance just as high as that of the wild type. This seemed to rule out a direct role for DPA in resistance to heat, although it is still quite often assumed by some researchers. The DPA-negative, heat-resistant spore mutants were not easy to work with. They reverted to DPA-positive or to heat-sensitivity readily and were therefore easily lost. As a result of these difficulties, doubts about their authenticity have crept into the literature, but a recent re-isolation has confirmed that they are fully heat-resistant and quite devoid of detectable DPA (Dring & Gould 1981; Fig. 1).

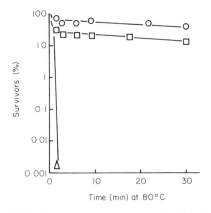

Fig. 1. Heat resistance of *Bacillus cereus* mutant spores: ○, dipicolinic acid (DPA)-positive 'wild type'; △, DPA-negative, heat-sensitive mutant; □, DPA-negative, heat-resistant revertant. From Dring & Gould (1981).

If DPA, or calcium dipicolinate, within spores is not then directly involved in heat resistance, what is? There is much evidence now, from mutants with well-defined biochemical defects that influence the quantity or structure of peptidoglycan in the spore (Pearce & Fitz-James 1971; Imae & Strominger 1976), from experiments on wild-type cells with lysozyme and other lytic enzymes that hydrolyse peptidoglycan (Gould & King 1969; Cassier & Ryter 1971) and from the effects of metal cations and other positively charged molecules that form complexes with it (Gould & Dring, 1975*a*), that

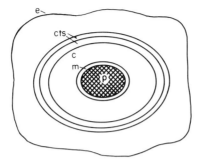

Fig. 2. Diagrammatic representation of the structure of a typical bacterial endospore: *p*, protoplast, containing cytoplasm and site of dipicolinic acid and calcium; *m*, plasma membrane, which becomes the membrane of the new vegetative cell following germination; *c*, cortex, consisting of loosely cross-linked, specially electronegative peptidoglycan; *cts*, proteinaceous, enzyme-resistant spore coats; *e*, exosporium, present on spores of some species, function unknown.

together strongly implicates the cortex as an important structure in the spore controlling heat-resistance and probably dormancy. The cortex is the region beneath the protein-rich spore coats and surrounds the central protoplast or 'core' (Fig. 2). A most interesting aspect of the involvement of the cortex in heat resistance is that the genetic material and most of the other protected macromolecules within the spore are in the central protoplast and are therefore physically separated from the cortex that surrounds and protects them.

It is thought currently that the specially electronegative and loosely cross-linked nature of peptidoglycan in the cortex is important in resistance, acting by controlling the water content of the enclosed protoplast and particularly by maintaining its water content at a low level. However, there is still doubt about how this actually occurs and about what the water content in the spore protoplast actually is. Topical ideas range from a cortex that contracts and squeezes the protoplast to partially dehydrate it (Lewis *et al.* 1960), to a cortex that expands following cation-exchange or limited hydrolysis to squeeze the protoplast by compression (Alderton & Snell 1963; Gould & Dring 1975*a*). It has been proposed that highly anisotropic expansion (Warth 1978) or even radial growth, could bring sufficient pressure to bear to dehydrate the protoplast essentially through reverse osmosis (Algie 1980).

Alternatively, simple osmotic equilibrium between protoplast and surrounding cortex, and the counter-ions associated with the highly electronegative polymer, could ensure relative dehydration as long as the level of low molecular mass, and therefore osmotically active, molecules in solution within the protoplast is low (Gould & Dring 1975*b*). Dielectric studies (Carstensen & Marquis 1975) indicated that the levels of mobile soluble ions in spores is indeed low when compared with vegetative cells. These studies have also detected mobile (counter-) ions in the cortex of coat-defective spores, but not in spores of the wild type coated normally (Carstensen *et al.* 1979), so there is certainly more understanding to be gained concerning the actual conditions within the spore.

However, whatever the mechanism, it is clear that the electronegative, loosely cross-linked peptidoglycan in the cortex is involved and this is especially interesting from the point of view of preservation because a number of factors that interfere with the properties of this polymer also affect heat resistance of the spore. For example, acid conditions, which will be expected to protonate peptidoglycan, reduce heat resistance, presumably by altering its surface charge, by reducing its expansion, by reducing its osmotic pressure or by destroying anisotropy. Simple osmotic dehydration, achieved by suspending the spores in high concentrations of non-permeant solutes, can dramatically reimpose heat resistance (Gould 1978). A number of multivalent cations including calcium, magnesium, lanthanum and cationic dyes like Alcian blue, will sensitize spores to heat as long as the spore coats are first made permeable to them by chemical treatment or by the use of ionophores such as A23187 that allow cations like calcium to readily cross anisotropic lipophilic barriers (Gould & Dring 1975*b*). Turning to combination preservation techniques, Stegeman *et al.* (1977) have suggested that the heat-sensitizing effect of prior ionizing radiation may possibly be explained by radiation-induced breaks in, or decarboxylation of, the peptidoglycan and consequent loss of its expansive or osmotic properties. Grecz *et al.* (1982) found that heat plus radiation greatly increased the numbers of single strand breaks in spore DNA and suggested that the synergism resulted from heat-inactivation of repair enzymes.

The picture of the spore now emerging is therefore of a cell *whose resistance depends on very effective homeostasis with respect to the maintenance of a lowered water content in the central protoplast*, but also of a cell that is far more flexible and manipulatable than thought hitherto. The homeostatic mechanism is built into the structure of the cell and requires no energy for its maintenance. The flexibility includes easy modification of the spore's heat resistance, and even extends to its physical structure, which can be 'collapsed', for instance, by exposure to HCl gas at low controlled water vapour pressures (Lelieveld 1980). Even suspension of certain spores in strong solutions of sucrose (Bothipaksa & Busta 1978) can cause obvious physical collapse and 'wrinkling' of the coat, presumably through osmotic dehydration of the underlying cortex and consequent plasmolysis (Fig. 3).

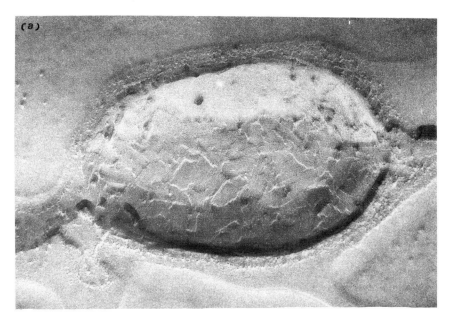

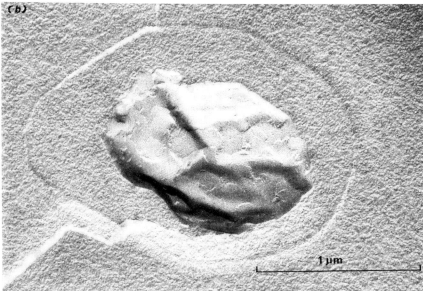

Fig. 3. Freeze etch electron micrographs of spores of *Bacillus cereus* showing shrinkage and collapse of surface structures in solutions of high osmolality. (*a*) *Bacillus cereus* wild-type spores in 6 mol/l glycerol, a permeant solute which does not cause shrinkage. (*b*) *Bacillus cereus* DPA-negative heat-sensitive spore (see Fig. 1) in 2·2 mol/l sucrose, which apparently does not easily permeate the spore and therefore brings about shrinkage osmotically (Bothipaska & Busta 1978). Data of G. J. Dring & G. W. Gould (unpublished)).

Since it has become more certain that the water content of the spore protoplast is an important factor in resistance, it is perhaps not surprising that resistance can be manipulated osmotically. Indeed, even germinated spores, whose heat resistance has fallen many thousand-fold, can be made fully as heat resistant as the ungerminated spore by rapid suspension in strong solutions of non-permeating solutes, like sucrose, that will osmotically dehydrate the just-rehydrated core (Dring & Gould 1975). Glycerol, at equivalent osmolality, will not bring about this protection, most probably because it readily permeates the cell and therefore does not cause osmotic dehydration. Permeability differences may also explain the relative effectiveness of different solutes, at equivalent water activities, as germination inhibitors (Jakobsen & Murrell 1977; Anagnostopoulos & Sidhu 1980).

3. Water Relations of Cells and the Tolerance of Vegetative Cells to Low Water Activity

A major preservation technique aimed at restricting the growth of the vegetative forms of bacteria, yeasts and moulds is the reduction of water activity, brought about by drying or by adding salts, sugars or other solutes. It is used in a wide variety of dried foods, cured products, conserves and 'intermediate moisture foods' (see review by Troller 1980).

A clue to the mechanism of inhibition by low water activity is best illustrated by the experiment described in Fig. 4, in which the water activity in a culture of exponentially growing *B. subtilis* was suddenly reduced by the addition of salt (sodium chloride). The immediate result was that the cells became plasmolysed and stopped growing. However, during the following hour or so, metabolism continued and brought about major changes, particularly in the concentrations of soluble amino acids in the cytoplasmic pool. In *B. subtilis* the proline level increased most of all. In other bacteria, glutamic acid, γ-amino-butyric acid and proline levels rise singly or in concert (Measures 1975) and intracellular potassium and glucose levels may rise as well (Roller & Anagnostopoulos 1982). Eventually, the intracellular concentrations of these low molecular mass solutes are increased sufficiently to balance the external osmolality. At this point (see Fig. 4) the cell rehydrates and growth recommences. However, there is now a considerable energy requirement for the synthesis or accumulation of the intracellular solutes and for their retention against a steep concentration gradient, so that the growth rate falls and the efficiency of conversion of substrate to biomass ('yield') is reduced also.

Whereas bacteria utilize predominantly amino acids, K^+ and glucose for 'osmoregulation', yeasts utilize polyhydric alcohols, probably in an anal-

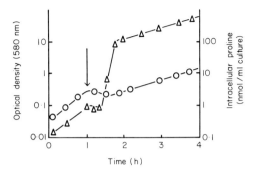

Fig. 4. Amino acid accumulation by *Bacillus subtilis* vegetative cells during 'osmoregulation'. The osmolality of an exponentially growing culture was suddenly raised by the addition of 1 mol/l NaCl (arrow) reducing the water activity to about 0·97. The figure shows the subsequent changes in optical density (○) and the large rise in intracellular proline level (△) which occurs during adaptation and prior to the recommencement of (slower) growth. From Gould & Measures (1977).

ogous manner (Brown & Simpson 1972; Brown 1974). In some yeasts, even intracellular ethanol concentrations as well as glycerol rise in response to osmotic stress, and may even reach levels that bring about a reduction in cell viability (Panchal & Stewart 1980). Some algae (e.g. the marine species *Dunaliella*; Borowitzka & Brown 1974) and some terrestrial lichens that endure dry environments (Lewis & Smith 1967), accumulate high levels of glycerol. Some halophytic plants, like the bacteria, accumulate proline (Stewart & Lee 1974) and high intracellullar proline levels accompany drought-resistance (Singh *et al.* 1972) and frost-resistance (Le Saint & Catesson 1966) of some plants. Many estuarine invertebrates react to changing salinity with changes in intracellular amino acid levels (Gilles 1974), and amino acids play a major (osmotic) role in cell volume regulation in fish (Forster & Goldstein 1979).

The groups of solutes that vary greatly in intracellular concentration in response to changes in external osmolality have the general property of interfering minimally with metabolic activities within the cell. They may have little effect on a cell's intracellular enzymes, for example, whereas an environmental, or food preservation, solute at the same osmolality may be very inhibitory should it be able to enter the cell. For this reason, the intracellular osmoregulatory solutes have aptly been termed 'compatible solutes' by Brown & Simpson (1972).

Solutes that readily permeate the cell (e.g. glycerol in the case of most bacteria) do not elicit the osmoregulatory response. Evidently, bacteria respond to the movement of water to or from the cytoplasmic space rather than to water activity or to the external solute itself. It has been suggested that the initial trigger, in vegetative cells, is a rise in intracellular K^+ level brought about by the loss of water accompanying plasmolysis (Gould & Measures 1977), but the detailed mechanism has not been worked out. The key point, however, is that *the cell osmoregulation mechanism operates to maintain homeostasis with respect to water content.*

A practical implication of the energy cost involved is that any restriction of energy supply will tend to be specially synergistic with lowered water activity when used as a method of preservation. Most obviously, this is seen in oxygen-free vacuum packs, in which the potential ATP-generation and growth yield by facultative anaerobes is further reduced.

If dried, or dehydrated by solutes at such concentrations as to exceed a cell's capacity to osmoregulate, then growth must, of course, cease. However, changes which are very important with respect to preservation may still occur, and particularly concerning the heat resistance of the inhibited cells. The most dramatic examples of this were given by Corry (1974, 1976*b*) who demonstrated increases in the heat resistance of *Salmonella typhimurium* exceeding 700-fold after suspending cells in strong solutions of solutes (sucrose, glucose, fructose, sorbitol) that did not readily permeate the cell membrane and hence dehydrated the cytoplasm. Equally strong solutions of glycerol, that did permeate, and therefore did not withdraw water, had a much smaller effect. It was clearly shown that with *Salmonella* species (Corry 1976*b*) and with yeasts (Corry, 1976*a*), the rise in heat resistance correlated well with plasmolysis of the cells. Evidently such treatments, in some respects, make the vegetative cell protoplasts 'spore-like'.

4. Mechanism of Vegetative Cell Resistance to Low pH Values

Acidification is a widely used food preservation procedure whose mode of action has, most surprisingly, received little attention until recently.

In contrast to their reaction to low water activity, vegetative cells of bacteria suddenly 'shifted' to a lower pH value do not undergo a long adaptation period but, unless the shift is very large, change growth rate immediately (Brown *et al.* 1980; Fig. 5). What happens within the cell during and after such a shift, or during growth at a suboptimal pH-value? *The most important principle operating again, as with cells growing at low water activities, seems to be homeostasis, in that the cells react generally to maintain internal pH values constant.* In order to do this, an organism needs to cope with an increasing leakage of protons into the cell cytoplasm as the pH is reduced, and these must be removed if internal acidification is to be avoided. Removal of protons, against what may be a steep concentration gradient of several hundred- or thousand-fold in many acid-preserved foods, is energy

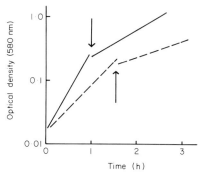

Fig. 5. Effect of abrupt pH reduction on growth rates of *Streptococcus faecalis* (———) and *Escherichia coli* (– – – –). Cultures growing exponentially at pH 7·2 and 7·0, respectively, were adjusted to pH 4·8 at the points indicated by arrows. From Brown *et al.* (1980).

demanding, so that as the external pH is reduced, more and more energy is utilized to maintain internal pH constant and less and less is therefore available for synthesis of cell material. A major consequence of this is that yield falls dramatically (Fig. 6*a*). The data in Fig. 6*a* were derived from chemostat cultures run at constant growth rate; however, as growth rate is reduced the fraction of the cell's energy utilized for maintenance (whether or not the external pH is reduced) rises (Fig. 6*b*), so that the energy available for additional proton-extrusion becomes even less.

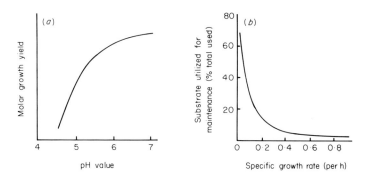

Fig. 6. Fall in molar growth yield with reduction in pH value and increase in the fraction of a cell's energy utilized for maintenance as growth rate falls. Together these determine the minimum pH for growth. (*a*) Change in molar growth yield of *Streptococcus faecalis* growing on glucose at various pH values. (*b*) Change in fraction of substrate used by *Escherichia coli* for 'maintenance' as growth rate is reduced. From Brown *et al.* (1980) and Förstel & Schlesser (1976).

Eventually, the low pH limit for growth is reached when the rate of energy generation is too low for maintenance *and* synthesis, and when this is exceeded, substantial acidification of the interior of the cell occurs. Again, being an energy-demanding homeostatic mechanism, restriction of energy supply will improve the effectiveness of preservation at low pH values. As with lowered water activity, vacuum or oxygen-free packaging are therefore well known as useful adjuncts, and any other energy-restricting procedure will add additional 'hurdles' (Leistner & Rödel 1976). There are certainly additional important, and sometimes poorly defined, hurdles operating in many foods to deliver some unplanned preservation. This may underly the recent demonstration that under certain (minimum hurdle?) conditions of low redox potential in protein-rich media with low organic acid content, growth and toxin formation by *Clostridium botulinum* can occur at pH values below 4·6, which has always been the standard delineating pH for safe acid canning and safety of some other low pH-preserved foods (Raatjes & Smelt 1979; Smelt *et al.* 1982).

5. Mechanism of Action of Preservatives

A particularly useful synergism that may operate on the pH-homeostatic mechanism is one first suggested by Freese *et al.* (1973) who proposed that some of the widely used weak lipophilic acid food preservatives, like sorbate (see Sofos & Busta 1981) benzoate and propionate, may owe part of their effectiveness to their solubility, in the undissociated form, in the cell membrane and to their consequent action as 'proton ionophores'. Such chemicals essentially allow protons to leak into cells more rapidly than they would in their absence and thereby increase the energy requirements of the cell to maintain its relatively alkaline internal pH. In effect the lipophilic acids increase the proton flux across the cell membrane, so that the leakage of protons becomes equivalent to that found at a much lower pH in their absence (Fig. 7). Interference with the proton gradient across the membrane disrupts many of the chemiosmotic-related functions of the cell (Mitchell 1972; Garland 1977), such as amino acid transport, as shown by Freese *et al.* (1973) and Eklund (1980). These effects were shown to occur with mammalian cells as well as with bacteria (Sheu *et al.* 1975) and correlated well with the partition coefficient, and therefore with the likely membrane-solubility, of the lipophilic acid (Freese *et al.* 1979). Sorbate reversibly inhibited germination of spores of *B. cereus* and *Cl. botulinum*, competitively with the L-alanine germinant (Smoot & Pierson 1981), and may act similarly. Many other lipophilic molecules are effective preservatives; examples include fatty acids, other lipids, some of the phenolic antioxidants

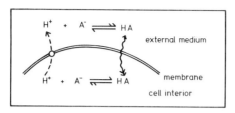

Fig. 7. Basis of acidification of the cell interior at low external pH values in the presence of weak lipophilic acids. The dashed line indicates the energy-requiring export of protons that normally keeps the cell interior more alkaline than the environment. HA is the undissociated form of the lipophilic acid, which tends to equilibrate to equal concentrations on each side of the membrane. The double headed arrows indicate the net flow of reactants when the exterior is more acid than the interior.

(Brannen *et al.* 1980), alcohols and glycols (Herman *et al.* 1980), esters (Conley & Kabara 1973; Mills *et al.* 1980), amides and amines (Kabara *et al.* 1972). The aliphatic members of these are often optimally active near the region of C_{12} in chain length, and are generally more effective against Gram negative than against Gram positive species (Kabara 1981). Their modes of action are most probably related to those suggested for the weak lipophilic acid food preservatives in that most of them probably act by interfering with membrane function. For example, the fact that the preservative activity of short chain alcohols and glycols rises with surface tension-reducing capacity has been interpreted as indicating action via the cell membrane (Herman *et al.* 1980), and a generally similar basic mechanism, on which will be superimposed additional specific effects (e.g. inhibition of yeast enolase by sorbate; Azukas *et al.* 1961), may operate for many of the lipophilic materials listed above.

In the face of such a wide range of low pH-potentiated, lipid-soluble antimicrobial substances, however, many of which occur naturally in the environment, evolution has not stood still. For example, some organisms, like the yeast *Saccharomyces bailii*, can adapt phenotypically to very high levels of some of these agents. *The adapted cells again seem to become resistant essentially through homeostasis, in this instance by developing a system that effectively excretes the acid out of the cell* (e.g. benzoic and sorbic acids; Warth 1977). Again, it was clearly shown that the excretion was dependent on a suitable energy source. For example, starved cells accumulated the preservatives, at low pH values, but on addition of glucose, there was rapid excretion (Warth 1977), so the principle of minimizing the availability of energy in order to obtain most effective preservation still holds.

The inorganic acid-potentiated preservatives, e.g. sulphite and nitrite, like the organic acids, are also more effective at reduced than at near-neutral pH values. This is thought to be because protonation influences the proportion of the antimicrobial forms of the ions.

However, action via chemiosmotic uncoupling or other membrane function targets is less likely than with the organic preservatives, and specific effects of the reactive chemicals are probably of far greater importance. For example, a recent study has shown that high concentrations of nitrite, heated at acid pH values, even cause germination-like change in spores of *Cl. perfringens* by hydrolysing peptidoglycan in the spore cortex (Ando 1980). Under conditions closer to food preservation practice, reaction with specific microbial enzymes is more probably responsible for activity, for example nitrite very effectively inhibits the phosphoroclastic system in *Cl. botulinum* (Woods & Wood 1982) and in *Cl. sporogenes* via the reaction of nitrous oxide with pyruvate: ferredoxin oxidoreductase (Woods *et al.* 1981). Nitrosothiols formed by reaction of nitrite with food sulphydryl compounds will inhibit outgrowth from *B. cereus* spores, it is thought by reaction with spore coat or membrane thiol groups (Morris & Hansen 1981). The possible involvement of iron nitrosyls, and other derivatives of nitrite heated in complex organic mixtures was recently reviewed by Benedict (1980).

6. Conclusions

In only a few of the preservation techniques listed earlier, are the mechanisms of action, or mechanisms of resistance reasonably understood. We have therefore been selective to illustrate an underlying theme: that many micro-organisms have developed very effective homeostatic mechanisms throughout evolution, and these must be overcome for effective food preservation. The homeostasis may, for instance, be 'passive', as in the spore (Table 2), acting to keep the central protoplast constantly low enough in water content to remain resistant and dormant for years, or even for hundreds or thousands of years, whatever the water status of the environment. In contrast, homeostasis may be active (Table 2), as in the vegetative cell, to keep its water content high, and thereby allow an organism to keep functioning no matter what happens to the water activity in its environment: or, to keep the internal pH value of a cell comfortably near neutral no matter how acid, or full of ionophores, the environment becomes.

Unfortunately for the food scientist, these mechanisms work against easy preservation. On the other hand, the improved understanding of preservation now becoming available slowly but steadily, is already helping to improve procedures and put them on a more rational basis and, one hopes, will continue to do so even more in the future

We thank G. J. Dring for previously unpublished data and J. M. Stubbs for electron microscopy.

TABLE 2

Key aspects of microbial homeostatic mechanisms

Cell	Spore	Vegetative cell
Basis of homeostasis	Passive, non-energetic ('potential energy basis')	Active; maintained energetically ('kinetic energy basis')
Factors interfering with homeostasis	Release from dormancy	Reduction in availability of energy	Increase in demand for energy
Means of interference	Initiation of germination with natural germinants or with 'false triggers'	Removal of oxygen; limitation of nutrients; reduction of temperature	Reduction of water activity: reduction of pH value: addition of ionophores and other membrane-active compounds
	Damage to key structures		
	Chemical or enzymic attack on coats, cortex etc.		

7. References

ALDERTON, G. & SNELL, N. S. 1963 Base exchange and heat resistance in bacterial spores. *Biochemical and Biophysical Research Communications* **10**, 139–143.

ALGIE, J. E. 1980 The heat resistance of bacterial spores due to their partial dehydration by reverse osmosis. *Current Microbiology* **3**, 287–290.

ANAGNOSTOPOULOS, G. D. & SIDHU, H. S. 1980 The effect of water activity and the a_w-controlling solute on spore germination of *Bacillus stearothermophilus*. *Journal of Applied Bacteriology* **50**, 335–349.

ANDO, Y. 1980 Mechanism of nitrite-induced germination of *Clostridium perfringens* spores. *Journal of Applied Bacteriology* **49**, 527–535.

AZUKAS, J. J., COSTILOW, R. N. & SADOFF, H. L. 1961 Inhibition of alcoholic fermentation by sorbic acid. *Journal of Bacteriology* **81**, 189–194.

BENEDICT, R. C. 1980 Biochemical basis for nitrite-inhibition of *Clostridium botulinum* in cured meat. *Journal of Food Protection* **43**, 877–891.

BOROWITZKA, L. J. & BROWN, A. D. 1974 The salt relations of marine and halophilic species of the unicellular green alga, *Dunaliella*. The role of glycerol as a compatible solute. *Archiv für Mikrobiologie* **96**, 37–52.

BOTHIPAKSA, K. & BUSTA, F. F. 1978 Osmotically induced increase in thermal resistance of heat-sensitive, dipicolinic acid-less spores of *Bacillus cereus* Ht-8. *Applied and Environmental Microbiology* **35**, 800–808.

BRANNEN, A. L., DAVIDSON, P. M. & KATZ, B. 1980 Antimicrobial properties of phenolic antioxidants and lipids. *Food Technology* **34**, 42–53.

BROWN, A. D. 1974 Microbial water relations: features of the intracellular composition of sugar-tolerant yeasts. *Journal of Bacteriology* **118**, 769–777.

BROWN, A. D. & SIMPSON, J. R. 1972 Water relations of sugar-tolerant yeasts: the role of intracellular polyols. *Journal of General Microbiology* **72**, 589–591.

BROWN, M. H., MAYES, T. & LELIEVELD, H. L. M. 1980 The growth of microbes at low pH. In *Microbial Growth and Survival in Extremes of Environment* ed. Gould, G. W. & Corry, J. E. L. pp. 71–98. Society for Applied Bacteriology Technical Series No. 15. London and New York: Academic Press.

CARSTENSEN, E. L. & MARQUIS, R. E. 1975. Dielectric and electrochemical properties of bacterial cells. In *Spores VI* ed. Gerhardt, P., Costilow, R. N. & Sadoff, H. L. pp. 563–571. Washington D.C.: American Society for Microbiology.

CARSTENSEN, E. L., MARQUIS, R. E., CHILD, S. Z. & BENDER, G. R. 1979 Dielectric properties of native and decoated spores of *Bacillus megaterium*. *Journal of Bacteriology* **140**, 917–928.

CASSIER, M. & RYTER, A. 1971 Sur un mutant de *Clostridium perfringens* donnant des spores sans tuniques à germination lysozyme-dépendante. *Annales de l'Institut Pasteur, Paris* **121**, 717–732.

CONLEY, A. J. & KABARA, J. J. 1973 Antimicrobial action of esters of polyhydric alcohols. *Antimicrobial Agents and Chemotherapy* **4**, 501–506.

CORRY, J. E. L. 1974 The effect of sugars and polyols on the heat resistance of salmonellae. *Journal of Applied Bacteriology* **37**, 31–43.

CORRY, J. E. L. 1976*a* The effect of sugars and polyols on the heat resistance and morphology of osmophilic yeasts. *Journal of Applied Bacteriology* **40**, 269–276.

CORRY, J. E. L. 1976*b* Sugar and polyol permeability of *Salmonella* and osmophilic yeast cell membranes measured by turbidimetry, and its relationship to heat resistance. *Journal of Applied Bacteriology* **40**, 277–284.

DRING, G. J. & GOULD, G. W. 1975 Reimposition of heat resistance on germinated spores of *Bacillus cereus* by osmotic manipulation. *Biochemical and Biophysical Research Communications* **66**, 202–208.

DRING, G. J. & GOULD, G. W. 1981 Reisolation of the *B. cereus* T DPA-negative heat resistant spore mutant of Hanson *et al*. *Spore Newsletter* **7**, 130–131.

EKLUND, T. 1980 Inhibition of growth and uptake processes in bacteria by some chemical food preservatives. *Journal of Applied Bacteriology* **48**, 423–432.

FREESE, E., LEVIN, B. C., PEARCE, R., SREEVALSAN, T., KAUFMAN, J. J., KOSKI, W. S. & SEMO, N. M. 1979 Correlation between the growth inhibitory effects, partition coefficients and teratogenic effects of lipophilic acids. *Teratology* **20**, 413–440.

FREESE, E., SHEU, C. W. & GALLIERS, E. 1973 Function of lipophilic acids as antimicrobial food additives. *Nature, London* **241**, 321–325.

FÒRSTEL, H. & SCHLESSER, G. 1976 Role of maintenance metabolism for biomass production. *Abstracts of the Fifth International Fermentation Symposium, Berlin.*

FORSTER, R. P. & GOLDSTEIN, L. 1979 Amino acids and cell volume regulation. *The Yale Journal of Biology and Medicine* **52**, 497–515.

GARLAND, P. B. 1977 Energy transduction and transmission in microbial systems. In *Microbial Energetics* ed. Haddock, B. A. & Hamilton, W. A. pp. 1–21. Symposium of the Society for General Microbiology No. 27. Cambridge: Cambridge University Press.

GILLES, R. 1974 Amino acid metabolism and control of cell volume. *Archives International de Physiologie et Biochimie* **82**, 423–483.

GOULD, G. W. 1978 Practical implications of compartmentalization and osmotic control of water distribution in spores. In *Spores VII* ed. Chambliss, G. & Vary, J. C. pp. 21–26. Washington D.C.: American Society for Microbiology.

GOULD, G. W. & DRING, G. J. 1975*a* Role of an expanded cortex in the resistance of bacterial endospores. In *Spores VI* ed. Gerhardt, P., Costilow, R. N. & Sadoff, H. L. pp. 541–546. Washington D.C.: American Society for Microbiology.

GOULD, G. W. & DRING, G. J. 1975*b* Heat resistance of bacterial endospores and the concept of an expanded osmoregulatory cortex. *Nature, London* **258**, 402–405.

GOULD, G. W. & KING, W. L. 1969 Action and properties of spore germination enzymes. In *Spores IV* ed. Campbell, L. L. pp. 276–286. Bethesda, Md: American Society for Microbiology.

GOULD, G. W. & MEASURES, J. C. 1977 Water relations in single cells. *Philosophical Transactions of the Royal Society, London, B.* **278**, 151–166.

GRECZ, N., BRUSZER, G. & AMIN, I. 1982 Effect of radiation and heat on bacterial spore DNA. In *Combination Processes in Food Irradiation.* IEAFE-SM-250/22. Vienna: International Atomic Energy Authority.

HALVORSON, H. O. & SWANSON, A. 1969 Role of dipicolinic acid in the physiology of bacterial spores. In *Spores IV* ed. Campbell, L. L. pp. 121–132. Bethesda, Md: American Society of Microbiology.

HANSON, R. S., CURRY, M. V., GARNER, J. V. & HALVORSON, H. O. 1972 Mutants of *B. cereus* strain T that produce thermoresistant spores lacking dipicolinic acid and have low levels of calcium. *Canadian Journal of Microbiology* **18**, 1139–1143.

HERMAN, E. B., HAAS, G. J., CROSBY, W. H. & CARTE, C. J. 1980 Antimicrobial action of short chain alcohols and glycols. *Journal of Food Safety* **2**, 131–139.

IMAE, Y. & STROMINGER, J. L. 1976 Relationship between cortex content and properties of *Bacillus sphaericus* spores. *Journal of Bacteriology* **126**, 907–913.

JAKOBSEN, M. & MURRELL, W. G. 1977 The effect of water activity and a_w-controlling solute on germination of bacterial spores. In *Spore Research 1976*, ed. Barker, A. N., Wolf, J., Ellar, D. J., Dring, G. J. & Gould, G. W. pp. 819–834. London & New York: Academic Press.

KABARA, J. J. 1981 Food grade chemicals for use in designing food preservation systems. *Journal of Food Protection* **44**, 633–647.

KABARA, J. J., CONLEY, A. J. & TRUANT, J. P. 1972 Relationship of chemical structure and antimicrobial activity of alkyl amides and amines. *Antimicrobial Agents and Chemotherapy* **2**, 492–498.

LEISTNER, L. & RÖDEL, W. 1976 Inhibition of micro-organisms by water activity. In *Inhibition and Inactivation of Vegetative Microbes* ed. Skinner, F. A. & Hugo, W. B. pp. 219–237. Society for Applied Bacteriology Symposium No. 5, London & New York: Academic Press.

LELIEVELD, H. L. M. 1980 A note on the effect of hydrogen chloride on the morphology of *Bacillus subtilis* spores. *Journal of Applied Bacteriology* **48**, 59–61.

LE SAINT, A. M. & CATESSON, A. M. 1966 Variations simultanées des teneurs en eau en sucres solubles en acides amines et de la pression osmotique dans le phloem et le cambium de sycamore pendant les périodes de repose apparent et de reprise de la croissance. *Comptes Rendus Academie Scientifique D* **263**, 1463–1475.

LEWIS, D. M. & SMITH, D. C. 1967 Sugar alcohols (polyols) in fungi and green plants. *New Phytology* **66**, 143–184.

LEWIS, J. C., SNELL, N. S. & BURR, H. K. 1960 Water permeability of bacterial spores and the concept of a contractile cortex. *Science, New York* **132**, 544–545.

MEASURES, J. C. 1975 Role of amino acids in osmoregulation of non-halophilic bacteria. *Nature, London* **257**, 398–400.

MILLS, J. C. J., RICHARDSON, T. & JASENSKY, R. D. 1980 Antimicrobial effects of N^α-palmitoyl-L-lysine-L-lysine ethyl ester dihydrochloride and its use to extend the shelf life of creamed cottage cheese. *Journal of Agricultural and Food Chemistry* **28**, 812–817.

MITCHELL, P. 1972 Chemiosmotic coupling in energy transduction: a logical development of biochemical knowledge. *Journal of Bioenergetics* **3**, 5–24.

MORRIS, S. L. & HANSON, J. N. 1981 Inhibition of *Bacillus cereus* spore outgrowth by covalent modification of a sulphydryl group by nitrosothiol and iodoacetate. *Journal of Bacteriology* **148**, 465–571.

PANCHAL, C. J. & STEWART, G. G. 1980 The effect of osmotic pressure on the production and excretion of ethanol and glycerol by a brewing yeast strain. *Journal of the Institute of Brewing* **86**, 207–210.

PEARCE, S. M. & FITZ-JAMES, P. C. 1971 Sporulation of a cortexless mutant of a variant of *Bacillus cereus*. *Journal of Bacteriology* **105**, 339–348.

RAATJES, G. J. M. & SMELT, J. P. P. M. 1979 *Clostridium botulinum* can grow and form toxin at pH values lower than 4·6. *Nature, London* **281**, 398–399.

ROLLER, S. D. & ANAGNOSTOPOULOS, G. D. 1982 Accumulation of carbohydrate by *Escherichia coli* B/r-1 during growth at low water activity. *Journal of Applied Bacteriology* **52**, in press.

SHEU, C. W., SALOMON, D., SIMMONS, J. L., SREEVALSAN, T. & FREESE, E. 1975 Inhibitory effect of lipophilic acids and related compounds on bacteria and mammalian cells. *Antimicrobial Agents and Chemotherapy* **7**, 349–363.

SINGH, T. N., ASPINALL, D. & PALEG, L. G. 1972 Proline accumulation and varietal adaptability to drought in barley; a potential metabolic measure of drought resistance. *Nature (London) New Biology* **236**, 188–190.

SMELT, J. P. P. M., RAATJES, G. J. M., CROWTHER, J. S. & VERRIPS, C. T. 1982 Growth and toxin formation by *Clostridium botulinum* at low pH values. *Journal of Applied Bacteriology* **52**, 75–82.

SMOOT, L. A. & PIERSON, M. D. 1981 Mechanisms of sorbate inhibition of *Bacillus cereus* T and *Clostridium botulinum* 62A spore germination. *Applied and Environmental Microbiology* **42**, 477–483.

SOFOS, J. N. & BUSTA, F. F. 1981 Antimicrobial activity of sorbate. *Journal of Food Protection* **44**, 614–622.

STEGEMAN, H., MOSSEL, D. A. A. & PILNICK, W. 1977 Studies on the sensitizing mechanism of pre-irradiation to a subsequent heat treatment on bacterial spores. In *Spore Research 1976* ed. Barker, A. N., Wolf, J., Ellar, D. J., Dring, G. J. & Gould, G. W. pp. 565–587. London & New York: Academic Press.

STEWART, G. R. & LEE, J. A. 1974 The role of proline accumulation in halophytes. *Planta* **120**, 279–289.

TROLLER, J. A. 1980 Influence of water activity on micro-organisms in foods. *Food Technology* **34**, 76–83.

WARTH, A. D. 1977 Mechanism of resistance of *Saccharomyces bailii* to benzoic, sorbic and other weak acids used as preservatives. *Journal of Applied Bacteriology* **43**, 215–230.

WARTH, A. D. 1978 Molecular structure of the bacterial spore. *Advances in Microbial Physiology* **7**, 1–45.

WOODS, L. F. J. & WOOD, J. M. 1982 A note on the effect of nitrite inhibition on the metabolism of *Clostridium botulinum*. *Journal of Applied Bacteriology* **52**, 109–110.

WOODS, L. F. J., WOOD, J. M. & GIBBS, P. A. 1981 The involvement of nitric oxide in the inhibition of the phosphoroclastic system in *Clostridium sporogenes*. *Journal of General Microbiology* **125**, 399–406.

Predictive Modelling of Food Safety with Particular Reference to *Clostridium botulinum* in Model Cured Meat Systems

T. A. ROBERTS

Agricultural Research Council, Meat Research Institute,
Langford, Bristol, UK

B. JARVIS

Leatherhead Food R.A., Leatherhead, Surrey, UK

Contents

1. Introduction

TRADITIONALLY bacteriologists have examined a product before and after storage, or after its probable involvement in an episode of food-borne illness, for numbers and types of bacteria deemed to be important. Such data have been accumulated over many years in different laboratories and under different circumstances, and related work under laboratory simulations has determined the conditions under which microbial growth is limited. From such data attempts are made to assess the shelf life and microbiological stability and safety of foods. The intention of this paper is to draw attention to the need for a timely review of attitudes in food microbiology. We do not presume to say that everything that has gone before is wrong; that what we now propose is right; or that all 'traditional' microbiology should be stopped. We have the greatest respect for scientists a generation or two ago who, with equipment and facilities inferior to today's, often performed better designed experiments than are done today and, in many cases, showed clearer lines of thought and reasoning. In all areas there are

FOOD MICROBIOLOGY
ISBN 0 12 589670 0

outstanding publications which laid the foundations of food microbiology as we now understand it. Nor is it our intention to criticize what others have done before; rather to suggest ways in which we, as food microbiologists, might develop predictive microbiology so that we obtain the best value from what is widely considered to be an expensive, yet largely negative, science.

Our intention, therefore, is simply to consider whether we should continue along current lines, or whether a modified approach is required. We should consider whether the duplicated effort, and the similar, but often uncomparable, experiments during the ten years since the last symposium (*Journal of Applied Bacteriology* 1971, **34**, 1–213) has been worthwhile in cost-effective terms and in relation to the progress made in understanding food microbiology, particularly with respect to predicting microbiological growth in foods.

2. Possible Changes in Attitude

Several changes of attitude are possible:

(1) traditional bacteriological methodology could be replaced by something more rapid, i.e. be mechanized, so that we obtain essentially the same information more quickly, or with less manpower, and therefore more cheaply;
(2) we could concern ourselves much less with the detailed identity of micro-organisms, and develop systems based upon monitoring their activities relevant to a particular food environment; or
(3) we might measure the growth response of microbes to as many as possible of the factors which influence the rate and type of microbial growth in food systems, and from those data attempt to predict what will happen when foods are stored. This is, to some extent, the application and extension of microbiological compositional analysis (Tuynenburg-Muys 1975): to predict the likely, the possible and the impossible.

Some will argue that the third option has been the overall target of food microbiologists for many years, and that everyone has been working towards this ultimate goal of being able to predict the nature and extent of microbiological changes. Certainly under particular storage conditions the type of spoilage and the hazard from microbes of public health significance are rather well understood. It has been possible to identify conditions in which survival or growth of microbes is unlikely ever to occur. We could not agree, however, that our range of knowledge extends sufficiently over the common conditions of food storage through distribution, retailing and domestic use. By way of example, there are few data on microbiological changes in red meats stored within the temperature range 5–20°C. If such information were more readily available, much routine bacteriological analysis would no longer be necessary.

In the absence of any stimulated change in direction, we feel that research over the next ten years will continue in much the same way as previously, i.e. an obsession with defining 'minima': the minimum temperature for growth, the minimum pH for growth, the minimum water activity for growth. Such values are relevant only to the experimental system in which they were obtained and have little or no relevance to foods where microbes grow in competition. They can also have most unfortunate implications when incorporated into legislation (e.g. categorizing meat products by water activity as in the EEC draft Council Directive on Meat Products (VI/2197/78-EN Rev 2); chilling carcasses to below 7°C as soon as possible, presumably to prevent the multiplication of *Salmonella*, as in the EEC Council Directive on health problems affecting intra-Community Trade In Fresh Meat (64/433/EEC; amended 66/601/EEC; 69/349/EEC).

In recent years several workers have investigated the effects of factors acting in combination (Baird-Parker & Freame 1967; Beuchat 1973; Emodi & Lechowich 1969; Matches & Liston 1972*a*,*b*; Ohye & Christian 1967; Ohye *et al.* 1967; Pivnick *et al.* 1969; Riemann *et al.* 1972; Segner *et al.* 1966) and this seems to us to be a self-evidently important step since there are always at least three factors controlling microbial growth; the pH value, the water activity and the temperature. However, if we list all the factors which *may* affect microbial growth (Table 1) the list becomes so long as to make impossible any attempt to test simultaneously all treatment combinations. Hence it will be necessary initially to exclude some variables to form a database of manageable proportions.

TABLE 1

Factors to be taken into account in laboratory studies of microbial growth responses

1. The pH value and buffering capacity of the system
2. The water activity and choice of humectant(s)
3. The temperature and duration of incubation
4. The nature and severity of any prior treatment likely to 'damage' cells
5. The effects of antimicrobial additives
6. Substrate limitation
7. (Anti)Microbial metabolites and competition
8. Choice of strain, e.g. from a culture collection or isolated from food
9. Work with single or mixed cultures
10. Initial numbers of microbes and, if more than one type, the initial proportions
11. Experimental variation
12. Laboratory to laboratory variation also important
13. Data formatting and the most appropriate statistical/mathematical analysis of the results

3. Alternative Investigational Approaches

Experiments in laboratory media are comparatively easier to perform than those in foods, but the concern is always how closely the microbial responses in media mimic those in foods. Our view is that data are currently so sparse that a degree of experimentation in media is useful, particularly to define the combinations of factors where growth of particular microbes cannot possibly occur, even after prolonged incubation (Roberts & Ingram 1973; Roberts *et al*. 1979).

There are examples where results in laboratory media clearly do not represent what happens in the food, e.g. the extent of the 'Perigo' effect in media (Perigo *et al*. 1967; Perigo & Roberts 1968) is much greater than it is in meat products (Jarvis *et al*. 1976). Nitrite is a good example where high expenditure on research over the years has led to prolific publications, yet a relatively poor understanding remains of its precise antimicrobial role in cured meats (Benedict 1980).

It is clearly impossible to test separately every food formulation (or even each food product). Experience some 20 years ago in the field of food irradiation may provide a salutory lesson. The intention was to clear for human consumption foods which had been irradiated and proven to be safe with respect to toxicology, wholesomeness and microbiology. The regulatory authorities at that time were advised by experts in the field of food irradiation that irradiation should be treated in much the same way as food additives, i.e. prohibited in general use but permitted in specific approved instances. One consequence was that each food would require to be tested separately for toxicology, wholesomeness and microbiological safety. The arguments then ranged over what constitutes different foods or different food products. The resultant confusion helped to prevent any significant application of food irradiation. With hindsight it would have been preferable to have designed a range of 'experimental foods' containing differing proportions of carbohydrates, protein and fats, and other components as necessary, and to make the relevant tests on a formulation that broadly represented a large number of commodities. Perhaps a similar approach is appropriate in food microbiology, where it is also obvious that study of individual commodities is unlikely to prove adequate. Such an approach remains to be developed and proven to be relevant.

4. Pasteurized Cured Meats as an Example

For some years the microbiological stability and safety of pasteurized cured meats has been in the minds of official bodies and industry. This has arisen because of changes in commercial practice which originated partly from reputed consumer demand for more bland foods and partly from changes in technological practice. Additionally there have been pressures to reduce, or even eliminate, nitrite and nitrate as curing salts because they may contribute to the formation of potentially carcinogenic nitrosamines both in the product and *in vivo* by reaction with secondary and tertiary amines (reviewed in Sofos *et al.* 1979 and Roberts *et al.* 1981*a*). The factors known to be important in the stability and bacteriological safety of pasteurized cured meats include sodium chloride, pH, sodium nitrite, storage temperature, the extent of the heat treatment and the numbers of bacteria (spores) present initially. This knowledge is a consequence of research by several groups (e.g. Silliker *et al.* 1958; Riemann 1963; Spencer 1966; Pivnick *et al.* 1969; Christiansen *et al.* 1973; Baird-Parker & Baillie 1974; Tompkin *et al.* 1977) and our own groups. However, because of differences in experimental design, these data cannot be bulked and used in a single mathematical analysis.

As it became obvious that very large experiments were necessary to take into account factors acting in combination, it also became clear that one laboratory could hardly hope to perform all the experimental work. So, using common methodology (Rhodes & Jarvis 1976), we shared two large experiments (Roberts *et al.* 1976). For the purposes of the present paper we are less concerned with the results *per se* than with the fact that we obtained much the same answers in the two different laboratories—thus we felt that sharing experiments was a real possibility. Subsequently, mainly for reasons of different systems of funding, our two lines of research have gone in slightly different directions.

At the Meat Research Institute we have attempted to obtain an overall database from a relatively enormous experiment. If the data proved suitable, this would lead to a single predictive equation for toxin production by *Clostridium botulinum* which would be relevant to product safety, particularly when formulations require modification. An example is given in Table 2.

At the Leatherhead Food R.A. studies were made on selected parts of the interactive system, but in such a way that some combinations of factors remained common. This approach led initially to development of a risk factor hypothesis for changes in preservative combinations (Jarvis *et al.* 1979). Subsequently a number of predictive equations (Table 3) was developed, none of which is complete in itself, but all of which can be handled by computers to give a predictive measure of safety and stability.

The objective of these studies has not been solely to develop a predictive model, but also to provide a database which can be evaluated against products in trade. We realized from the outset that the results from our

TABLE 2

The probability of toxin production by Clostridium botulinum *types A and B in pasteurized pork slurry in the pH range 5·5 to 6·3 **

Probability of toxin production $(P) = 1/(1 + e^{-\mu})$, where $\mu = 4\cdot679$	
$- (1\cdot47 \times N)$	where $N = $ NaNO$_2$, μg/g $\times 10^2$
$- (1\cdot104 \times S)$	where $S = $ NaCl, % w/v on the water
$+ (0\cdot1299 \times T)$	where $T = $ storage temperature, °C
$- (2\cdot09) + (0\cdot67 \times N)$	if 500 μg/g nitrate added
$- (6\cdot238) + (0\cdot8264 \times S)$	if 1000 μg/g isoascorbate added
$- (1\cdot7049) + (0\cdot3987 \times N)$	if 0·3% polyphosphate† added
$- (1\cdot771) + (0\cdot3997 \times N)$	if heat treatment *High*‡
$- (0\cdot01937 \times N \times T)$	
$- (1\cdot2824)$	if nitrate and polyphosphate added
$+ (0\cdot99)$	if nitrate added and heat treatment *High*

* From Roberts *et al.* (1981c).
† Curaphos 700 (Fibrisol Service Ltd., London W3 8TE, UK).
‡ *High* heat treatment = 80°C/7 min + 70°C/1 h.

TABLE 3

Examples of predictive equations from work in model cured meat slurries (LFRA)

	Salt (% w/v)	Nitrite (μg/g)	Process*	Storage temperature (°C)	Sorbate (% w/v)	Phosphate (% w/v)
A	3·3–5·5	40–300	L	20, 25	—	—
B	3·5–5·5	40–300	H	20, 25	—	—
C	2·0–3·5	75–175	M	20	—	0–0·5

Probability of toxin production $(P) = 1/(1 + e^{-\mu})$, where,

A $\mu = 0\cdot43T - 2\cdot01S + 0\cdot01N - 0\cdot0013NT + 1\cdot2$
B $\mu = 0\cdot43T - 1\cdot53S + 0\cdot0064N - 0\cdot001NT - 1\cdot698$
C $\mu = 7\cdot94 - 0\cdot045N - 0\cdot54S + 4\cdot87P - 2\cdot85SP$

where $S = $ salt (% w/v), $N = $ nitrite (μg/g), $T = $ storage temperature (°C), $P = $ phosphate (% w/v).

* Processes: L = low (20 to 70°C in 7 min); M = moderate (as L plus holding for 1 hour at 70°C); H = high (as L plus holding for 2 hours at 70°C).

experimental systems might not apply directly to particular products and might only be of use in a relative sense, i.e. it might only be possible to say that if certain factors were changed in a particular way the overall effect would be to reduce, or increase, the likelihood of toxin production. However, even this limited success would improve upon merely identifying factors and treatment combinations which had a statistically significant effect on toxin production and might allow the relative effects of the statistically important factors and combinations to be computed. With hindsight, the equation from the very large single experiment is easier to use than the multiple experiment equations, but both provide predictive data of a similar nature (Table 4.).

During the period of this research other groups have similarly continued their interest in the safety of pasteurized cured meats but because of differences in experimental design and in the formulation of the basal 'product' in which the microbes grow, it is difficult to compare our results directly with theirs. LFRA comparisons (Rhodes 1979) revealed less toxin production by *Cl. botulinum* in a 'thick' slurry than in a 'thin' slurry. If that trend continued to even thicker products the MRI formulae would over-estimate the likelihood of toxin production.

Within the joint experiment (Roberts *et al.* 1976) and in the last series of experiments (Jarvis, Rhodes & Patel 1979; Roberts, Gibson & Robinson 1981*a,b,c*, 1982) storage temperatures have been compared. Progressive

TABLE 4

Probability of toxin production (%) by Clostridium botulinum *types A and B in pasteurized cured meat slurries*

Salt (% w/v)	NaNO₂ (μg/g)	Probability of toxin production after 6 months at 20°C: low heat process (L)	
		MRI	LFRA
2·5	100	62	76
	200	16	34
	300	2	7
4·5	100	5	13
	200	1	2
	300	0*	0†

* MRI calculation based on the equation in Table 2 (see Roberts *et al.* 1981c). Results rounded to the nearest whole number, i.e. '0' = <0·49%.
† LFRA calculation based on equation B in Table 3. '0' = 'very much less than 1%'.
pH of pork slurry 5·5–6·3.
Inoculum 10 spores per bottle (replicate).

reduction in the storage temperature from 35°C initially merely slowed the growth rate of *Cl. botulinum*, i.e. the same proportion of samples contain toxin from 35 to 20°C or 17·5°C, provided that samples were incubated for a sufficient time. However, when the temperature was reduced to 15°C there was a marked reduction in the proportion of replicates containing toxin. Hence incubation at 35°C does *not* mimic what will happen slowly at 15°C, and the relevance of accelerated storage tests to chill storage must be questioned. We wished also to store replicates at 12·5 and 10°C, but previous experience suggested that the number of replicates containing toxin would have been too small to allow any mathematical interpretation without vast replication—perhaps a thousand or more replicates per treatment combination.

5. What of the Future?

We have mentioned above the impossibility of comparing data from different workers using different experimental models. We feel that the time has come to suggest that experiments be shared between appropriate laboratories using standard methodology, adequate replication, and with adequate account being taken of inter-laboratory variation. Additionally, those data must from the outset be in a form to facilitate mathematical analysis—the use of computer compatible data acquisition, formatting and storage would permit comparison to be made readily between laboratories and enable workers in the future to build upon the original data banks.

We believe it is time to suggest further that standardized systems should be developed for data acquisition and storage in *all* areas of microbiology so that results from one laboratory can readily be made available to another and incorporated in subsequent analyses.

There are numerous imaginary objections to these suggestions and changes in attitude will be required of the experimenter, editors of journals and even by those who evaluate the performance of scientists to judge promotability. Not least it will require a change in attitude of funding authorities.

With the availability of satellite communications and the enormous increase in computer facilities, one might even contemplate the establishment of a national, or international, data bank. This could be consulted and further data added, provided that standard procedures had been followed to acquire those data. Of course, adequate standard methods are not available, and few of us have the experience of recording and storing large quantities of data in computer compatible format and of analysing such large amounts of data. We must move away from the ideas that replication is a substitute for precise standard methods, and must accept also that in the absence of adequately standardized procedures we shall not obtain anything better than trends. However, such trends can be very useful.

In the absence of such a change in philosophy we seem likely to continue to collect vast amounts of data, at enormous expense, and constantly to rediscover perfectly well-established principles and yet be unable to evaluate fully those data. We believe that after 5–10 years co-ordinated research effort a considerable proportion, but not all, 'traditional' microbiology would become unnecessary, and that the response of many products to storage abuse and to prescribed storage conditions would be largely predictable.

6. References

BAIRD-PARKER, A. C. & BAILLIE, M. A. H. 1974 The inhibition of *Clostridium botulinum* by nitrite and sodium chloride. In *Nitrite in Meat Products* ed. Krol, B. & Tinbergen, B. J. pp. 77–90. Proceedings of an International Symposium, Zeist, 1973, Wageningen: PUDOC.

BAIRD-PARKER, A. C. & FREAME, B. 1967 Combined effect of water activity, pH and temperature on the growth of *Clostridium botulinum* from spore and vegetative cell inocula. *Journal of Applied Bacteriology* **30**, 420–429.

BEAN, P. G. & ROBERTS, T. A. 1974 Effects of pH, sodium chloride and sodium nitrite on heat resistance of *Staphylococcus aureus* and the growth and recovery of damaged cells in laboratory media. *Proceedings of the 4th International Congress of Food Science and Technology (IUFOST), Madrid, September 1974*, **III**, 93–102.

BENEDICT, R. C. 1980 Biochemical basis for nitrite-inhibition of *Clostridium botulinum* in cured meat. *Journal of Food Protection* **43**, 877–981.

BEUCHAT, L. R. 1973 Interacting effects of pH, temperature and salt concentration on growth and survival of *Vibrio parahaemolyticus*. *Applied Microbiology* **25**, 844–846.

CHRISTIANSEN, L. N., JOHNSTON, R. W., KAUTTER, D. A., HOWARD, J. W. & AUNAN, W. J. 1973 Effect of nitrite and nitrate on toxin production by *Clostridium botulinum* and on nitrosamine formation in perishable canned comminuted cured meat. *Applied Microbiology* **25**, 357–362 (erratum 26, 653).

EMODI, A. S. & LECHOWICH, R. V. 1969 Low temperature growth of type E *Clostridium botulinum* spores, I. Effects of sodium chloride, sodium nitrite and pH. *Journal of Food Science* **34**, 78–81.

GIBSON, A. M., ROBERTS, T. A. & ROBINSON, A. 1982 Factors controlling the growth of *Clostridium botulinum* types A and B in pasteurized cured meats. IV. The effect of pig breed, cut and batch of pork. *Journal of Food Technology* **17**, 471–482.

INGRAM, M. 1971 Microbiological changes in foods—general considerations. *Journal of Applied Bacteriology* **34**, 1–8.

INGRAM, M. 1976 The microbiological role of nitrite in meat products. In *Microbiology in Agriculture, Fisheries and Food* ed. Skinner, F. A. & Carr, J. G. pp. 1–18. Society for Applied Bacteriology Symposium Series No. 4, London & New York: Academic Press.

JARVIS, B., RHODES, A. C., KING, S. E. & PATEL, M. 1976 Sensitization of heat damaged spores of *Clostridium botulinum* type B to sodium chloride and sodium nitrite. *Journal of Food Technology* **11**, 41–50.

JARVIS, B., RHODES, A. C. & PATEL, M. 1979 Microbiological safety of pasteurized cured meats: inhibition of *Clostridium botulinum* by curing salts and other additives. In *Food Microbiology and Technology* ed. Jarvis, B., Christian, J. H. B. & Michener, H. D. pp. 251–266. Parma, Italy: Medicina Viva.

MATCHES , J. R. & LISTON, J. 1972*a* Effects of incubation temperature on the salt tolerance of *Salmonella. Journal of Milk and Food Technology* 35, 39–44.

MATCHES, J. R. & LISTON, J. 1972*b* Effect of pH on low temperature growth of *Salmonella. Journal of Milk and Food Technology* 35, 49–52.

OHYE, D. F. & CHRISTIAN, J. H. B. 1967 Combined effects of temperature, pH and water activity on growth and toxin production by *Clostridium botulinum* types A, B and E. In *Botulism 1966* ed. Ingram, M. & Roberts, T. A. pp. 217–223, Proceedings of the 5th International Symposium on Food Microbiology, Moscow, July 1966. London: Chapman & Hall.

OHYE, D. F., CHRISTIAN, J. H. B. & SCOTT, W. J. 1967 Influence of temperature on the water relations of growth of *Clostridium botulinum* type E. In *Botulism 1966* ed. Ingram, M. & Roberts, T. A. pp. 136–143, Proceedings of the 5th International Symposium on Food Microbiology, Moscow, July 1966. London: Chapman & Hall.

PERIGO, J. A. & ROBERTS, T. A. 1968 Inhibition of clostridia by nitrite. *Journal of Food Technology* 3, 91–94.

PERIGO, J. A., WHITING, E. & BASHFORD, T. E. 1967 Observations on the inhibition of vegetative cells of *Clostridium sporogenes* by nitrite, which has been autoclaved in a laboratory medium, discussed in the context of sublethally processed cured meats. *Journal of Food Technology* 2, 377–397.

PIVNICK, H., BARNETT, H. W., NORDIN, H. & RUBIN, J. L. 1969 Factors affecting the safety of canned, cured, shelf-stable luncheon meat inoculated with *Clostridium botulinum. Canadian Institute of Food Technology Journal* 2, 141–148.

RHODES, A. C. 1979 *The effect of curing and processing parameters on the survival and growth of* Clostridium botulinum *in cured meat and fish.* Ph.D. Thesis, University of Reading.

RHODES, A. C. & JARVIS, B. 1976 A pork slurry system for studying inhibition of *Clostridium botulinum* by curing salts. *Journal of Food Technology* 11, 13–23.

RIEMANN, H. 1963 Safe heat processing of canned cured meats with regard to bacterial spores. *Food Technology, Champaign* 17, 39–42, 45–46, 49.

RIEMANN, H., LEE, W. H. & GENIGEORGIS, C. 1972 Control of *Clostridium botulinum* and *Staphylococcus aureus* in semi-preserved meat products. *Journal of Milk and Food Technology* 35, 514–523.

ROBERTS, T. A. & INGRAM, M. 1973 Inhibition of growth of *Clostridium botulinum* at different pH values by sodium chloride and sodium nitrite. *Journal of Food Technology* 8, 467–475.

ROBERTS, T. A., JARVIS, B. & RHODES, A. C. 1976 Inhibition of *Clostridium botulinum* by curing salts in pasteurized pork slurry. *Journal of Food Technology* 11, 25–40.

ROBERTS, T. A., BRITTON, C. R. & SHROFF, N. N. 1979 The effect of pH, water activity, sodium nitrite and incubation temperature on growth of bacteria isolated from meat. In *Food Microbiology and Technology* ed. Jarvis, B., Christian, J. H. B. & Michener, H. D. pp. 57–71. Parma, Italy: Medicina Viva.

ROBERTS, T. A., GIBSON, A. M. & ROBINSON, A. 1981*a* Factors controlling the growth of *Clostridium botulinum* types A and B in pasteurized cured meats. I. In pork slurry in the pH range 5·54–6·36. *Journal of Food Technology* 16, 239–266.

ROBERTS, T. A., GIBSON, A. M. & ROBINSON, A. 1981*b* Factors controlling the growth of *Clostridium botulinum* types A and B in pasteurized cured meats. II. In pork slurry in the pH range 6·27–6·72. *Journal of Food Technology* 16, 267–281.

ROBERTS, T. A., GIBSON, A. M. & ROBINSON, A. 1981*c* Prediction of toxin production by *Clostridium botulinum* in pasteurized pork slurry. *Journal of Food Technology* 16, 337–355.

ROBERTS, T. A., GIBSON, A. M. & ROBINSON, A. 1982 Factors controlling the growth of *Clostridium botulinum* types A and B in pasteurized cured meats. III. The effect of potassium sorbate. *Journal of Food Technology* 17, 307–326.

SEGNER, W. P., SCHMIDT, C. F. & BOLTZ, J. K. 1966 Effect of sodium chloride and pH on the outgrowth of spores of type E *Clostridium botulinum* at optimal and suboptimal temperatures. *Applied Microbiology* 14, 49–54.

SILLIKER, J. H., GREENBERG, R. A. & SCHACK, W. R. 1958 Effect of individual curing ingredients on the shelf stability of canned comminuted meats. *Food Technology, Champaign* 12, 551–554.

SOFOS, J. N., BUSTA, F. F. & ALLEN, C. E. 1979 Botulism control by nitrite and sorbate in cured meats: a review. *Journal of Food Protection* **42**, 739–770.

SPENCER, R. 1966 Processing factors affecting stability and safety of non-sterile canned cured meats. *Food Manufacture* **41**, 1–4.

TOMPKIN, R. B., CHRISTIANSEN, L. N. & SHAPARIS, A. B. 1977 Variation in inhibition of *Clostridium botulinum* by nitrite in perishable canned comminuted cured meats. *Journal of Food Science* **42**, 1046–1048.

TUYNENBURG-MUYS, G. 1975 Microbial safety and stability of food products. *Antonie van Leeuwenhoek* **41**, 369–371.

Developments in Heat Treatment Processes for Shelf-stable products

P. G. BEAN

Metal Box p.l.c., Wantage, Oxfordshire, UK

Contents

1. Introduction

In 1810 Nicolas Appert, a skilled French chef, published his patent for the preservation of food by heat. The foods were placed in glass bottles closed with corks secured by wire, and then heated in boiling water baths. Appert's imitators in the UK, Donkin and Hall, followed the French patent, but finding glass too fragile and corks too porous, began to use soldered, tin-plated iron containers. Although the new long-life foods found immediate favour with the military, especially sailors, production was slow and only gradual advance was made into the domestic retail market. By the beginning of the twentieth century, however, the picture was rapidly changing partly due to the introduction of the now familiar open-top can which was then capable of being sealed at a rate of 50 cans/min. By comparison, today's higher speed canning lines are capable of filling and sealing cans at speeds up to 1200 cans/min.

During the present century, shelf-stable heat-preserved foods have become a familiar part of national food supplies all over the world. Containers, equipment and products have developed to take advantage of new technology and new markets. Container materials now include aluminium, tin-free steel, aluminium foil, plastics and foil/plastic combinations in addition to glass and tin-plate. Extrusion, injection moulding, drawing and

FOOD MICROBIOLOGY
ISBN 0 12 589670 0

wall-ironing are manufacturing techniques which have been added to traditional glass blowing and tin-plate soldering for container manufacture. With the introduction of plastics heat sealing was adopted for some container closures and welding has recently been introduced in the manufacture of some tin-plate cans. Food preparation, filling, sealing and processing equipment have been developed to suit the needs of new containers, and to increase line speeds by the introduction of continuous processing and automation. Innovation in food ingredients has led to a variety of recipe and formulated heat-preserved foods taking advantage of blended spices, structurally modified starches and other developments.

Although Appert and his contemporaries were not entirely correct in believing that exclusion and 'heat fixation' of air was the basis of successful heat preservation, nevertheless Appert's two principal tenets apply equally well to prevention of microbial spoilage, i.e. (1) enclose the food in an hermetically sealed container; (b) apply heat to the food and container for an appropriate period.

Shelf-stable heat-preserved foods are commonly classified according to their pH value into low-acid foods with pH values greater than 4·5 and medium or high-acid foods with pH values less than 4·5. This division reflects the inability of *Clostridium botulinum* and other more heat-resistant mesophilic spore-formers to grow out from spores in food at pH levels below 4·5. Hence, low-acid foods (pH >4·5) are subjected to the most severe heat preservation treatments since, theoretically, outgrowth from surviving spores is unrestricted. Two areas applicable to low acid foods representing technology currently being exploited and potential future technology are considered: (a) the use of alternative methods of heating food which reduce the exposure time (*per se* or by the use of high temperatures), while still relying upon heat as the sole preserving agent; (b) alternative methods of preservation which although including heat as a part of the procedure do not rely on heat as the sole preserving agency.

2. Alternative Methods of Heating

Possible alternatives to traditional steam heating in pressure vessels or retorts include the use of thin cross-section packs, aseptic packaging, heating by direct application of gas flame to the container, and heating by microwave energy or electrical current. In all cases the criterion for preservation remains the same as for traditional steam processing, i.e. destruction of bacterial spores by heat.

A. Thin cross-section packs

Processing times can be reduced with thin packs, such as shallow trays or flexible pouches, containing viscous or solid products which heat by conduction, since in comparison with the traditional tin can or glass jar, the distance from the external heat source to the centre of the product is reduced (Table 1). Microbiologists have been concerned principally with the integrity of heat-sealed semi-rigid or flexible thin packs, resulting in appropriate recommendations for filled pack handling and overwrapping to obviate the risks of recontamination of the processed product during handling and distribution (Turtle & Alderson 1971).

TABLE 1

Comparison of process time/temperature for open-top can and retort pouch

Product	Open-top can*	Retort pouch†	F_0 Value
Stewed steak in gravy	75 min at 121°C (16 oz)	24 min at 121°C (16 oz)	8–10
Chocolate pudding	85 min at 113°C (8 oz)	16 min at 121°C (16 oz)	6
Chicken in cream sauce	69 min at 121°C (16 oz)	24 min at 121°C (16 oz)	8
Roast beef in gravy	60 min at 115·6°C (7 oz)	22 min at 121°C (8 oz)	8–10

* 16 oz Open-top can, 73 mm diam. × 117 mm high.
† 16 oz Retort pouch, *ca.* 151 mm wide × 202 mm long × 25 mm deep.

B. Aseptic packaging

Aseptic packaging takes the principle of rapid heating a stage further by reducing product thickness to a thin film passed through a heat exchange system where product heating is very rapid. Because temperatures above 120°C have a greater destructive effect on bacterial spores than on food quality, high temperatures and short holding times can be used without detriment to product quality. Product and container, however, are treated separately and then brought together in a sterile environment.

The microbiologist's concern with aseptic packaging has principally been with spore heat resistance at ultra high temperatures and methods used to sterilize empty containers. The question is whether the rate of spore death (usually expressed as decimal reduction times or D-values) which changes in a regular fashion through the ultra-high temperature processing range (130–160°C), affects the validity of extrapolation of heat resistance data from lower temperatures. Some workers have reported linear relationships between log D-value and temperature in the UHT range (Pflug & Esselen 1953; Busta 1967; Adams 1973) whereas other workers have reported non-linear relationships (e.g. Edwards *et al.* 1965; Miller & Kandler 1967; Neaves & Jarvis 1978). Although, in my laboratories, we showed a linear relationship between log D and temperature for *Bacillus stearothermophilus* Th 24 and *Cl. thermosaccharolyticum* 'O' spores when heated in phosphate buffer in capillary tubes over the range 120–145°C (Bean *et al.* 1979), at higher temperatures difficulties in accurately controlling temperature and exposure time in seconds resulted in a high proportion of spurious results (Fig. 1). Experimental difficulties of this nature may in part explain the

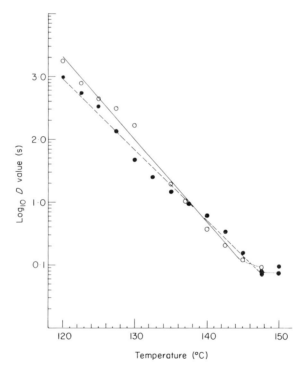

Fig. 1. Heat resistance of *Clostridium thermosaccharolyticum* strain 'O' spores (○——○) and *Bacillus stearothermophilus* TH24 spores (●— —●) in 0·007 mol/l phosphate buffer, pH 6·6.

TABLE 2

The resistance of spores* of
Clostridium botulinum *type B to*
hot air

Temperature (°C)	D-Value (min)
148	1·92
154·4	1·02
160	0·38
165·6	0·22

*Spores of strain 34B dried on aluminium
strips (Denny *et al.* 1979). $z = 16\cdot4°C$.

TABLE 3

The resistance of spores* of
Clostridium botulinum *type B to*
35% hydrogen peroxide

Temperature (°C)	D-Value (min)
28	2·52
54·4	0·12
71·1	0·025
87·8	0·0165

*Spores of strain 196B dried on plastic
strips (Denny *et al.* (1979). $z = 24\cdot4°C$.

different results obtained by different workers and supports the view of
Hermier *et al.* (1975) that UHT resistance should preferably be determined
experimentally in the UHT processing plant.

Container sterilization using superheated steam at temperatures around
160°C has been used successfully for many years, but Denny *et al.* (1979)
have recently published hot-air resistance figures for *Cl. botulinum* 34B
spores dried on aluminium (Table 2) which suggests that this may be an
equally effective, and less energy consuming, method. The most widely used
sterilant for heat-sensitive materials, such as plastics, is probably hot
hydrogen peroxide although u.v. irradiation alone or in combination with
peroxide has been used to sterilize web material formed into containers
immediately prior to filling. Denny *et al.* (1979) have also published data on
the resistance of *Cl. botulinum* 196B to hot hydrogen peroxide, the spores
having been dried on plastic strips (Table 3).

C. Other methods

Direct flame heating and microwave heating may be considered as alternative methods of utilizing rapid heating and ultra-high temperatures to effect
heat processing. Flame heating is applied to the outside of rotating metal
containers, the rotation dispersing the applied heat rapidly throughout the
product (Leonard *et al.* 1975). Microwave heating possibly has greatest
potential as a means of continuous in-line heating of liquids as an alternative
to heat exchangers. The uneven heating effects sometimes experienced
when microwaves are used to heat heterogeneous products is a possible
limitation on this form of heating (Sale 1976).

A recent development as an alternative heating method to steam or water under pressure, and one which is currently attracting considerable attention from food technologists, is the direct application to food of an alternating electrical current. This technique, known as Ohmic heating, is said to result in very rapid and uniform heating (Anon. 1976).

3. Modified Heat Processing

Preservation techniques which utilize heat as a part of, but not the sole, preserving agent and aim at the destruction of *Cl. botulinum* spores and inhibition of outgrowth from the more resistant spores, may be considered as an intermediate step between traditional heat preservation and stabilization by physical or chemical methods alone.

Some heat-preserved foods are stabilized by heat processes significantly lower than classic heat resistance data on spoilage organisms would indicate to be necessary. Those data have generally been established in phosphate buffer at pH 7·0 and optimal conditions for spore outgrowth. In practice, the pH of heat-preserved foods is usually less than pH 7·0 and conditions for outgrowth may be sub-optimal at least in terms of water activity. Cameron *et al.* (1980) recently published data on the heat resistance of the spoilage organism *Cl. sporogenes* PA3679 as influenced by pH in both phosphate

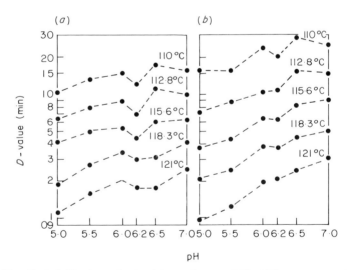

Fig. 2. The effect of pH value on heat resistance of spores of *Clostridium sporogenes* PA 3679 in (*a*) phosphate buffer and (*b*) buffered pea purée. (From Cameron *et al.* 1980.)

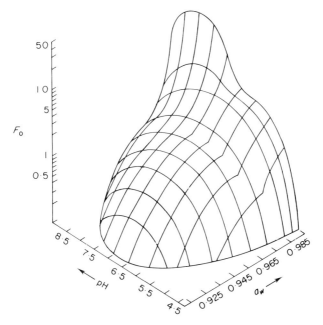

Fig. 3. Isometric illustration of the combined effect of pH and a_w on the limiting F_0 value for combined suspension of culture collection strains of *Bacillus* spp. (From Braithwaite & Perigo 1971.)

buffer and buffered pea purée (Fig. 2). These results showed the well-established reduction of heat resistance with decreasing pH. However, this reduction was greater in the food base than in buffer and perhaps even more significantly, was greater at 121°C than at lower temperatures.

Braithwaite & Perigo (1971) published the results of preliminary experiments on the survival and outgrowth of spores heated at various pH values and water activity levels, using a nutrient agar system and mixed spore suspension containing 10^5 spores of each of 151 culture collection strains of *Bacillus* spp. The results were presented as a three-dimensional diagram (Fig. 3) showing the most severe heat treatment, as F_0 value (equivalent time in min at 121°C, assuming $z = 10°C$), at which survivors were observed at any of four incubation temperatures, viz. 25, 35, 45 or 55°C.

Figure 4 shows a plan view of the diagram with limiting F_0 values drawn as contours, illustrating the effect of substrate pH and water activity on the severity of heat treatment required to stabilize the model system. The authors recognized the limitations of these data since only *Bacillus* spp. were investigated and all the strains were from culture collections.

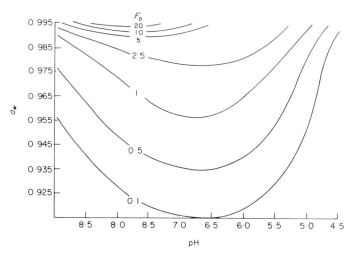

Fig. 4. Combinations of pH and a_w which result in various limiting F_0 values for culture collection strains of *Bacillus* spp. (From Braithwaite & Perigo 1971.)

4. Experimental Results

The experiments of Braithwaite & Perigo (1971) were repeated using mixed suspensions of naturally occurring spores obtained from untreated powdered spices, well-rotted animal manure or soil samples collected around Calcutta. Slurries of 1000 g of the different materials were initially strained through muslin and then subjected to a centrifugation–resuspension programme starting at 500 **g** and increasing to 10,000 **g**. The resulting suspensions of spores were stored separately at 5°C after dilution with distilled water to a final concentration of 10^6 to 10^7 spores/0·5 ml inoculum each of aerobic, as well as anaerobic, mesophilic and thermophilic types. It was necessary to combine spores from spices and manure to obtain a sufficiently dense suspension. In addition, a third spore suspension was prepared containing *ca.* 10^5 spores/0·5 ml of each of 13 clostridial species previously isolated from spoiled canned foods.

These spore suspensions and the original mixed culture collection *Bacillus* spp. spore suspension were heated separately in Oxoid Reinforced Clostridial Agar (RCA), adjusted to different pH and water activity levels to determine F_0 values required to prevent outgrowth on subsequent incubation, as described previously by Braithwaite & Perigo (1971). In this latter study RCA was used as the heating/culturing substrate rather than Oxoid Nutrient Agar, and 5-ml screw-capped instead of 28-ml screw-capped bottles. Each heated bottle was incubated *in toto* instead of dispensing the contents into Petri dishes, and only three incubation temperatures, 35, 45 and 55°C, were used. The medium was dispensed in 11·5 ml amounts and after autoclaving and cooling to 50°C, 0·5 ml of spore suspension was added and the medium mixed thoroughly before commencing heating.

The data from each spore suspension have been expressed in the form of three-dimensional models (Fig. 5) initially drawn on isometric paper, as described by Braithwaite & Perigo (1971), in which the curved surface represents the limiting F_0 value at a variety of pH and a_w combinations for a given storage temperature. This information is more conveniently expressed as a series of transverse sections through the three-dimensional model at specified F_0 values, thereby producing a contour plan of each model (Fig. 6).

Comparison of the plan diagrams (Fig. 6) with the original plan diagram of Braithwaite & Perigo (1971) (Fig. 4) showed that the original model did not adequately represent the limiting parameters obtained from naturally occurring spores. A new composite diagram was therefore constructed (Fig. 7) based on data from a total of 10,342 individual experimental determinations made throughout this series of experiments.

Examination of this new composite diagram shows it to be made up of a complex of curves, each part resulting from the survival and growth of spores from one or more sources. Although outgrowing spore species were not identified distinctions could be drawn between those growing aerobically or anaerobically and at thermophilic or mesophilic temperatures. In this way three regions of the diagram can be distinguished although some overlap inevitably exists between them.

Region I. This is a high resistance peak similar to, but more extensive than, that occurring in the original diagram (Fig. 4). This peak is bounded by pH values 5·8–8·0 and a_w values 0·970–0·995. Outgrowth was observed only at 45 and 55°C, gas was not produced, but growth occurred in both the presence and absence of air. It is therefore reasonable to conclude that survival and outgrowth of typical flat-sour organisms of the *B. stearothermophilus* type was responsible for this high resistance peak.

Region II. This region is adjacent to the high resistance peak and appears as a bulge or shoulder on the acidic side of the peak. It is bounded by pH 4·5–5·8 and a_w values 0·980–0·995. Again, outgrowth was only observed at 45 or 55°C incubation temperatures but in this case gas was produced and growth only occurred in the absence of air. It is reasonable to conclude, therefore, that survival and outgrowth of *Cl. thermosaccharolyticum*-type organisms were responsible for the shape of this region.

Region III. This region is the whole of the remainder of the diagram where process values significantly less than those observed in the other two regions were sufficient to prevent outgrowth. This region lies within pH values 5·8–8·0 and water activity 0·880–0·970, and pH 5·0–5·8 and a_w 0·94–0·98.

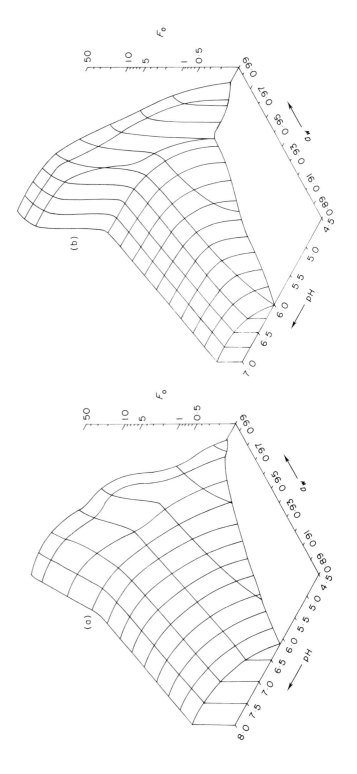

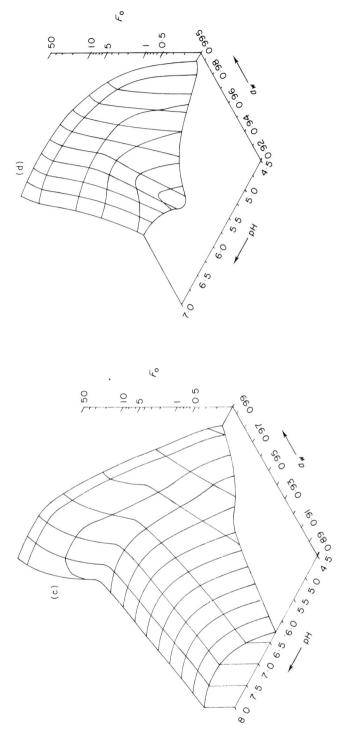

Fig. 5. Isometric illustration of the combined effect of pH and a_w on limiting F_0 value. The heating/culturing medium was Reinforced Clostridial Agar (Oxoid). (a) Spores from manure and untreated spices; (b) spores from Calcutta soil; (c) culture collection *Bacillus* spp. spores; (d) spoilage *Clostridium* spp. spores.

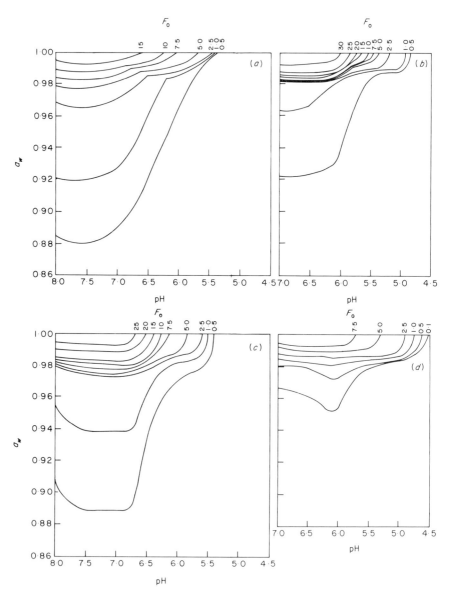

Fig. 6. Combinations of pH and a_w which result in limiting F_0 values. The heating/culturing medium was Reinforced Clostridial Medium (Oxoid). (a) Spores from manure and untreated spices; (b) spores from Calcutta soil; (c) culture collection *Bacillus* spp. spores; (d) spoilage *Clostridium* spp. spores.

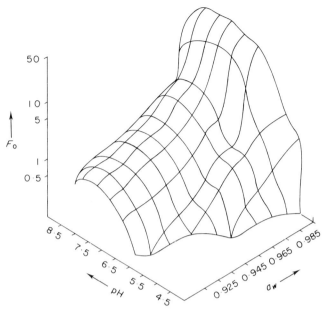

Fig. 7. New isometric illustration of the combined effect of pH and a_w on the limiting F_0 values. Results bulked from all experiments on naturally occurring and culture collection spores.

Growth occurred in this region at 35°C in both the presence and absence of air, and gas production was not observed. It is reasonable, therefore, to conclude that survival and outgrowth of mesophilic *Bacillus* spp. was principally occurring in this region.

Comparison of the new diagram with the original derived from culture collection strains of *Bacillus* spp. show general similarity in shape although considerable differences from the original findings occur in some areas (Fig. 8).

Such differences were not unexpected since culture collection strains may be expected sometimes to become attenuated with respect to resistance, and the thermophilic anaerobes not present in the original experiments are known to be heat-resistant under relatively acidic conditions. An interesting detail of the later work is that the clostridial spores used failed to grow out, even at optimum pH values, when subjected to sub-botulinal heat treatments at water activity levels below 0·950 (Fig. 5).

A plan view of the final composite diagram with limiting F_0 values drawn as contours shows the effects of substrate pH and water activity on the severity of heat treatment required to stabilize the system (Fig. 9).

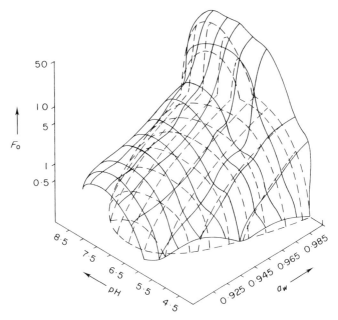

Fig. 8. Comparison of the original and new composite isometric illustrations of pH and a_w effects on limiting F_0 values. — — —, original data from *Bacillus* spp. spores in nutrient agar; ————, combined data from naturally occurring spores, *Clostridium* spp. and *Bacillus* spp. spores.

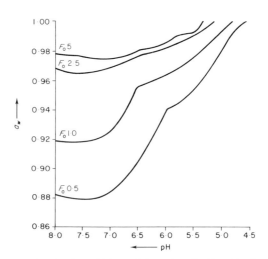

Fig. 9. Combinations of pH and a_w which result in various limiting F_0 values. Combined data from naturally occurring spores, *Bacillus* spp. and *Clostridium* spp. spores.

5. Future Developments

These later results are currently being challenged using a food substrate and different acidulant and humectant systems. Application of the information obtained so far will require further investigation of the effects of heating to determine the process levels at which inhibition of outgrowth, as opposed to thermal destruction, of spores occurs. If inhibition of outgrowth is a significant controlling factor, the effect of processing and storage conditions on subsequent pH or water activity levels in the food in relation to spore outgrowth must be thoroughly investigated. At the present time, therefore, it would be unrealistic to rely upon pH and water activity control as a means of reducing the severity of heat processes, and irresponsible to reduce those processes below the accepted minimum for control of *Cl. botulinum*.

However, the results presented here give sufficient indication of the possible benefits in modifying traditional heat processes to warrant further exploration of control of pH and water activity as adjuncts to heat processing in the stabilization of low-acid foods.

6. References

ADAMS, D. M. 1973 Inactivation of *Clostridium perfringens* Type A spores at ultra high temperatures. *Applied Microbiology* **26**, 282–287.

ANON. 1976 Lectrofood Corporation. Electrical resistance cooking appliance. US Patent 4099-454 (A37).

BEAN, P. G., DALLYN, H. & RANJITH, H. M. P. 1979 The use of alginate spore beads in an investigation of UHT processing. In *Proceedings of the International Meeting on Food Microbiology and Technology, April 1978*, ed. Jarvis, B. *et al.* pp. 281–294. Parma: Medicina Viva.

BRAITHWAITE. P. J. & PERIGO, J. A. 1971 The influence of pH, water activity and recovery temperature on the heat resistance and outgrowth of *Bacillus* spores. In *Spore Research 1971* ed. Barker, A. M., Gould, G. W. & Wolf, J. pp. 289–302. London & New York: Academic Press.

BUSTA, F. E. 1967 Thermal inactivation characteristics of bacterial spores at ultra high temperatures. *Applied Microbiology* **15**, 640–645.

CAMERON, M. S., LEONARD, S. J. & BARRETT, E. L. 1980 Effect of moderately acidic pH on heat resistance of *Clostridium sporogenes* in phosphate buffer and in buffered pea purée. *Applied and Environmental Microbiology* **39**, 943–949.

DENNY, C. B., SHAFER, B. & ITO, K. 1979 Inactivation of bacterial spores in products and on container surfaces. *International Conference on UHT Processing and Aseptic Packaging of Milk and Milk Products, North Carolina State University, Raleigh*.

EDWARDS, J. L., BUSTA, F. F. & SPECK, M. L. 1965 Heat injury of *Bacillus subtilis* spores at ultra high temperatures. *Applied Microbiology* **13**, 858–864.

HERMIER, J., BEGUE, P. & CERF, O. 1975 Relationship between temperature and sterilising efficiency of heat treatments of equal duration. Experimental testing with suspensions of spores in milk heated in an ultra high temperature steriliser. *Journal of Dairy Research* **42**, 437–444.

LEONARD, S., MERSON, R. L., MARSH, G. L., YORK, G. K., HEIL, J. R. & WALCOTT, T. 1975 Flame sterilization of canned foods: an overview. *Journal of Food Science* **40**, 246–249.

MILLER, I. & KANDLER, D. 1967 Temperatur-und Zeit-Abhängigkeit der Sporenablotung in Bereich der Ultrahocherhitzung. *Milchwissenschaft* **22**, 686–691.

NEAVES, P. & JARVIS, B. 1978 Thermal inactivation kinetics of bacterial spores at ultra high temperatures with particular reference to *Clostridium botulinum*—second report. *Leatherhead Food R.A. Research Report No. 286.*

PFLUG, I. J. & ESSELEN, W. B. 1953 Development and application of an apparatus for study of thermal resistance of bacterial spores and thiamine at temperatures above 250°F. *Food Technology* **7**, 237–241.

SALE, A. J. H. 1976 A review of microwave for food processing. *Journal of Food Technology* **11**, 319–329.

TURTLE, B. I. & ALDERSON, M. G. 1971 Sterilisable flexible packaging. *Food Manufacturer* **45**, 23, 48.

New Interest in the Use of Irradiation in the Food Industry

F. J. LEY

Irradiated Products Ltd., Swindon, Wiltshire, UK

Contents

1. Units

RECOMMENDATIONS have been made by the British Committee on Radiation Units and Measurements (BCRU) to change the units at present used in measuring radioactivity and ionizing radiations. The proposal is intended to bring the units in line with the International System of Units (SI) and is summarized in Table 1 for those units of particular interest. It has been

TABLE 1

Summary of old and new (SI) irradiation units

	Absorbed dose units		Radioactivity units	
	old	new	old	new
Name	rad	gray	curie	becquerel
Symbol	rad	Gy	Ci	Bq
Units	100 erg/g	1 J/kg	$3 \cdot 7 \times 10^{10}$ dis/s	1 dis/s
Relationship	1	0·01	1	$3 \cdot 7 \times 10^{10}$

FOOD MICROBIOLOGY
ISBN 0 12 589670 0

recommended that the date on which the old units are abandoned for official use should not be before 1982 and should not be delayed later than 1985. The old units are used throughout this review.

2. Introduction

The possibility of using ionizing radiation for the treatment of food for preservation or other purposes has attracted the serious attention of scientists in many countries during the last 30 years. Of particular interest are gamma rays as emitted by the radioisotope cobalt 60, or X-rays and electrons generated by electrical machines such as the linear accelerator. The use of these types of radiation has been long familiar in medical practice, in both diagnosis and therapy; in the latter case in relation to inactivation of viable tumour cells. Not so widely appreciated is the remarkable impact of radiation sterilization as applied to medical devices, particularly plastic disposables, surgical dressings, sutures and pharmaceuticals. In contrast, practical application in the food industry has been small scale and spasmodic and often confined to pilot-scale processing and market trials, largely due to the question of safety for consumption.

Food processes such as those based on heat, drying or freezing have been accepted as safe largely because of traditional use, but irradiation was conceived at a time when such treatment was associated with the atom bomb and with induced radioactivity, and at a time of increasing awareness of hazards to man through dietary changes and the use of chemical additives. National health authorities took the view that irradiation should be regarded as hazardous until proved safe. This unique situation for a food process presented experimental problems in demonstration of safety which have proved both long-term and expensive. Much effort has been provided by the United States through government agencies and the army, whilst in Europe many countries contributed to an international programme of research beginning in 1970, devoted entirely to this question.

The revival of industrial interest in the food area was stimulated in 1981 by a recommendation from relevant United Nations organizations that the process should now be accepted by national health authorities up to a certain radiation dose level. In addition, guidance is offered in the control of the process with respect to the licensing and operation of radiation facilities. Many countries are actively examining the recommendations, including the UK. Broad acceptance will lead to the implementation of a number of applications including several aimed at foods traded internationally.

3. Choice of Source

The general potential of irradiation for the treatment of food is seen in Table 2, alongside other processes currently in commercial use in chemical and general biological areas. This brings into context the dose levels required to bring about desired effects. The level of dose is directly related to the size of the radioisotope source required or the power of the electron machine to be installed. In the chemical area, where high doses are required to be delivered to high throughputs of wire and cable or fast moving film, the electron machine has been the choice. Such machines produce millions of rads in fractions of a second. Because of the limiting electron penetration of foods, machines are most suitable for in-line operations involving the treatment of individual packs no more than a few centimetres thick. However, there have been recent advances in the design of such machines and progress in the more efficient production of X-rays which give the same penetration as gamma-rays at the same energy level. Gamma-rays are particularly suitable for use in the medical sterilization area where there is a considerable attraction in processing medical devices in their final shipping carton, or sacks of powder such as kaolin talc in 25 kg sacks.

The radioisotope cobalt 60 with a half-life of 5·3 years, is the most attractive gamma emitter for industrial use. It can be produced in large

TABLE 2

Doses required for chemical and various biological applications of radiation

Application	Dose range (Mrad)
Modification of the properties of polymers, e.g. polyethylene or PVC	5·0–25·0
Long-term ambient storage, e.g. meat	4·0–6·0
Inactivation of *Bacillus anthracis* spores in hair or fur	2·0–2·5
Sterilization of medical devices and pharmaceuticals	1·5–2·5
Sterilization of packaging materials for medical or food use	1·0–2·5
Rendering laboratory animal diets 'pathogen-free'	1·0–2·5
Decontamination of food ingredients	0·7–1·0
Inactivation of *Salmonella* sp.	0·3–1·0
Reduction of micro organisms in cosmetics	0·3–1·0
Extended storage of meat or fish (0–4°C)	0·2–0·5
Prolongation of fruit storage	0·2–0·5
Control of parasites	0·01–0·2
Control of insects	0·01–0·2
Inhibition of sprouting, e.g. in potatoes	0·01–0·02
Increase in mutation rate in seeds and plants	0·001–0·01

quantities in certain nuclear reactors by neutron bombardment of the inactive cobalt 59 for periods usually in excess of 6 months. Canada is the main source of supply with smaller quantities available from France and the UK. Altogether some 90 plants are operating world wide. These plants currently house about 50 million curies between them with a capacity to build up to 100 million and there are several more plants now under construction. Following the commissioning in the UK of the first ever large-scale cobalt 60 plant in 1960, eight are now in operation. There could be heavy demands to be met if food irradiation is implemented even in limited areas, and no doubt there will be a place for both radioisotope and machine sources.

4. International Findings on Food Safety

In a review of the practical problems associated with animal feeding studies in the 1960s (Ley 1970), a change in objective away from clearances of specific foods to a broad clearance was strongly advised at the commencement of the international research programme on the subject initiated in 1970. Twenty-six countries, including the UK, contributed to a programme with its directorate centred at Karlsruhe, Federal Republic of Germany, and with most of the budget spent on toxicological research contracted out to established laboratories.

The results obtained from animal feeding studies were examined in 1976 by an international committee of experts (Anon. 1977a), brought together by the Food and Agricultural Organization (FAO), the International Atomic Energy Agency (IAEA) and the World Health Organization (WHO), all UN agencies. Irradiated wheat, potatoes, papayas, strawberries and chicken received 'unconditional acceptance'. Onions, rice, fresh cod and redfish received 'provisional acceptance', meaning that further information was deemed necessary. Of more importance, the committee stressed that irradiation should be regarded as a process akin to heating and freezing and should not be considered as an additive to food in relation to toxicological evaluation. Following this the scientific committee of the Karlsruhe project placed much greater emphasis on research aimed at obtaining a broad clearance of irradiated foods.

In 1980 the compiled data was again submitted to an FAO/IAEA/WHO Expert Committee this time requesting general clearance up to a dose level of 1·0 Mrad. The findings gave a clear recommendation for acceptance of this proposal (Anon. 1981a). It was agreed that there was no toxicological hazard and no general nutritional problems associated with the process. The approved types and sources of ionizing radiation are not sufficiently energetic to induce radioactivity in foods. Further, it was agreed that no unique microbiological hazards are likely to be introduced; considerations included the question of induced radiation resistance in micro-organisms.

A comprehensive account of the microbiological principles in food irradiation was given by Ingram & Roberts (1966), and specific problems in this area have been reviewed in special meetings (Anon. 1967). By the use of suitable radiation doses and good commercial practice, both in radiation plant operation and in the storage and distribution of food, no new problems are likely to arise. Each individual process will require scrutiny with respect to microbiological aspects, but this is true of any proposed process whether based on existing or new technology.

5. Applications

A. Microbiological considerations

A range of applications involving doses up to 1·0 Mrad are summarized in Table 3 and this is wide enough to arouse commercial interest in a number of

TABLE 3

Food applications of particular interest below the upper limit of the recommended dose* of 1·0 Mrad

General application	Specific examples	Dose (Mrad)
Decontamination of food ingredients	Various spices Onion powder Dyes Mineral supplements	1·0
Inactivation of *Salmonella*	Meat and poultry Egg products Shrimps and frog legs Meat and fish meal	0·3–1·0
Extended storage of fruit etc.	Strawberries Mangoes Papayas Dates Cocoa beans	0·2–0·5
Inhibition of sprouting or growth	Potatoes Onions Garlic Mushrooms	0·01–0·3

* With respect to the acceptability of any foods—Report of a Joint FAO/IAEA/WHO Committee (Anon 1981a).

areas. Those involving the control of micro-organisms require doses greater than 0·25 Mrad, the particular dose being specific to each application depending on the species involved, their numbers, the environmental conditions during irradiation, and the level of inactivation required. In general, it is accepted that viruses are more resistant than bacterial spores, resistance increasing with decreasing particle size, and in turn spores are more resistant than vegetative organisms, yeasts and moulds. However, there are some obvious exceptions as with *Micrococcus radiodurans* which fortunately is non-pathogenic and uncommon. Figure 1 illustrates the difference in the shapes of some dose/survival curves for different species irradiated in the same environment, plotted in a semi-logarithmic manner to convenience mathematical manipulation. The difference in resistance between species can be compared conveniently by observing that a 10^5 reduction in initial population covers a dose range of 0·05–1·6 Mrad.

The influence of different factors on the resistance of a given species may be measured using the D_{10} value or decimal reduction dose which can be obtained from the linear part of the dose/survival curve. The linear plot is expressed mathematically as

$$\log_{10} \frac{N}{N_0} = -kD,$$

where N is the number of cells surviving a treatment D, N_0 is the initial number of viable cells, and k is a constant equal to the slope of the curve. The slope can be seen to be $1/D_{10}$ and the number of log cycles or powers of 10 by which a population is reduced by a given treatment is obtained by dividing the treatment dose by the appropriate D_{10} value. If the quotient is designated x, then the term 10^x is referred to as the inactivation factor. A knowledge of this factor for a given species in a particular food, together with a figure for the likely initial level allows a definition of effectiveness of a given treatment.

The role of oxygen in enhancing radiation damage is well established in radiobiology. D_{10} values for bacteria may be increased by between 2 and 4 times under anaerobic conditions. Temperature during irradiation has a striking effect which is noteworthy in relation to the possibility of the treatment of frozen foods for control of pathogens. Microbes are more resistance to irradiation in the frozen state. Table 4 also shows that the very nature of the food itself is important in modifying radiation resistance. The large differences are accounted for not only in terms of gaseous environment and temperature, but also by the presence of natural sensitizing and protective agents and even by the water content of the organisms. A review of the influence of such factors, as well as those involving the radiation itself, such as dose rate and dose fractionation, is available (Ley 1973).

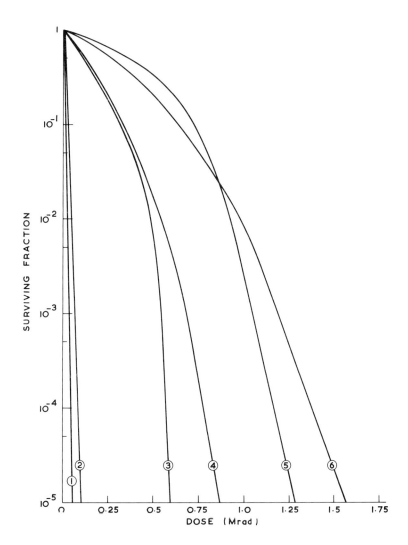

Fig. 1. Typical dose/survival curves for a number of bacterial species irradiated in buffer suspension in the presence of air. 1. *Pseudomonas* sp. 2. *Salmonella typhimurium*. 3. *Streptococcus faecium* $A_2 1$. 4. *Bacillus pumilis* spores. 5. *Clostridium botulinum* type A, spores. 6. *Micrococcus radiodurans*.

TABLE 4

The influence of media and temperature on the
radiation resistance of Salmonella typhimurium
expressed in D_{10} *values (krad)*

	D_{10} value (krad)	
Medium	At ambient temperature	When frozen ($-15°C$)
Phosphate buffer	20·8	39·1
Kaolin powder	21·0	. . .
Meat	55·8	96·3
Whole egg	63·2	68·0
Corned beef	80·0	. . .
Bone meal	91·0	. . .
Desiccated coconut	158·0	. . .

Data from Ley *et al.* (1963, 1970).

Excluded from Table 3 is the whole area of sterilization of meats and seafoods where inactivation of spores of *Clostridium botulinum* determines the dose requirement. A vast research and development programme with this objective was maintained in the United States under the auspices of the US Army and recently transferred to the Research Centre of the US Department of Agriculture in Philadelphia. Sliced beef, pork, ham and chicken of good quality has been produced in plastic pouches with long shelf life at ambient temperature. However, the process involves preliminary thermal inactivation of enzymes followed by high-dose irradiation at about $-30°C$ to protect against organoleptic changes. Even if such a high dose of between 4 and 5 Mrad is eventually approved the commercial value of such a process is somewhat doubtful. More details are given in a review by Urbain (1978).

B. Individual proposed processes

(i) Reduction of contaminants in spices

Contamination of spices with bacteria, moulds and yeasts presents a serious problem since their addition to various foods can be the cause of spoilage. Irradiation at the recommended maximum dose of 1·0 Mrad gives a marked reduction in the high populations encountered without causing any change in odour or taste (Tjaberg *et al.* 1972; Vajdi and Pereira 1973). Similar results have been obtained with onion powder showing initial microbial numbers of 10^6 to 10^7/g, to be reduced to less than 25,000/g with complete absence of yeasts, moulds and coliforms (Silberstein *et al.* 1979). Commercial application of the process in this area is very likely since the product is expensive enough to tolerate the additional cost and the annual tonnage involved is within the capacity of existing facilities, certainly in the UK. Furthermore, the current use of ethylene oxide gas for this treatment is in question, although irradiation will be more expensive, probably in the order of £120/tonne.

(ii) *Salmonella control*

A number of thorough investigations have been made into the efficacy of radiation for the treatment of some foods. Ley *et al.* (1970) studied the use of gamma radiation for the elimination of salmonellas from frozen meat intended for pet food, and recommended a minimum dose of 0·6 Mrad, resulting in an inactivation factor of 10^5. This dose would also result in a considerable reduction in numbers of other pathogens of public health significance. The process applied in practice to frozen blocks of meat as routinely handled has obvious advantages over the alternative heat processes which results in a cooked product. In Canada, attention has focused on the inactivation of *Salmonella* in poultry (Ouwerkerk 1981). A dose of 0·3 Mrad is recommended for chickens and turkeys at 'on ice' temperature. Health and Welfare of Canada have already approved this process subject to commercial runs to prove its effectiveness under practical conditions. Approximately the same dose is used in the Netherlands for treatment of frozen shellfish and frogs' legs.

A commercial-scale trial irradiation of pre-cooked, peeled, frozen shrimps was undertaken in Australia in 1978 (Wills 1981). About 65 tonnes were treated with 0·6–0·8 Mrad, resulting in at least a hundredfold reduction in total count and reduction of *Staphylococcus aureus, Salmonella* spp. and *Escherichia coli* to below the levels of detection. The treatment had no effect on organoleptic properties; the shrimps were sold through restaurants with no public reaction following wide publicity through the communications media.

(iii) *Extended storage of meats and fish*

A dose within the range of 0·1–0·2 Mrad is sufficient to control microbial spoilage which would otherwise occur during storage of fresh meat at below 5°C. This treatment is part of a procedure suggested by Urbain (1973) who claims that satisfactory quality of beef can be maintained for up to 21 days. Irradiation of meats and poultry effectively inactivates the radiation-sensitive gram negative rods, including *Pseudomonas*, *Achromobacter* and *Flavobacterium*. Drip control and colour preservation are obtained by polyphosphate treatment and protection against oxidation by vacuum packaging.

Whilst white fish will tolerate low doses, the fatty species such as mackerel, salmon or herring are severely affected organoleptically. In the UK a detailed investigation was made of the use of a dose of 0·35 Mrad to improve the distribution and retail sale of white fish. The microbiological and toxicological research was concentrated on cod. It was envisaged that frozen-at-sea fish would be held in buffer store to supply continuously production lines involved in thawing, filleting, phosphate dipping and vacuum packaging in plastic pouches. These packs would be irradiated immediately, resulting in a storage life improvement at temperatures not exceeding 3°C. Particular emphasis was given to the question of health hazards due to *Clostridium botulinum* type E. It was concluded that good commercial handling practice, including tight control of temperature, would obviate any danger (Shewan & Hobbs 1970). An interesting combination process is envisaged for shrimp by Savagaon *et al.* (1972), the relatively heat-sensitive *Cl. botulinum* type E spores being inactivated by a heat treatment prior to packaging, followed by low-dose irradiation to control the adventitious contamination.

(iv) *Preservation of fruits and vegetables*

The high acidity in fruit is sufficient to prevent bacterial spoilage; the decay which can be controlled by irradiation is associated with yeasts and moulds. Elimination of moulds and considerable reduction in yeast contamination can be achieved with comparatively low doses. Detrimental changes are apparent at doses exceeding 0·3 Mrad, particularly texture loss, partly due to pectin degradation, pitting of citrus fruit, colour change and loss of natural flavour. However, there has been considerable success with strawberries, with improved storage life at both ambient and refrigerator temperatures, as demonstrated in the Netherlands some years ago. The combination of dipping in hot water at 50°C followed by irradiation of sub-tropical fruits such as mangoes and papayas using 75 krad, produced excellent results with respect to the control of post-harvest diseases (Brodrick & Thomas 1977). The irradiation also controls insect infestation and delays ripening. These fruits could become more common in European markets with the solution to shipping and quarantine problems.

Vegetables are damaged by radiation in much the same way as fruits, leafy vegetables losing crispness and flavour whilst peas, beans and carrots become soft and discoloured. Radiation does not inhibit metabolic activity in fresh produce and even if sterility could be achieved without damage, a blanching treatment would be necessary for long storage. Mushrooms were successfully irradiated at 0·25 Mrad and marketed in large-scale trials in the Netherlands (de Zeeuw 1975). This treatment keeps the caps tight and maintains whiteness for considerable periods.

(v) *Miscellaneous*

Insects and parasites at all growth stages succumb to comparatively low doses. However, unlike the use of chemical agents, irradiation confers no control against reinfestation and is not, therefore, seen as being applicable to grain handled and stored in bulk in up-country areas. More attractive is the application to products such as pre-packed dried dates. Also of interest is irradiation of cocoa beans for both insect control and mould inhibition.

The radiation inhibition of sprouting in potatoes, onions, garlic and shallots is well established. With potatoes, the most important root crop, radiation causes an immediate increase in sugar content and a decrease in vitamin C, but these changes become insignificant after several weeks' storage and do not detract from the usefulness of the process. There are indications, however, that the treatment increases the susceptibility of the tubers to mould attack, apparently due to prevention of periderm formation in tubers damaged by handling. Furthermore, the gradual increase in sweetening, which normally occurs in potatoes during prolonged storage, is accelerated. It is necessary, therefore, apart from economic considerations, to keep the dose level as low as possible to minimize these side effects whilst achieving an acceptable degree of sprout suppression.

A feasibility study carried out in the UK in the 1960s was not encouraging because of variation in harvest size and hence crop value, and the varying need for long-term storage coupled with the high cost of transport to a radiation facility. However, the process is being tried on a commercial scale in several countries.

C. Commercial use

The first regular commercial use of irradiation occurred in Japan, being applied to sprout control in potatoes. A purpose-built cobalt 60 plant was commissioned in 1974 at Hokkaido commencing with 5000 tonnes per annum and reaching 30,000 tonnes in 1978, and is still continuing. Other countries, in particular Italy, Hungary and Israel, have undertaken the distribution and sale of pilot quantities of irradiated potatoes, onions and garlic.

The Netherlands has been at the forefront of practical application. The market trials with mushrooms and strawberries undertaken during 1969–70 were not followed up for economic reasons. However, a number of foods or food ingredients for human consumption are irradiated regularly and currently marketed very successfully. These include spices, cocoa bean powder, shrimps and frogs' legs. Spice irradiation is extending rapidly; many tonnes are also irradiated in Belgium and Hungary and authorization has been given in France. Tonne quantities of strawberries, mangoes and papayas have recently been irradiated and marketed in South Africa in consumer acceptance trials. Up to 50 tonnes of irradiated mangoes were shipped to the Netherlands and found to be in excellent condition compared with controls. Small quantities of irradiated foods have also been distributed in Bulgaria, Czechoslovakia and the USSR.

It is of significance to record that gamma radiation has been used in the UK for the sterilization of laboratory animal diet since 1962, although no food for human consumption has been distributed. Other countries now use irradiated diets bringing the total quantity in use to about 1500 tonnes per annum. The diet is intended for specified-pathogen-free and germ-free breeding colonies, and useful experience of the effectiveness of irradiation for microbiological control has been gained, as well as evidence of the absence of toxic effect and minimal change in nutritive value (Ley *et al.* 1969). A detailed microbiological investigation of a number of batches of diet showed a D_{10} value of 172 krad for natural spore contaminants (Halls & Tallentire 1978). Based on an initial count of 10^5/g a radiation dose of 2·5 Mrad would give a calculated survival of 1 organism/10^{10} g of diet, thus giving a very high margin of microbiological safety.

6. Control of the Process

A. Legislation

By 1981 most countries had special regulations referring to the irradiation of food for human consumption, either incorporated into already existing regulations governing food additives, as in the USA and Canada, or especially introduced, as in the UK. In practical terms, irradiation of food for sale for human consumption is prohibited and an exemption from this control is given only after scrutinizing evidence of the safety of each process proposed. Exceptionally, Dutch legislation allows limited uses without such demanding evidence which is the reason for such wide experience of the process in the Netherlands with respect to both market trials and now full commercial use.

The UK introduced the Food (Control of Irradiation) Regulations in 1967 followed by an amendment in 1969, allowing the irradiation of food for patients in hospital who need sterile diets as an essential factor in their treatment. These regulations followed a detailed official review of the subject (Anon. 1964). Another review is expected in 1983 following the FAO/IAEA/WHO Expert Committee recommendations of 1980 (Anon. 1981a) regarding overall clearance up to a dose of 1·0 Mrad as discussed earlier.

A significant development in the USA has been the issue of a notice inviting comment on a proposed modification of the rules governing irradiation under the 1958 amendment to the Federal Food, Drug and Cosmetic Act. The notice (Anon. 1981b) proposes broad clearance of food up to 100 krad. The use of doses higher than this would still require toxicological studies on specific foods unless such foods or ingredients represent a very small proportion of the diet. This proposed broad clearance of the use of low doses has been stimulated by the desire to replace ethylene bromide fumigation as a means of insect control. A serious situation faces states such as Florida and California with regard to the spread of Mediterranean fruit fly. A number of comments regarding the notice has been made to the USFDA in an attempt to elevate the level of broad clearance to 1·0 Mrad in line with the international recommendations. A decision is expected in early 1983.

B. General standard and code of practice

The Codex Alimentarius Commission which acts under the FAO/WHO Food Standards Programme is developing general standards for irradiated foods. The object is to remove barriers to international trade by incorporating such standards into the national laws of the 120 member states. A draft general standard and code of practice is currently moving through the various stages of approval which, at the earliest, could be complete in 1983.

The standard refers to the upper dose limit of 1·0 Mrad and the range of radiation sources which may be used. Overall control of the process will be exerted by the licensing and registration of facilities by competent national authorities. The facilities themselves should be operated by suitably qualified and trained personnel in accordance with the recommended code of practice. These recommendations follow the procedures already operative for facilities involved in the sterilization of medical devices and pharmaceutical products. The standard also stresses that foods must be handled in accordance with acceptable hygienic practice. Labelling is an area of contention and it is suggested that requirements should accord with the standards generally accepted for food in general. The acceptance of irradiation as a process should obviate the need for labelling—there is no scientific reason for it.

The recommended code of practice refers to the various radiation plant parameters which influence dose, including source strength, conveyor speed or dwell time of the product in the radiation cell, the distance of product from the source and the product density. Control of such parameters is straightforward, but varies for radioisotope or machine source. Detailed records are expected. This control is backed up with dosimetry measurements. Measurement of absorbed dose can be made on the product itself and there is a choice of techniques available (Anon. 1977b). The recommended upper limit of dose for food is referred to as an overall average dose of 1·0 Mrad, which is suggested to be the mean of minimum and maximum doses as measured normally at the centre and on the outside, respectively, of the product.

7. Discussion

New interest by the food industry in irradiation is timely. It is envisaged that many countries will accept the process at least within the limits recommended by the relevant UN organizations. Faced with increasing concern over the safe use of such chemicals as ethylene oxide for inactivating micro-organisms or ethylene dibromide for insect control, health authorities are seeing the attraction of the process. Increase in international food trade is another factor since irradiation might be the answer to quarantine problems, as for example with the import of exotic fruits from Africa or for the treatment of food in the developing countries where so much spoilage occurs during storage.

Official approval of the process will stimulate industry to undertake research and development studies in specific areas and these will reveal the technical feasibility of the many applications proposed. The fact that chemical changes can occur in food is evident at high doses in terms of organoleptic properties. The tolerable dose varies from food to food and in some instances can be very low, e.g. milk, and therefore no application can be foreseen. It is quite impracticable, if not impossible, to characterize and quantitate by direct analysis all the chemical changes that could occur in irradiated foods. However, considerable effort has been made, the results of which form a substantial part of the information available for scrutiny with respect to safety considerations (Elias & Cohen 1977). Freedom from legislative controls will also allow market trials to proceed in order to test public consumer reaction and reveal technical and economic feasibility.

The radiation industry is also interested in this new potential already having behind it the experience of commercial processing in the medical and chemical industries. There is already sufficient capacity in the UK to see the food industry through development and pilot plant stages or even to undertake irradiation of certain food ingredients where the total throughput is relatively small. Large-scale applications will need in-line facilities and even low-dose processes demand large cobalt 60 sources. Referring to the poultry industry in Canada and the potential for *Salmonella* control, Ouwerkerk (1981) quotes a figure of 7·5 tonnes/h needing 1 MCi for a dose of 0·3 Mrad. Such a facility, including source, shield and handling equipment, would cost between £750,000 and £1,000,000. Many such facilities would be required to satisfy the whole of the Canadian poultry industry if irradiation became accepted practice. This is just one example and the question arises as to the availability of this radioisotope in such large quantities. It is doubtful whether such a demand could be met over a period of a few years. By far the main supplier of cobalt 60 is Atomic Energy of Canada Limited, followed by Commissariat a L'Energie Atomique in France. The UK Atomic Energy Authority is unlikely to continue any supply after 1982. Caesium 137 appears to be an attractive alternative source of gamma radiation: it is present in significant quantities in fission-product liquor from separation plants, but the cost of such plants is very high and there are other disadvantages.

Electron machine manufacturers will be keeping abreast of developments since machines can operate at a high rate, although penetration is limited. However, new machines of high beam-energy will compete with gamma rays and so too will X-rays produced when the electron accelerator is aimed at water-cooled tungsten plate. The perfecting of such equipment will depend on the response of the food industry to the new outlook.

The various organizations of the United Nations already mentioned, as well as the Organization for Economic Co-operation and Development (OECD) which sponsored the Karlsruhe project, have played important roles in the development of the technology. Much of the science and technology of the subject is published in the proceedings of a series of international meetings already referred to in the text. The next few years might see the implementation of many of the ideas which have been put forward. The first applications will be modest in size, largely using existing service facilities, but even large food manufacturers will hesitate over the high capital investment incurred for high throughputs. The many market trials with irradiated foods already carried out in a number of countries have not initiated any adverse public comment, indeed the idea is attractive if it replaces some of the many toxic chemicals now in use or if the process is seen to have a significant impact in the area of food hygiene or in minimizing food losses incurred through spoilage.

8. References

ANON. 1964. *Report of the Working Party on Irradiation of Food.* London: HMSO.

ANON. 1967 *Microbiological Problems in Food Preservation by Irradiation.* IAEA Publication No. STI/PUB/168. London: HMSO.

ANON. 1977a *Wholesomeness of Irradiated Food.* Report of Joint FAO/IAEA/WHO Expert Committee. WHO Technical Report Serial No. 604. London: HMSO.

ANON. 1977b. *Manual of Food Irradiation Dosimetry.* IAEA Technical Report Serial No. 178. London: HMSO.

ANON. 1981a *Wholesomeness of Irradiated Food.* Report of a Joint FAO/IAEA/WHO Expert Committee. WHO Technical Report Serial No. 659. London: HMSO.

ANON. 1981b Policy for irradiated foods, advance notice of proposed procedures for the regulation of irradiated foods for human consumption. US Federal Register Vol. 46, No. 59, March 27th.

BRODRICK, H. T. & THOMAS, A. C. 1977 Radiation preservation of subtropical fruits in South Africa. In *Food Preservation by Irradiation* Vol. I. Proceedings of a symposium, Wageningen, The Netherlands, IAEA Publication NO. STI/PUB/470 pp. 167–178. London: HMSO.

DE ZEEUW, D. 1975. *Commercialisation of Irradiated Potatoes, Mushrooms, Onions and Spices in the Netherlands.* Proceedings of a meeting on Requirements for the Irradiation of Foodstuffs on a Commercial Scale, Vienna. IAEA Publication No. STI/PUB/394, pp. 133–139. London: HMSO.

ELIAS, P. S. & COHEN, A. J. (eds) 1977 *Radiation Chemistry of Major Food Components.* Amsterdam & New York: Elsevier/North-Holland Inc.

HALLS, N. A. & TALLENTIRE, A. 1978 Effects of processing and gamma irradiation on the microbiological contaminants of a laboratory animal diet. *Laboratory Animals* **12**, 5–10.

INGRAM, M. & ROBERTS, T. A. 1966 Microbiological principles in food irradiation. In *Food Irradiation.* Proceedings of symposium, Karlsruhe, Germany. IAEA Publication No. STI/PUB/127, pp. 267–285. London: HMSO.

LEY, F. J. 1970 Safety of irradiated food-biological aspects. *Proceedings 3rd International Congress of Food Science and Technology*, Washington, D.C. pp. 806–815.

LEY, F. J. 1973 The effect of ionizing radiation on bacteria. In *Manual on Radiation Sterilization of Medical and Biological Materials.* IAEA Technical Report Serial No. 149, pp. 37–63. London: HMSO.

LEY, F. J., FREEMAN, B. H. & HOBBS, B. C. 1963 The use of gamma radiation for the elimination of *Salmonella* from various foods. *Journal of Hygiene, Cambridge* **61**, 515–529.

LEY, F. J., BLEBY, J., COATES, M. E. & PATERSON, J. S. 1969 Sterilization of laboratory animal diets using gamma radiation. *Laboratory Animals* **3**, 221–254.

LEY, F. J., KENNEDY, T. S., KAWASHIMA, K., ROBERTS, D. & HOBBS, B. C. 1970 The use of gamma radiation for the elimination of *Salmonella* from frozen meat. *Journal of Hygiene, Cambridge* **68**, 293–311.

OUWERKERK, T. 1981 Salmonella control in poultry through the use of gamma irradiation. *Proceedings of a Meeting on Combination Processes for Food Irradiation.* Colombo, Sri Lanka, IAEA Publication No. STI/PUB/568, pp. 335–345. London: HMSO.

SAVAGAON, K. A., VENUGOPAL, V., KEMAT, S. V., KUMTA, U. S. & SREENIVASEN, A. 1972 Radiation preservation of tropical shrimp for ambient temperature storage. 1. Development of heat-radiation combination process. *Journal of Food Science* **37**, 148–150.

SHEWAN, J. M. & HOBBS, G. 1970 The botulism hazard in the proposed use of irradiation of fish and fishery products in the UK. In *Preservation of Fish by Irradiation.* IAEA Panel Proceedings Series, pp. 117–124. London: HMSO.

SILBERSTEIN, O., GALETTO, W. & HENZI, W. 1979 Irradiation of onion powder: effect on microbiology. *Journal of Food Science* **44**, 975–981.

TJABERG, T. B., UNDERDAL, B. & LUNDE, G. 1972 The effect of ionizing radiation on the microbiological content and volatile constituents of spices. *Journal of Applied Bacteriology* **35**, 473–478.

URBAIN, W. M. 1973 The low-dose preservation of retail cuts of meat. In *Proceedings of Symposium of Radiation Preservation of Food, Bombay, India.* IAEA Publication No. STI/PUB/317, pp. 155–227. London: HMSO.

URBAIN, W. M. 1978 Food irradiation. *Advances in Food Research* **24**, 155–216.

VAJDI, M. & PEREIRA, R. R. 1973 Comparative effects of ethylene oxide, gamma radiation and microwave treatment on selected spices. *Journal of Food Science* **38**, 893–896.

WILLS, P. A. 1981. Commercial application of freezing–irradiation combination process for pasteurisation of two specific batches of cooked, peeled shrimps. *Proceedings of a Meeting on Combination Processes for Food Irradiation.* Colombo, Sri Lanka, IAEA Publication No. STI/PUB/568, pp. 291–304. London: HMSO.

New Methods for Controlling the Spoilage of Milk and Milk Products

B. A. LAW AND L. A. MABBITT

Microbiology Department, National Institute for Research in Dairying,
Shinfield, Reading, Berkshire, UK

1. Introduction

IT IS NOT surprising that cow's milk is an extremely good food for calves—it is also an excellent food for man. A glance at its chemical composition reveals why this is so (Table 1) and also shows clearly why it is a very satisfactory medium for microbial growth, a fact which enormously complicates the harvesting of the product and its processing. Those familiar with a farm environment will appreciate the impossibility of excluding bacteria from the milk supply during milking; the consequent problem of controlling bacterial multiplication in the raw milk prior to processing is obvious.

The total numbers of bacteria contaminating the milk at the farm varies from about 10^3/ml when hygiene is very good, to 10^6/ml when it is poor. With such inocula in such a good growth medium, and bearing in mind the need in most cases to transport the milk and store it at manufacturing creameries for

FOOD MICROBIOLOGY
ISBN 0 12 589670 0

TABLE 1

Approximate composition of cows' milk

Water	*ca.* 87% (w/v)
Carbohydrates	*ca.* 4·9% (w/v)
	α-lactose 37% of total lactose
	β-lactose 63%
	(minor) Glucose, galactose, polysaccharides, sugar phosphates
Lipids	*ca.* 3·7% (w/v)
	Triglycerides 98% of total lipid
	(minor) Diglycerides, monoglycerides, phospholipids, sterols, carotenoids, vitamins A, D, E. K
Proteins	*ca.* 3·5% (w/v)

	% of total protein
Casein	
α-Casein	53–70
β-Casein	25–35
κ-Casein	8–15
γ-Casein	3–7
Non-casein	14–24
β-Lactoglobulin	7–12
α-Lactalbumin	2–5
Blood serum albumin	0·7–1·3
Euglobulin	0·8–1·7
Pseudoglobulin	0·6–1·4
Immunoglobulins	1·3–2·8
Proteose-peptone	2–6

(minor) Lipases, esterases, alkaline phosphatase, acid phosphatase, xanthine oxidase, lactoperoxidase, proteases, catalase, sulphydryl oxidase, amylase, aldolase, ribonuclease, carbonic anhydrase, saldase, rhodonase, lactase, lactoferrin.

Vitamins	A, B_1, B_2, Nicotinic acid, B_6, Biotin, Folic acid, B_{12}, C, D, E, K
	Choline, Inositol
Ash	0·7% (w/v)

appreciable periods before processing, urgent action must be taken to restrain bacterial growth and prevent spoilage of this important and valuable foodstuff (at doorstep prices the value of milk for the UK is around £2500M per annum). Fortunately, freshly drawn milk contains several natural antibacterial systems. As well as the immunoglobulins, it contains two other proteins, lactoferrin (LF) and lactoperoxidase (LP), whose action can suppress bacterial growth. Lactoferrin binds Fe^{3+} very strongly (association constant $= 10^{36}$) and prevents its uptake by bacteria (Reiter *et al.* 1975; Griffiths & Humphreys 1977). Its potential usefulness lies in its capacity to inhibit selectively those bacteria whose Fe^{3+} requirement is high (e.g. *Escherichia coli*) while allowing the growth of those with low Fe^{3+} requirements (e.g. the lactic acid bacteria used for dairy product manufacture). However, LF is not present at sufficiently high concentrations to be effective in bovine milk and the milk citrate is antagonistic to its Fe^{3+} binding capacity (Masson 1970; Reiter *et al.* 1975). It can be isolated easily from bovine milk whey (Law & Reiter 1977) but is not available commercially in sufficient quantities to justify milk preservation trials by supplementation.

Lactoperoxidase is also present in only small amounts in bovine milk but it acts catalytically and is therefore a more potent and potentially useful natural antibacterial agent. This enzyme uses hydrogen peroxide to oxidase thiocyanate ions (SCN^-) to an unstable intermediate product (probably hypothiocyanate, $OSCN^-$) which kills Gram negative bacteria (e.g. *E. coli*, *Salmonella* spp., *Pseudomonas* spp.; Björck *et al.* 1975). The complete mode of action of the lactoperoxidase bactericidal systems (LPS) is not understood but it prevents amino acid uptake (Marshall 1978) and interferes with membrane energization in *E. coli* by inhibiting the membrane bound D-lactate dehydrogenase (Law & John 1981). Other workers have reported the oxidation of protein-bound sulphydryl groups by LPS but were unable to relate this directly to cellular function (Thomas & Aune 1978), though earlier studies by Oram & Reiter (1966) suggested that glycolytic enzymes may be involved.

There is sufficient LP in raw bovine milk to kill the Gram negative bacteria with spoilage potential but the activity of the LPS as a whole is restricted by the amounts of SCN^- and H_2O_2 normally present in milk (Reiter & Marshall 1979) and essential for activity of the LPS.

Thiocyanate concentrations can be made adequate by adjusting the cows' diet (Korhonen 1980) but H_2O_2 must be supplied exogenously.

2. Sources of Contamination of Milk

As previously implied, the types and efficiencies of cow management and washing techniques for milking systems are varied and it is impossible to provide generalized data about the various sources and numbers of the different species of bacteria which get into the milk supply. The numbers due to aerial contamination of the milk are usually very small but are nevertheless important in special circumstances; for example, even low numbers of clostridial spores in milk for Emmental and Dutch-type cheese may cause detrimental gas formation in the product. Such defects are uncommon in our more acidic hard cheeses, so aerial contamination of milk in the UK is unlikely to be important except in special cases. The three main important sources of bacteria are the cows' teats, milking and storage equipment, and the interior of the udder—the last sometimes being a more important source than is generally realized. Table 2 indicates the relative contributions of these sources.

TABLE 2

Sources of the microflora of ex-farm milk and their estimated contribution to total numbers

	Number of bacteria per ml of milk in or from				
Production hygiene	Total	Udder interior	Teats	Equipment	Somatic cells per ml udder interior
Good	10,000	200	4800	5000	200,000
Average	50,000	1000	20,000	29,000	400,000
Poor	1,000,000	2000	100,000	898,000	1,000,000

A. Udder interior

The bacteria in milk taken aseptically from a healthy udder are derived mainly from the streak canal of the teat and are unlikely to exceed 10^2–10^3/ml. However, udder disease (mastitis) is present in all dairy herds in varying degrees. In England and Wales about 1–2% of all cows have clinical infections and about 35% are subclinically infected (Wilson & Richards 1980). The number of pathogens in the milk of clinical cases may reach *ca.* 10^8/ml but such milk must by law be excluded from the main supply. In subclinical cases, the number of pathogens may be about 10^5/ml, but the effect of dilution with milk from uninfected cows will bring the numbers down to a few thousand per ml of the bulked milk. Because of the severity of its effect *E. coli* is an important mastitis pathogen at present, and consequently any contribution of coliforms to milk from this pathological source makes a nonsense of their use as indicators of faecal contamination.

B. Teat surfaces

Concentrated housing of cows combined with wet conditions, as may occur in winter, introduces an almost insoluble problem of controlling contamination of the udders. A similar problem presents itself when cows are outdoors in ill-drained and dirty areas. Heavily contaminated teats may contribute about 100,000 bacteria/ml of milk. For this and aesthetic reasons husbandry advisers and relevant codes of practice recommend that the teats are washed before milking. To do this effectively under farm conditions and production pressures is no mean task. A particular difficulty is that unless considerable care is taken, dirt and bacteria on the udder may be washed down to the more important teat surfaces from which they gain easy access to the milk supply.

An average herd size of, say, 50 cows may produce about 1000 litres of milk at one milking so that any contamination from equipment is considerably diluted by the large volume of fluid produced. At this level of production milk with a final count of only 10,000 bacteria/ml derived from the equipment must have washed off 10^{10} bacteria from the contact surfaces (area, *ca.* 10 m^2), indicating a surface contamination level of *ca.* 10^9/m^2. The equipment has to be very dirty to reach this level and reference to Table 2 indicates that there must therefore be considerable scope for improvement in the efficiency of cleaning equipment on the average farm.

C. Important types of bacteria in the milk supply

The species of bacteria contaminating milk during production can be surmised from the main sources, i.e. soil, faeces, udder, skin and equipment. These contaminants are important for two main reasons: the justified aesthetic and health objection to any gross contamination during production, and the undoubted ability of some of the degradative bacterial contaminants to multiply in the milk during transport or storage. The resulting deterioration in milk quality before processing may be due to the ensuing change in composition of the substrate or, more probably, to heat-tolerant lipases and proteases of these contaminants which can survive pasteurization and continue their activity during storage of the product. This latter problem increases as the developed countries move rapidly to milk handling systems which involve refrigerated storage at the farm at 4°C or below and subsequent transport and storage at low temperatures at the creamery. Under such conditions the microflora of the milk at the point of processing is almost entirely made up of psychrotrophic bacteria, mainly pseudomonads but also aeromonads, *Acinetobacter* spp. (Table 3) and even psychrotrophic spore formers (Law *et al.* 1979). It is unfortunate that some of these species,

TABLE 3

Psychrotrophic bacteria commonly isolated from refrigerated raw market milk

---

Pseudomonas: fluorescens, fragi, putida, putrefaciens,
    aureofaciens
Acinetobacter spp.
Flavobacterium spp.
Cytophaga spp.
Aeromonas spp.
Bacillus: sphaericus, circulans

---

particularly pseudomonads, frequently produce heat-resistant proteases and lipases referred to above; the ability to measure these enzymes at the very low concentrations usually present would clearly be advantageous.

3. Methods for Controlling Bacterial Growth in Raw Milk

A. Prevention of contamination before processing

Prevention is always better than cure but under commercial farm conditions the best that can be achieved is to minimize contamination without unduly affecting production costs.

Despite decades of research and major advances in techniques and treatment of the disease, the control of mastitis only slowly reduces the number of infected udders. When, as in the case of mastitis, there are a number of potential pathogens, their balance tends to change under the impact of applied control measures while the disease maintains its grip. In England and Wales a survey has indicated that about 32% of cows' udders are infected with streptococci, staphylococci or coliforms (Wilson & Richards 1980). The measures to control mastitis resulting from collabora-tive work at the National Institute for Research in Dairying and the Central Veterinary Laboratory, Weybridge, UK, have been adopted by most developed countries. They consist of antibiotic infusion of the udders of all cows at 'drying off' and of udders showing clinical symptoms during lactation together with the treatment of all teats *after* milking with disinfectant to control cow–cow transmission and teat–teat transmission of pathogens. Unfortunately this teat disinfection does not, and could not, control patho-gens which contaminate the teat between milking—the so-called environ-mental pathogens such as *Streptococcus uberis* and *E. coli*—precisely those now causing trouble when the control measures are exerting a significant effect. The teat dips now being used, mainly iodophors, remain active on the skin surface for only a few hours and the present hope is that a teat disinfectant with persistent activity will become available for application after milking. A naïve approach would suggest that persistence would best be achieved by treating the negatively charged skin surface with a positively charged disinfectant and clearly neither iodine- nor chlorine-based products are in this category. The persistence on skin of positively charged agents, e.g. chlorhexidine, is already appreciated in medicine and we have some evidence that future progress in mastitis control may well be in the use of such materials. For the time being, however, the numbers of bacteria entering milk from infected udders are unlikely to decrease markedly in the short term.

TABLE 4

Contamination of milk with bacteria from surfaces of cows' teats after different hose washing treatments

Treatment	Total colony count/ml milk (geometric mean)* ($\times 10^{-3}$)
Unwashed	7·5
Washed with water, left wet	7·9
Washed with water, dried	4·2
Washed with NaOCl,† left wet	4·1
Washed with NaOCl,† dried	1·5

* Data from Cousins (1978).
† NaOCl solution contained 600 mg/l available chlorine.

In preparing the cow for milking the importance of drying the teats after washing has been emphasized recently by work at NIRD (Table 4). As indicated earlier, washing the teats and leaving them wet may not significantly reduce the contamination of milk from the teat surface. Whether this is due to the numbers of bacteria suspended in the aqueous film on the teat or whether a moist teat releases bacteria more easily from deeper layers of the epithelium is not clear. Unfortunately drying of the teat with paper towels introduces another operation with a financial penalty into a busy milking routine and whether it can be adopted as a general procedure depends perhaps on further innovation at the farm. Nevertheless its use may reduce the contribution from heavily contaminated teats, but only two- to five-fold.

Improved cleaning systems for milking machine pipelines would also have a very beneficial effect on the hygienic quality of our milk. If they are to be used by the farmer they must be cheap and simple to operate. The acid boiling water (ABW) system devised at NIRD falls into this category and has been used extensively in practice. Boiling water is sucked through the plant by the milking vacuum, entraining acid in the first 4 min of flow and washing acid-free in the remaining 2 min (Cousins & McKinnon 1979). Its success probably depends on the effective combination of heat and low pH on the surfaces contacted by the solution and on the conduction of heat by metal to those difficult crevices the solution does not reach. Nevertheless, greater attention to the design and lay-out of milking machine pipelines in order to facilitate cleaning would reduce markedly the contamination of milk derived from milking equipment.

B. Monitoring contamination

There is little doubt that an accurate monitoring scheme combined with a financial incentive is the most effective way of ensuring product quality. For this reason a search for good, cheap, preferably rapid methods for estimating bacterial numbers or metabolites in food has been in progress for decades. Over the past 50 years the UK dairy industry has used Breed smears, poured plates, roll tubes, stability tests and dye reduction tests (methylene blue, resazurin). The scientific basis for the current reduction tests, never very convincing, has dissolved progressively as refrigeration of milk at the farm increased; in general, the psychrotrophic flora has low reductive capacity and the correlation between numbers of bacteria in refrigerated milk and the time of reduction is scientifically unacceptable (correlation coefficients of only about 0·5 are usual). More recently research has revealed other possible indirect tests; calorimetry, impedance, ATP, pyruvate. They are all either too time-consuming for industrial application, lack reproducibility or the results are poorly correlated with bacterial numbers in milk.

An accurate, direct method of counting the bacteria in milk would give the least ambiguous result. For application at the creamery the method should be able to produce a result in less than 30 min so that it could be used to reject unsatisfactory tanker milk on arrival at the creamery or to divert milk in store to the most suitable product. Furthermore, it should be sufficiently sensitive to enable the lower numbers of bacteria in milk at the farm to be counted accurately; thereby allowing milk at both ends of the transport chain to be assessed by the same method and the causes of changes in hygienic quality during transport and storage to be located with certainty. In developing such a method, the Breed smear method (0·01 ml milk smeared over 1 cm^2 of slide, stained and counted microscopically) was taken as the starting point because of its simplicity and rapidity. Three main criticisms of this method needed attention: the very small sample is unrepresentative; the bacteria are not uniformly distributed over the area due to surface tension and other effects; the method is very insensitive because of the large microscopic factor—a mean count of 1 bacterium/field might represent about 10^6/ml of milk. These problems could be overcome by using a larger volume of milk and by concentrating the numbers of bacteria before counting.

The direct epifluorescent filter technique (DEFT) which emerged finally (Pettipher *et al.* 1980) is summarized in Table 5. Results obtained with this method correlated well with the plate count in a number of laboratories (Table 6). The manual version of the test is cheap to operate (capital costs £3000; running costs <50p per sample) but an automated version for centralized testing is being developed and is estimated to cost about £15,000.

TABLE 5

Summary of procedures for direct epifluorescent (DEFT)
test for counting bacteria in milk

A. *Membrane filtration*
   1. Add 2 ml milk to 0·5 ml crude trypsin
   2. Add 2 ml Triton X-100 (0·5%)
   3. Incubate at 50°C for 10 min
   4. Pre-warm 0·6 μm pore size Nucleopore membrane filter and holder
   5. Filter-treated milk
   6. Rinse reaction tube and filter tower

B. *Staining*
   7. Overlay filter with 2·5 ml acridine orange/Tinopal (0·025%) for 5 min
   8. Rinse with 2·5 ml pH 3 buffer
   9. Rinse with 2·5 ml ethanol (95%)
   10. Mount membrane in immersion oil
   11. Examine by epifluorescent microscopy

TABLE 6

Relationship between log_{10} *DEFT count* (y) *and* log_{10} *plate*
count (x) *for raw milk in six laboratories*

Laboratory No.	a	b	r
1	0·81	0·83	0·92
2	1·28	0·68	0·86
3	0·04	0·98	0·93
4	0·92	0·76	0·94
5	0·11	0·78	0·90
6	0·99	0·78	0·96
OVERALL	0·61	0·86	0·93

$$y = a + bx$$
where a = intercept on y axis, b = slope, r = correlation coefficient.

A great advantage of the technique is that a preservative may be added to the milk sample (Pettipher & Rodrigues 1982) which greatly facilitates transport to testing centres. Also the membranes are stable after staining and may be sent to a central laboratory for counting later if this is advantageous. Finally, slight modifications of the technique enabled the method to be used for counting somatic cells (Pettipher & Rodrigues 1981a). The dairy industry now has, therefore, the prospect of using a relatively cheap and flexible technique capable of measuring all the current hygiene parameters. Whether it will be used routinely in practice will depend on organizational, economic and political factors.

The possibility of using DEFT for counting bacteria in food generally is being explored. In preliminary work it has been successful for pasteurized milk, dried milk and cream (Pettipher & Rodrigues 1981b). There seems no reason why its use could not be extended into medical areas where rapid enumeration of bacteria is required.

C. Monitoring enzyme levels

Psychrotrophic bacteria multiplying in refrigerated raw milk dominate its flora during storage. They also produce extracellular enzymes which survive heat treatment and may subsequently spoil milk products. Lipases can cause rancid defects in cheese and proteinases can cause gelling of UHT-treated milk and produce bitter flavour defects. These are economically the two most important psychrotrophic enzymes and efforts are now being made to develop rapid, sensitive assays for their detection as part of routine milk quality control. Such measurements may be more relevant to product quality than are bacterial counts because equal numbers of different species and strains of psychrotrophs growing in different milks do not necessarily produce critical amounts of enzymes. For example, some raw milk floras produce detectable proteinases when they multiply to only 1.5×10^6 c.f.u./ml, whereas others reach $>10^7$ c.f.u./ml before activity is detected (Table 7).

The reliable detection and measurement of low levels of microbial lipases and proteinases in ageing milk is extremely difficult using simple measurements of soluble nitrogen or free fatty acids because basal levels vary widely according to the source of the milk (Law 1979). The most recent attempts to devise new assays have been based on the use of exogenous substrates which are unrelated to milk proteins or triglycerides. For example, the hydrolysis of insoluble dye-labelled Hide Powder Azure (HPA) by psychrotrophic proteinases can be detected in milk by the appearance of soluble blue dye-labelled peptides (Cliffe & Law 1982). The detection level (Table 7) is low enough to distinguish UHT-treated milk which will gel after several weeks, from that which can safely be stored for several months without

TABLE 7

Minimum populations of raw milk psychrotrophic floras which produce detectable proteinase activity with the Hide Powder Azure assay carried out in different milk samples*

Milk sample	Population (c.f.u./ml) of		
	Ps. fluorescens PM1	*Ps. fluorescens* AR11	Undefined mixed flora
A	1.5×10^6	1.4×10^7	1.2×10^6
B	6.0×10^6	Nt	7.9×10^6
C	3.8×10^6	Nt	2.8×10^6
D	1.1×10^8	Nt	2.7×10^7

Nt, not tested.
* Cliffe & Law (1982).

deterioration. The hide powder azure assay can be used quantitatively but this requires two ether emulsification stages followed by ultracentrifugation to clarify the milk samples so that the solubilized blue dye can be estimated spectrophotometrically. This procedure is too complex and time-consuming for commercial quality control application but it will make possible direct investigation of the factors controlling enzyme production by psychrotrophs growing in milk, rather than in synthetic media.

The situation regarding lipase assays is analogous to that of proteinase assays (Law 1979) and satisfactory methods are not available at present. However, a sensitive fluorimetric method, involving the release of 4-methylumberlipherone (4-MU) from its oleic acid ester by lipases, is under investigation. The fluorescent reaction product, 4-MU can be detected directly in milk without the need for complicated extraction procedures, but the assay is not yet sufficiently sensitive to detect the very low levels of lipases which can impart rancidity to cheese after prolonged storage.

4. Preservation of Milk and Milk Products

The production of fermented milk products such as cheese, yoghurt and fermented milks has developed over the centuries as a means of preserving the nutritive value of milk and providing a safe, attractive food. The products of fermentation not only contribute to the characteristic flavour but may also prevent the growth of pathogens and of other bacteria whose metabolism would cause undesirable flavour or textural changes in the foodstuff. Of equal importance under modern manufacturing conditions is the preservation of raw milk quality before processing. It is an unfortunate symptom of 'progress' that when an improvement in technique is made in one area the advantage is often used in another (often unintended) area and so the advantage gained from the former is diminished. Thus the present practice of refrigerated storage of milk at the farm undoubtedly extends the keeping quality of the milk and could result in better quality milk at the point of processing. However, manufacturers often prefer to extend the time between production and processing to obtain greater flexibility in plant operation at the sacrifice of milk quality. Indeed at present they would often like to increase this time further, if possible to 3, 4, or even 5 days. A search for the most practical technique to achieve this is constantly in progress.

Refrigerated milk is stored at the farm at about 4°C until collected in insulated road tankers; in England and Wales the temperature of the milk at pick-up must not exceed 7°C. At the manufacturing creamery this milk is transferred to large insulated, cylindrical silos (10,000–30,000 gal, i.e. 45,000–135,000 litres) whose surface area/volume ratio is very low and the temperature of the milk, therefore, rises only very slowly to, say, 10°C at the most before processing. Nevertheless, the doubling time of the psychrotrophic bacteria at 7°C may be about 4 h and after 3 days at this temperature an original count of 10,000 bacteria/ml (good quality milk) becomes 4×10^7/ml at processing. Since it is at this level of bacterial number that the psychrotrophic metabolism generally has a marked deteriorative effect on product quality, it is clear that some further control of psychrotrophic growth is desirable. A number of methods may be considered; refrigeration below 4°C is the most obvious but other possibilities include an additional heat treatment (thermization) or the activation of the antibacterial system in milk. A more speculative approach is the addition to milk of a suitable preservative.

A. Refrigeration

The doubling time of psychrotrophs, although depending on species and strain, usually increases rapidly with fall in temperature below 10°C. At 4°C it may be *ca.* 8 h whereas at 1°C it may be 18 h or more. If milk could be kept at 1°C during the whole of the time up to processing, first-class milk with an original count of 10,000/ml would have a count of *ca.* only 100,000/ml after 3 days, a number unlikely to cause any decrease in product quality. Some countries, such as the Netherlands, are using lower refrigeration temperatures in this way. This increased attention to cooling means not only lower temperatures in the farm bulk tanks but also preferably the use of refrigerated milk tankers for transport and possibly further cooling during filling of the creamery silo. Clearly the economics of this approach must involve consideration of the capital outlay and energy costs.

B. Thermization

This process is energy-intensive because it involves two heat treatments, at the farm or creamery before storage, and again before manufacture or packaging. Thermization is used in the Netherlands but in the UK new regulations would be required before the process could be accepted. There is little published data on the process but Zall (1980) showed that treatment of farm milks at 60–65°C for 10 s reduced total counts of bacteria from 10,000 to 1000/ml. Also, little subsequent bacterial growth occurred in 3 days at 7°C, presumably due to thermal shocking of the cells. After 3 days the rate of bacterial growth was normal. However the process appears to give a useful lag period during which the bacteriological quality of the milk does not deteriorate.

C. Activation of inhibitory systems in milk

Although the lactoperoxidase system (LPS) in raw milk requires an exogenous source of H_2O_2 to kill bacteria, this cannot be legally added directly in many countries, and its effects are short-lived. Xanthine oxidase (Björck & Claesson 1979) and glucose oxidase (Björck & Rosen 1976) have been used to generate H_2O_2 *in situ* at such a rate that it does not accumulate, but is used by the LPS as it is produced. The LPS kills Gram negative psychrotrophs, and trials at NIRD showed that one initial treatment of raw milk at 5°C with 0·1 u of glucose oxidase/ml and 0·3 mg of glucose/ml prevented the multiplication of a lipase-producing strain of *Pseudomonas fluorescens* to numbers which, in untreated milk, resulted in the subsequent development of fatty acid rancidity in Cheddar cheese made from that milk (Table 8). Although

TABLE 8

Preservation of milk for cheesemaking using the lactoperoxidase system

| Milk treatment | Raw milk* counts (c.f.u./ml × 10⁴) | | Cheddar cheese (4-month-old) | |
	day 0	day 3	FFA† (μmol/10 g)	Mean rancidity score (0–4 scale)
None	15	1400	248	2·6
Milk + glucose/glucose oxidase	21‡	0·1	50	0·2

* Inoculated with *Pseudomonas fluorescens* strain AR11 at day 0.
† Free fatty acids.
‡ Count made before LPS activation by glucose/glucose oxidase.

this trial was successful in preserving milk for cheesemaking, the excess glucose resulted in a fast, over-acid cheese and the expensive glucose oxidase was irretrievable. Björck & Rosen (1976) avoided direct additions to the milk by immobilizing the glucose oxidase on glass beads and generating the glucose in small amounts from the milk lactose using immobilized β-galactosidase. When the mixed enzyme-associated glass beads are packed in a column, the system forms the basis of a continuous flow, cold sterilization unit (Fig. 1).

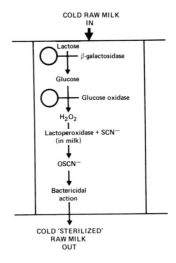

Fig. 1. Flow diagram of 'cold sterilization' unit using the lactoperoxidase system.

The preservation of milk in the developing countries presents a different problem in that ambient temperatures tend to be high (25–35°C), milking hygiene is poor and refrigeration is rarely available. Hydrogen peroxide is a permitted additive in these cases but, when used as the preservative *per se* is required at 300–500 mg/l. At these concentrations it has non-specific detrimental oxidative effects on vitamins and proteins which influence its nutritional value and manufacturing properties. Two reviews (Björck 1980; Korhonen 1980) have described how the LPS has been exploited in field trials in Africa so that only about 10 parts/10^6 H_2O_2 need to be added to prevent bacterial multiplication in warm milk during storage and transport. This requirement is so low that solid peroxides (e.g. alkali metal peroxide) might be used without causing health problems, greatly facilitating the handling and control of the system.

The antibacterial spectrum of the LPS is such that it not only kills Gram negative pathogens and spoilage bacteria, but also inhibits the multiplication of Gram positive mastitis organisms and lactic acid bacteria (Korhonen 1980). The African trials resulted in an increase from 26 to 88% in the numbers of milk samples which pass a 10 min resazurin test at one particular creamery (Björck 1980).

D. Addition of a preservative

Even to consider adding a preservative to a natural food like milk with its image of purity and wholesomeness is almost unthinkable. One would need to be convinced that the additive was completely innocuous and at the same time so cheap as to have no effect on the milk price. If there were promise of improvement in product quality and, if it could be shown that the preservative could be removed at the point of processing when its job was completed, that might be decisive. Such a compound, effective in controlling the multiplication of psychrotrophs, is carbon dioxide. The gas, or rather the HCO_3^- ion, is inhibitory to a number of food spoilage organisms, particularly psychrotrophs (Valley & Rettger 1927; Neiderhardt $et\ al.$ 1974) and this has been exploited for the preservation of fruit and vegetables (Smith 1963) and meat (Ogilvy & Ayres 1951; Enfors $et\ al.$ 1979).

In a typical experiment, the atmospheric head space of a closed vessel, containing half its volume of milk, is replaced by CO_2. After stirring for 30 min the milk pH falls to about 6·0 and the CO_2 content increases to about 30 mmol/l. When the milk is held at 4 or 7°C the inhibitory effect on the growth of psychrotrophs is pronounced (Fig. 2; King & Mabbitt, unpublished results). If 10^6 bacteria/ml is taken as the point where degradative changes in the milk become unacceptable the CO_2 treatment extends the

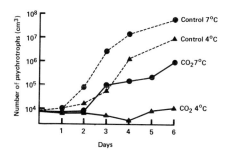

Fig. 2. The effect of CO_2 on psychrotrophs in raw milk.

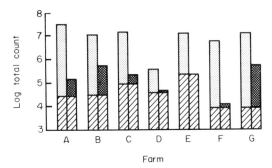

Fig. 3. The effect of CO_2 on bacterial growth in ex-farm milk stored for 3 days at 7°C.
▨, counts on receipt; ☐, stored without CO_2 treatment; ▓, stored with 30 mmol/litre of CO_2.

possible storage time by about 3 days (at 4°C) or 2 days (at 7°C) even for poor quality milk and by considerably longer times for good quality milk. The inhibitory phenomenon appears to be a general one for farm milk supplies (Fig. 3; King & Mabbitt, unpublished results).

The carbon dioxide is easily removed before pasteurization by warming under vacuum but this is not necessary if the milk is being converted to cheese or yoghurt. The point in the production/transport chain at which the gas can be most easily added to the milk still needs to be decided. The obvious points are during transfer to the milk tanker or during filling of the creamery silo. Development work and large scale trials will be necessary before the concept of CO_2 preservation is attractive to the dairy industry.

A second, perhaps more natural, preservative has gained some prominence, namely the use of lactic cultures. The disadvantage of storage of milk at low temperature is that the controlling effect of the lactic acid microflora, which multiplies at ambient temperatures, is lost. The inhibitory actions of these micro-organisms has already been cited in connection with the resistance of fermented dairy products to spoilage and the growth of pathogens. Although the main basis of this action is the effect of lactic acid at low pH, other factors are also involved: production of antibiotics, hydrogen peroxide and low reduction potential (Babel 1977). Whatever the mechanisms by which they function, the addition of lactic cultures as a preservative for refrigerated raw milk particularly if the milk is to be subsequently converted to cheese or other fermented products, has some attraction. Mixed streptococcal cultures (also containing *Leuconostoc cremoris*) are inoculated at 0·1–0·5% and although they do not multiply at say 4°C, they produce sufficient inhibitors (Juffs & Babel 1975) to reduce the numbers of psychrotrophs in stored raw milk by approximately 90%, compared with untreated control milk. The pH of the milk is not altered significantly and provided that it is heat treated before further processing, the added lactic acid bacteria do not alter its manufacturing properties.

E. Preservation of milk products

Processed liquid milk has to depend for a reasonable shelf life on well-controlled heat treatments and the elimination of post-heat treatment contamination, since preservatives are not at present permitted. When spoilage occurs in pasteurized milk it is usually due to spore formers which survive 72°C for 15 s. For example, the bitty cream defect results from the growth of *Bacillus cereus*. Fermented products on the other hand have a degree of natural resistance to spoilage conferred on them by the lactic acid bacteria.

Both mesophilic and thermophilic starter strains normally produce enough lactic acid to suppress the growth of the pathogens most commonly found in milk and curds (Reiter *et al.* 1964; Gilliland & Speck 1972; Haines & Harmon 1973; Pulusani *et al.* 1979). The pathogens isolated from cheese include *Staphylococcus*, *Salmonella*, enteropathogenic strains of *E. coli*, *Shigella*, *Brucella*, group D streptococci, *Clostridium* and *Mycobacterium tuberculosis*. Most of these organisms can only multiply if starter failure or slowness results in the production of a high pH, low acid cheese, though *M. tuberculosis* can survive in normal cheese and is best avoided by primary eradication. Viral infections from cheeses are rare but not unknown (e.g. encephalitis from sheeps' milk cheese) and several types of virus have been shown to survive in cheese (Coxsackie, Echovirus, and Foot & Mouth Disease virus). A comprehensive literature review of pathogens in milk and cheese was prepared recently under the auspices of the International Dairy Federation (Anon. 1980). The low moisture, high acid cheeses are particularly stable to internal spoilage, but they are susceptible to undesirable surface mould and yeast growth which is traditionally controlled with potassium sorbate. More recently, the mould inhibitor pimaricin has been gaining acceptance as an alternative. It is a colourless, odourless complex tetraene, produced by *Streptomyces natalensis* and selective against yeasts and moulds (Klis 1960). Depending on the method of application, pimaricin can be used at concentrations between 80 and 99% lower than sorbate to achieve similar mould control. Minimum inhibitory concentrations for pimaricin and sorbate are compared in Table 9. It is applied most commonly to hard cheese as an emulsion incorporated into the plastic coat or wrapping, though it can be included in brine vats or painted onto the cheese surface.

The so-called 'sweet curd' cheese (e.g. Gouda) are susceptible to late blowing defects detrimental to texture due to the growth of spore-formers (notably *Cl. tyrobutyricum*). It is normal for manufacturers to include 0·02%

TABLE 9

*Activities of pimaricin and sorbate against food spoil-
age moulds and yeasts (surface grown)*

	Minimum inhibitory concentration (μg/ml) of	
	Pimaricin	Sorbate
Aspergillus niger	5	1000
Fusarium sp.	10	500
Mucor mucedo	5	500
Oidium lactis	10	500
Penicillium digitatum	5	500
Rhizopus sp.	5	500
Saccharomyces cerevisiae	1–5	500
Torulospora rosei	10	500
Zygosaccharomyces barketi	1	500

* Data from Klis (1970).

nitrate into the cheese-milk as an inhibitor but there is increasing speculation concerning the role of nitrate and nitrite in the formation of carcinogenic nitrosamines in protein-containing foods under some manufacturing conditions or after consumption. Devoyod (1976) concluded that Edam and Gouda cheese made with nitrate-treated milk was not a hazard but that the nitrate content of the whey should be controlled because of its own toxicity to infants at high concentrations and also because it may be involved in nitrosamine formation during whey processing in the presence of other food ingredients. For these reasons, alternatives to nitrate, of which lysozyme has received the most attention, are under investigation. Both pure egg-white lysozyme and fresh egg-white itself are very effective inhibitors of late blowing bacteria in cheese, but dried egg-white is ineffective (Wasserfall *et al.* 1976). At present, therefore, there is no cheap, convenient method for using lysozyme (treatment with the pure enzyme would cost approximately £250/tonne of cheese) but circumstances may arise in which its use may be favoured instead of nitrate.

5. References

ANON. 1980 Behaviour of pathogens in cheese. *International Dairy Federation Bulletin* **122**, 3–23.
BABEL, F. J. 1977 Antibiosis by lactic culture bacteria. *Journal of Dairy Science* **60**, 815–821.
BJÖRCK, L. 1980 Enzymatic stabilization of milk—Utilization of the milk peroxidase for the preservation of raw milk. *International Dairy Federation Bulletin* **126**, 5–7.

BJÖRCK, L. & CLAESSON, O. 1979 Xanthine oxidase as a source of hydrogen peroxide for the lactoperoxidase system in milk. *Journal of Dairy Science* **62**, 1211–1215.

BJÖRCK, L. & ROSEN, C.-G. 1976 An immobilized two-enzyme system for the activation of the lactoperoxidase antibacterial system in milk. *Biotechnology and Bioengineering* **18**, 1463–1472.

BJÖRCK, L., ROSEN, C.-G., MARSHALL, V. M. & REITER, B. 1975 Antibacterial activity of the lactoperoxidase system in milk against pseudomonads and other Gram-negative bacteria. *Applied Microbiology* **30**, 199–204.

CLIFFE, A. J. & LAW, B. A. 1982 A new method for the detection of microbial proteolytic enzymes in milk. *Journal of Dairy Research* **49**, 209–219.

COUSINS, C. M. 1978 Milking techniques and the microbial flora of milk. In *Proceedings of the 20th International Dairy Congress*, Paris, p. 60T. Brussels: IDF.

COUSINS, C. M. & McKINNON, C. H. 1979 Cleaning and disinfection in milk production. In *Machine Milking* ed. Thiel, C. C. & Dodd, F. H. Reading: NIRD Publications.

DEVOYOD, J.-J. 1976 The use of nitrates in cheesemaking. *Annales de la Nutrition et de l'Alimentation* **30**, 789–792.

ENFORS, S. O., MOLIN, G. & TERNSTROM, A. 1979 Effect of packaging under carbon dioxide, nitrogen or air on the microbial flora of pork stored at 4°C. *Journal of Applied Bacteriology* **47**, 197–208.

GILLILAND, S. E. & SPECK, M. L. 1972 Interactions of food starter cultures and food-borne pathogens: lactic streptococci *versus* staphylococci and salmonella. *Journal of Milk and Food Technology* **35**, 307–310.

GRIFFITHS, E. & HUMPHREYS, J. 1977 Bacteriostatic effect of human milk and bovine colostrum on *E. coli*: Importance of bicarbonate. *Infection and Immunity* **15**, 396–401.

HAINES, W. C. & HARMON, E. G. 1973 Effect of variations in conditions of incubation upon inhibition of *Staphylococcus aureus* by *Pediococcus cerevisiae* and *Streptococcus lactis*. *Applied Microbiology* **25**, 169–172.

JUFFS, H. S. & BABEL, F. J. 1975 Inhibition of psychrotrophic bacteria by lactic cultures in milk stored at low temperatures. *Journal of Dairy Science* **58**, 1612–1619.

KLIS, W. S. 1960 Effectiveness of pimaricin as an alternative to sorbate for the inhibition of yeasts and moulds in food. *Food Technology* **13**, 124–132.

KORHONEN, H. 1980 A new method for preserving raw milk. The lactoperoxidase antibacterial system. *World Animal Review* **35**, 23–29.

LAW, B. A. 1979 Review of the progress of dairy science: Enzymes of psychrotrophic bacteria and their effects on milk and milk products. *Journal of Dairy Research* **46**, 573–588.

LAW, B. A. & JOHN, P. 1981 Effect of the lactoperoxidase bactericidal system on the formation of the electrochemical proton gradient in *E. coli*. *Federation of European Microbiological Societies Microbiology Letters* **10**, 67–70.

LAW, B. A. & REITER, B. 1977 The isolation and bacteriostatic properties of lactoferrin from bovine milk whey. *Journal of Dairy Research* **44**, 595–599.

LAW, B. A., COUSINS, C. M., SHARPE, M. E. & DAVIES, F. L. 1979 Psychrotrophs and their effects on milk and dairy products. In *Cold Tolerant Microbes in Spoilage and the Environment* ed. Russell, A. D. & Fuller, R. pp. 137–152. Society for Applied Bacteriology Technical Series No. 13. London & New York: Academic Press.

MARSHALL, V. M. 1978 *In vitro* and *in vivo* studies on the effect of the lactoperoxidase–thiocyanate–hydrogen peroxide system on *Escherichia coli*. Ph.D. Thesis, University of Reading.

MASSON, P. 1970 In *La Lactoferrine* ed. Arsscia SA, Bruxelles. Paris IV: Libraire Maloine SA.

NEIDERHARDT, F. C., BLOCH, P. L. & SMITH, D. F. 1974. Culture media for Enterobacteria. *Journal of Bacteriology* **119**, 736–747.

OGILVY, W. S. & AYRES, J. C. 1951 Post-mortem changes in stored meats. II. The effect of atmospheres containing CO_2 in prolonging the storage life of cut-up chickens. *Food Technology* **5**, 97–103.

ORAM, J. D. & REITER, B. 1966 The inhibition of streptococci by lactoperoxidase thiocyanate and hydrogen peroxide. The effect of the inhibitory system on susceptible and resistant strains of group N streptococci. *Biochemical Journal* **100**, 373–381.

PETTIPHER, G. L. & RODRIGUES, U. M. 1981a Rapid membrane filtration epifluorescent microscopic technique for the direct enumeration of somatic cells in fresh and formalin-preserved milk. *Journal of Dairy Research* **48**, 239–246.

PETTIPHER, G. L. & RODRIGUES, U. M. 1981b Rapid enumeration of bacteria in heat-treated milk and milk products using a membrane filtration–epifluorescent microscopy technique. *Journal of Applied Bacteriology* **50**, 157–166.

PETTIPHER, G. L. & RODRIGUES, U. M. 1982 A bacteriostatic mixture for milk samples and its effect on the bacteriological, cytological and chemical compositional analysis. *Journal of Applied Bacteriology* **52**, 259–262.

PETTIPHER, G. L., MANSELL, R., MCKINNON, C. H. & COUSINS, C. M. 1980 Rapid membrane filtration-epifluorescent microscopy technique for the direct enumeration of bacteria in raw milk. *Applied and Environmental Microbiology* **39**, 423–429.

PULUSANI, S. R., RAO, D. R. & SUNKI, G. R. 1979 Antimicrobial activity of lactic cultures: partial purification and characterization of antimicrobial compounds, produced by *Streptococcus thermophilus*. *Journal of Food Science* **44**, 575–578.

REITER, B. & MARSHALL, V. M. 1979 Bactericidal activity of the lactoperoxidase system against psychrotrophic *Pseudomonas* spp. in raw milk. In *Cold Tolerant Microbes in Spoilage and the Environment* ed. Russell, A. D. & Fuller, R. pp. 153–164. Society for Applied Bacteriology Technical Series No. 13. London & New York: Academic Press.

REITER, B., FEWINS, G. F., FRYER, T. F. & SHARPE, M. E. 1964 Factors affecting the multiplication and survival of coagulase-positive staphylococci in Cheddar cheese. *Journal of Dairy Research* **31**, 261–272.

REITER, B., BROCK, J. H. & STEEL, E. D. 1975 Inhibition of *E. coli* by bovine colostrum and post colostral milk. II. The bacteriostatic effect of lactoferrin on a serum susceptible and serum-resistant strain of *E. coli*. *Immunology* **28**, 83–95.

SMITH, W. H. 1963 The use of carbon dioxide in the transport and storage of fruit and vegetables. *Advances in Food Research* **12**, 95–146.

THOMAS, E. L. & AUNE, T. M. 1978 Lactoperoxidase, peroxide, thiocyanate antimicrobial system: correlation of sulphydryl oxidation with antimicrobial action. *Infection and Immunity* **20**, 456–463.

VALLEY, G. & RETTGER, L. F. 1927 The influence of carbon dixoide on bacteria. *Journal of Bacteriology* **14**, 101–137.

WASSERFALL, F., VOSS, E. & PROKOPEK, D. 1976 Studies on cheese ripening. The use of lysozyme instead of nitrate to inhibit late blowing of cheese. *Kieler Milchwirtschaftliche Forschungsberichte* **28**, 3–16.

WILSON, C. D. & RICHARDS, M. S. 1980 A survey of mastitis in the British Dairy Herd. *Veterinary Record* **106**, 431–435.

ZALL, R. R. 1980 Can cheesemaking be improved by heat treating milk on the farm? *Dairy Industries International* **45**, 25–31, 48.

Microbial and Chemical Changes in Chill-stored Red Meats

R. H. DAINTY, B. G. SHAW AND T. A. ROBERTS

*Agricultural Research Council, Meat Research Institute,
Langford, Bristol, UK*

Contents

1. Introduction

THE PURPOSE of this paper is to review developments, mainly in the past ten years, in the study of bacteriological and chemical changes that take place when red meat is stored under refrigeration.

There is a long history of research into the nature of bacteriological changes occurring on meat during storage to determine the identity of the bacteria that are the likely cause of spoilage. The level at which this can be done depends upon the sophistication of the classification systems available. There is now a better understanding of the taxonomy of the *Pseudomonas* strains which dominate the flora of meat stored in air, but this is a recent development, and most reports in the last decade on bacterial changes during aerobic storage have continued to describe the flora at genus level. These reports have therefore not contributed any major new information, but are reviewed with the earlier literature to summarize our current state of knowledge. Research into the microbiological consequences of vacuum packaging began about twenty years ago when it was shown that it extended shelf life by inhibiting pseudomonads and allowing a flora dominated by lactic acid bacteria to develop. Subsequent studies have provided more detailed

FOOD MICROBIOLOGY
ISBN 0 12 589670 0

information on the microflora and have revealed important effects of meat pH on microbial changes occurring during storage. These will be reviewed and an account given of recent studies on the microbial flora of meat stored in modified gas atmospheres.

Interest in defining the chemical changes associated with bacterial growth on meat stems from the possibility of using them as an index of microbial quality, shelf life or consumer acceptability. Prior to 1970, few systematic chemical studies had been undertaken and discussions of the subject relied heavily on analogies with other proteinaceous foods, particularly fish, and on data obtained with the classical strains of bacteria typically used for biochemical studies. Experiments completed in the last few years, however, are beginning to rectify this position. A definite order of substrate utilization during growth of representative strains of many of the different types of bacteria commonly associated with the spoilage of meat stored in air or vacuum packs has been demonstrated and its relevance to shelf life established. In addition, data on the chemical nature of the odours which often signal spoilage is accumulating, thus opening the way to more meaningful investigations of alternative methods for assessing acceptability.

2. The Bacteriology of Spoilage

A. Aerobic storage

For the purposes of this paper aerobic storage includes storage in air and in gas-permeable plastic films.

There are very few bacteria within the musculature of carcasses from healthy animals (Gill 1979) and microbial growth during chilled storage is largely a surface phenomenon. During aerobic storage of lean meat off-odours become evident when microbial numbers reach *ca.* $10^7/cm^2$, and slime can be seen when numbers reach *ca.* $10^8/cm^2$ (Ayres 1960).

Ten years ago (Ingram & Dainty 1971) earlier studies on the bacteria on aerobically stored meat (Kirsch *et al.* 1952; Ayres 1960; Gardner 1965*a*) had led to the conclusion that *Pseudomonas* spp. were responsible for spoilage. There was some confusion over their identity because most had been described as *Achromobacter* in early studies (Haines 1933*a*; Empey & Vickery 1933), the consequence of using a different classification system (Brown & Weidemann 1958). Attempts to identify these *Pseudomonas* strains (Kirsch *et al.* 1952; Wolin *et al.* 1957; Ayres 1960) failed because of inadequate descriptions of species in the literature. Gram negative non-motile bacteria had been detected in some studies (Ayres 1960; Gardner 1965*a*) but their role in spoilage was uncertain, as was their identity (Ingram & Dainty 1971). *Brochothrix thermosphacta* had been detected in the aerobic spoilage flora (Gardner *et al.* 1967) but was not thought to be important in spoilage except possibly on lamb (Barlow & Kitchell 1966). Most studies had been made on sliced or minced lean beef stored in containers or packages with little detailed information on the nature of bacterial growth on hanging carcasses or on fat surfaces. It was uncertain whether meat from animals of different species developed the same spoilage flora (Ingram & Dainty 1971). Knowledge in some, but not all, of these areas has improved in the last ten years.

(i) *Classification of the bacteria*

Most bacteria detected on refrigerated, aerobically stored meat are Gram negative rods, the majority of which can be identified to generic or family level with relatively few tests (Table 1). *Alteromonas putrefaciens* has not been reported on meat of normal pH ($<6\cdot0$) but must be included in the scheme to distinguish *Pseudomonas* strains from it.

One problem remaining is the grouping and nomenclature applied to the non-motile aerobic strains. The Gram negative non-motile aerobes on meat are non-pigmented and were for a long time called *Achromobacter* spp. This was never wholly satisfactory because the type species of *Achromobacter* was described as motile with peritrichous flagella. Ten years ago it was considered appropriate to propose classifying them in the genus *Acinetobacter* (Thornley 1967). However, this genus is now described as oxidase negative (Lautrop 1974) and cannot, therefore, accommodate oxidase positive strains, which form the majority on meat. These strains are not identifiable with any described species, but resemble members of the genus *Moraxella* which are non-motile, oxidase positive, non-saccharolytic coc-

TABLE 1

Differentiation of Gram negative bacteria found on chilled meat

	Motility	Oxidase	Glucose metabolism	Ornithine decarboxylase
Pseudomonas spp.	+	+	Oxidative	−
Alteromonas putrefaciens	+	+	Oxidative or inert	+
Acinetobacter spp.	−	−	Inert or oxidative	NA
Moraxella spp.	−	+	Inert	NA
Moraxella-like spp.	−	+	Oxidative	NA
Aeromonas spp.	+	+	Fermentative	NA
Enterobacteriaceae	+ or −	−	Fermentative	NA

NA, not applicable.

cobacilli. It is our practice to group oxidase positive, non-saccharolytic meat strains as *Moraxella* spp. and saccharolytic strains as *Moraxella*-like spp. Even this is not wholly satisfactory, and there is a clear need for a taxonomic study of the oxidase positive strains to establish their generic status and to determine whether or not different types exist amongst them.

Considerable attention has been paid to speciating and grouping the *Pseudomonas* strains because of their obvious importance in spoilage. Many problems in identifying members of this genus were solved when Stanier *et al.* (1966) demonstrated that species could be distinguished by examining their ability to use different carbon compounds as energy sources. However, psychrotrophic strains from foods were not included in that study and Davidson *et al.* (1973) found that most meat strains would not identify with the described species and biotypes. Shaw & Latty (1981, 1982), using the tests of Stanier *et al.* (1966), therefore employed numerical taxonomic methods to establish the relationship of meat strains to their described species and to named species which had previously been associated with the spoilage of chilled foods (*Ps. fragi*, *Ps. taetrolens*, *Ps. perolens*). The majority were contained in two closely related clusters of non-fluorescent strains and were identified with *Ps. fragi*. Two clusters of fluorescent strains were also detected which were distinct from all the species. An identification scheme developed from this work has been used on 789 strains from a range of red meats and successfully identified 90% of the isolates, confirming members of the two *Ps. fragi* clusters as the predominant types of pseudomonad on meat (Table 2).

Brochothrix thermosphacta, formerly known as *Microbacterium thermosphactum* but now reclassified (Sneath & Jones 1976), is the only Gram positive organism found in high numbers on refrigerated, aerobically stored meat. It was not described in detail when first isolated (McLean & Sulzbacher 1953) and was not easy to recognize until Davidson *et al.* (1968) described its properties more fully. Gardner (1966) developed a selective medium (STAA) for *Br. thermosphacta* and today any Gram positive, catalase positive rod or coccobacillus which forms chains and grows on STAA is identified conclusively with this organism.

(ii) *Changes in the bacterial flora during storage*

The composition of the initial microflora on meat is influenced by the source of most of the contamination, which could in different instances be soil from the hide, faecal material, or surfaces which have been in contact with the meat, e.g. cutting boards and knives. It is not surprising, therefore, that reports on the incidence of different genera on fresh meat vary widely (Table 3). Although the composition of the initial flora will influence the

TABLE 2

The identity of 789 Pseudomonas *strains from lean pork, beef, and lamb stored refrigerated under a gas-permeable plastic film*

Group/species	Shewan *et al.* [*] Group	%
Cluster 1 *Ps. fragi*†	II	25
Cluster 2 *Ps. fragi*†	II	50
Cluster 3 (fluorescent)†	I	10
Cluster 4 (fluorescent)†	I	ND
Ps. fluorescens biotype A‡	I	4
Ps. fluorescens biotype C‡	I	<1
Ps. putida	I	<1
Unidentified	—	10

*Shewan *et al.* (1960).
†Shaw & Latty (1982).
‡Stanier *et al.* (1966).
ND, not detected.

TABLE 3

Groups of organisms detected on fresh meat

	Beef*	Pork†	Beef‡	Pork§	Lamb‖	Pork¶
Micrococcus spp.	—	23**	48	2	71	13
Pseudomonas spp.	—	36	28	24	—	14
Moraxella/Acinetobacter spp.	44	—	—	54	—	29
Lactobacillus spp.	—	4	—	—	—	—
Flavobacterium spp.	28	8	—	8	—	22
Coryneforms	6	7	—	—	13	8
Yeasts	—	5	—	—	13	5
Enterobacteriaceae	—	8	—	3	—	1
Staphylococcus spp.	—	—	—	1	3	1
Kurthia spp.	—	3	—	2	—	—
Streptococcus spp.	—	—	—	—	—	6
Bacillus spp.	—	—	13	—	—	—
Brochothrix thermosphacta	—	4	—	—	—	1
Others and unclassified	22	2	10	7	—	1

*Gardner (1965a).　†Gardner *et al.* (1967).　‡Stringer *et al.* (1969).　§Enfors *et al.* (1979).　‖Shaw *et al.* (1980).　¶Blickstad *et al.* (1981).
**Percentage of isolates from the total aerobic viable count identified with the group.

changes taking place during storage, the over-riding factors are the conditions of storage.

Most meat is stored initially as hung carcasses and subsequently as unwrapped, loosely wrapped, or packaged joints. Empey & Vickery (1933) examined fat and lean surfaces of beef carcasses stored at −1°C and reported the predominant groups to be *Achromobacter* and *Pseudomonas*. The exact identity of those *Achromobacter* strains is uncertain because that group would have included strains of both *Pseudomonas* and *Moraxella/ Acinetobacter*. The detailed composition of the flora was not presented and no distinction was made between the flora on fat and lean meat. Stringer *et al.* (1969) presented more detail on the nature of the flora on refrigerated beef carcasses and attempted to speciate *Pseudomonas* strains. The complete breakdown of the predominant flora was *Ps. fragi* (54%), *Ps. geniculata* (31%), *Ps. fluorescens* (5%), *Moraxella/Acinetobacter* (9%). Although fat and lean surfaces were sampled separately, only a composite flora was presented. Selective media were not employed in either study and it is impossible to comment on the occurrence and growth of other minor groups of micro-organisms. Despite this lack of detailed information on bacterial changes on carcasses during storage no further studies have been reported in the literature.

Storage in unsealed containers best represents storage of meat joints unwrapped in conditions of high humidity. Gardner (1965a) has provided the most comprehensive study of microbial growth on lean meat stored refrigerated in air in this manner. The composition of the flora was determined by identifying isolates from total viable count plates and several selective media were used to monitor the growth of different groups of organisms. Minced beef was used with an initial total viable count of 10^4–10^5/g composed of *Acinetobacter/Moraxella* spp. (22%), *Corynebacterium* spp. (14%), *Flavobacterium* spp. (10%) and unclassified strains (54%). *Pseudomonas* strains became detectable amongst the dominant flora after 4 days, and at 7 days, when the total viable count reached 10^7/g, they formed more than 95% of the population, the remainder being mostly *Acinetobacter/Moraxella* strains. Presumptive coli-aerogenes organisms and yeasts grew, but comprised only a small fraction of the flora. *Brochothrix thermosphacta* was not detected amongst the dominant population after storage and its growth could not be monitored because of the lack of a selective medium at that time. This information has not been improved upon in the last decade.

Gardner *et al.* (1967) identified the predominant bacteria on lean pork after storage at 2°C in air under a perforated polyethylene film, finding *Pseudomonas/Acinetobacter/Moraxella* (96%), *Br. thermosphacta* (2%) and lactobacilli (2%). Patterson (1970) also reported that the *Pseudomonas/*

Acinetobacter/Moraxella group predominated in minced lamb stored refrigerated in air in Petri dishes. This suggests that there is little difference between the spoilage flora of lean beef, pork and lamb stored in air in conditions of high humidity.

At retail outlets a considerable amount of meat is displayed in trays covered by plastic films with high permeability to gases (O_2-permeability *ca*. 10,000 $cm^3/m^2/day/atm$ O_2). Ayres (1960) identified the predominant bacteria on sliced lean beef stored under a gas-permeable film as *Pseudomonas*, confirming an earlier report by Halleck *et al*. (1958) on minced lamb. Roth & Clark (1972) employed selective media to monitor the growth of fluorescent pseudomonads, *Br. thermosphacta* and lactobacilli on sliced beef stored under a gas-permeable film at 5°C. After 12 days fluorescent pseudomonads and *Br. thermosphacta* accounted collectively for 60% of the total aerobic count ($8 \times 10^9/cm^2$), the remainder being non-pigmented pseudomonads and *Acinetobacter/Moraxella* strains. Lactobacilli grew, but accounted for less than 1% of the flora at 12 days. Gardner *et al*. (1967) detected organisms belonging to the *Pseudomonas/Acinetobacter/Moraxella* group predominating on pork in gas-permeable packs and also detected *Br. thermosphacta*. Its incidence amongst the dominant population averaged 11% in gas-permeable packs and 2% on samples stored in air. Despite the high permeability of the packaging materials used for this type of storage, some restriction of gaseous diffusion to and from the pack occurs, resulting in a slight accumulation of CO_2 (Gardner *et al*. 1967). Carbon dioxide is inhibitory to *Pseudomonas* strains and its presence might account for the slightly higher incidence of *Br. thermosphacta* on meat stored in gas-permeable films than on meat stored in air. Carbon dioxide accumulation may also partially account for the unusual observation of Barlow & Kitchell (1966) that 100% of isolates identified from the predominant flora of lamb were *Br. thermosphacta* after storage under a gas-permeable film.

We have studied the spoilage flora of beef, pork and lamb stored under a gas-permeable film (polyvinyl chloride; O_2-permeability 10,000 $cm^3/m^2/day/atm$ O_2). *Pseudomonas* strains were the most common group (42–60%) on all three types of meat, while *Moraxella/Moraxella*-like strains accounted for 16–23% of the flora. *Brochothrix thermosphacta* was considerably more common on lamb (22%) and pork (26%) than on beef (4%).

Some data have been presented within the last 10 years on microbial growth on meat stored in large bags of gas-impermeable plastic films and containing initially 10–15 litres of air. This technique has been used to represent storage in air in comparison with storage in various gas atmospheres. The predominant flora on pork loins stored in air at 4°C in this way (Enfors *et al*. 1979) was composed of non-fluorescent *Pseudomonas* spp.

(92%), *Ps. fluorescens* (5%) and *Bacillus subtilis* (3%). Lean beef stored in air-filled bags at 4°C had a flora composed of non-fluorescent *Pseudomonas* spp. (60%), *Ps. fluorescens* (16%), *Br. thermosphacta* (16%), lactic acid bacteria (4%) and *Micrococcus* spp. (4%) (Erichsen & Molin 1981). The predominant bacteria on lamb stored at −1°C were *Pseudomonas* spp. and *Br. thermosphacta* (Newton *et al.* 1977). Taken as a whole, these data are in general agreement with those in earlier studies on storage in air. The slightly higher incidence of *Br. thermosphacta* may be the result of some inhibition of *Pseudomonas* strains by CO_2 which gradually accumulates in the bags (Enfors *et al.* 1979; Erichsen & Molin 1981).

The aerobic spoilage of dry, firm, dark (DFD) meat has been given some attention in recent years. DFD meat has a high ultimate pH (>6·0) and low glucose content. The condition arises from a depletion of glycogen reserves in the musculature of the live animal, e.g. as a result of stress or cold followed by insufficient time to recover before slaughter, or from fright. Consequently, post-slaughter, there is a limited build-up of lactate, which determines the ultimate pH, and of glucose and glucose-6-phosphate which are intermediates in lactate formation from glycogen. Such meat has long been known to spoil more rapidly than that of normal pH but the reasons were not clear. Observations that the growth rates of fluorescent and non-fluorescent pseudomonads from meat are unaffected by pH in the range 5·5–7·0 (Gill & Newton 1977) suggested that the reduced shelf life is not caused by more rapid growth of spoilage bacteria. Newton & Gill (1978*a*) inoculated meat of different pH values with a fluorescent pseudomonad and detected no difference in growth rate on a DFD sample (pH 6·3) and a control sample (pH 5·85), although the lag phase was extended at the lower pH. Rey *et al.* (1976) monitored changes in numbers of psychrotrophic bacteria on naturally contaminated normal and DFD meat and also detected a reduced lag phase on DFD samples. While this is undoubtedly a factor in reducing shelf life, the current view (Newton & Gill 1981) is that DFD meat spoils more rapidly because it contains little or no glucose (see section below on chemical changes for a more detailed account). Erichsen & Molin (1981) detected only small differences in the composition of the microbial flora on normal and DFD meat after aerobic storage.

B. Vacuum-packed storage

It is common practice to butcher beef carcasses at the abattoir into 4–20 lb (1·8–9 kg) primal joints which are then stored vacuum-packed in bags of plastic materials with low permeability to gases (O_2 permeability 5–90 $cm^3/m^2/day/atm$ O_2; CO_2 permeability 20–300 $cm^3/m^2/day/atm$ CO_2). These materials restrict the flow of gases to such an extent that the atmosphere around the meat becomes depleted in O_2 (often <1% v/v) and enriched in CO_2 (>20% v/v) resulting in microbial changes quite different from those observed during aerobic storage.

The major differences between microbial growth on meat stored aerobically and in gas-permeable films were established nearly 20 years ago (Jaye et al. 1962). Under gas-impermeable films pseudomonads are inhibited by the carbon dioxide (Ingram 1962) and the predominant bacteria after storage are mainly lactic acid bacteria. Subsequent studies have provided a more detailed account of microbial changes during storage.

The maximum aerobic total count of bacteria detected on vacuum-packed beef is ca. $10^8/cm^2$ which is one-tenth the maximum detectable with aerobic storage under a gas-permeable film (Roth & Clark 1972). Comparisons of aerobic counts with those on selective media, and identification of isolates from aerobic count plates, consistently reveal lactic acid bacteria as the predominant group after storage (Roth & Clark 1972; Beebe et al. 1976; Seidemann et al. 1976; Dainty et al. 1979; Erichsen & Molin 1981). These lactic acid bacteria have not been thoroughly classified. Hitchener et al. (1982) identified 18 of 177 isolates with Leuconostoc mesenteroides but the remainder were not identifiable with species and could only be described as atypical streptobacteria or atypical betabacteria.

Reports on the growth of other groups of bacteria on vacuum-packed beef are inconsistent. Growth of Br. thermosphacta to ca. $10^6/cm^2$ has been reported by Sutherland et al. (1975b), Dainty et al. (1979) and Erichsen & Molin (1981), whereas Pierson et al. (1970) and Roth & Clark (1972) did not detect its growth. Growth of Pseudomonas strains has been reported (to 10^3–$10^6/cm^2$) in some studies (Sutherland et al. 1975b; Seidemann et al. 1976) but not in others (Pierson et al. 1970; Roth & Clark 1972). Enterobacteriaceae have not been enumerated in all studies but can constitute 10% of the flora after 4 weeks or longer (Beebe et al. 1976; Seidemann et al. 1976) when numbers up to $10^6/cm^2$ have been reported (Dainty et al. 1979). Dainty et al. (1979) reported that the Enterobacteriaceae were mostly Serratia liquefaciens.

Campbell et al. (1979) have shown that Br. thermosphacta will not grow anaerobically on meat of pH 5·8 or lower. They therefore suggested that the varied reports on the growth of Br. thermosphacta are due to differences in pH of the meat and/or the permeability of the packaging materials used. Growth of Pseudomonas strains can be influenced by film permeability (Newton & Rigg 1979) and by the degree of vacuum used to evacuate the bags (Seidemann et al. 1976). Increasing the storage temperature, even to 5°C, increases the growth of Enterobacteriaceae (Beebe et al. 1976).

When vacuum-packaged meat is repackaged under a gas-permeable film, as happens before retail sale, the flora which develops resembles that on meat held under aerobic conditions throughout storage (Roth & Clark 1972). However, the odour and appearance of that meat at spoilage may differ from that normally associated with aerobically stored meat (Sutherland *et al.* 1975*b*).

The spoilage of vacuum-packed DFD meat differs considerably from that of normal pH meat. The latter spoils with a sour or cheesy off-odour which develops slowly after about 8 weeks at 0°C, whereas DFD meat spoils more rapidly (3–6 weeks shelf life) often with an offensive putrid odour (Bem *et al.* 1976; Taylor & Shaw 1977). Hydrogen sulphide is often produced in vacuum packs of DFD meat resulting in the formation of the pigment sulphmyoglobin, giving exudate in the packs an undesirable, green appearance (Nicol *et al.* 1970). The most important H_2S-producing organism is *Alteromonas putrefaciens*, which is unable to grow on meat of normal pH (Seelye & Yearbury 1979; Gill & Newton 1979). It does not, however, predominate on the spoiled meat and the numbers of H_2S-producers, which also include some *Aeromonas* strains, form *ca.* 20% of the population after storage (Dainty *et al.* 1979). Lactic acid bacteria were detected as the most numerous group on vacuum-packed DFD meat by Dainty *et al.* (1979) whilst Patterson & Gibbs (1977) and Erichsen & Molin (1981) reported that the predominant flora was composed of similar numbers of lactic acid bacteria and *Br. thermosphacta*. Enterobacteriaceae grew to higher numbers on vacuum-packed DFD meat than on normal meat (Dainty *et al.* 1979).

Patterson & Gibbs (1977) described most psychrotrophic Enterobacteriaceae on vacuum-packed DFD meat as being similar to *Serratia liquefaciens*. Gill & Newton (1979) identified the most common Enterobacteriaceae with *Ser. liquefaciens* and *Yersinia enterocolitica* and showed that the former could produce spoilage odours. The reported growth of *Y. enterocolitica* is of concern because some strains produce enteritis in humans. However, Hanna *et al.* (1976) serotyped 8 *Y. enterocolitica*-like strains growing on vacuum-packed meat and found that none was 0:3, 0:8 or 0:9 strains, which are the only serotypes of proven pathogenicity to humans (Wauters 1981).

C. Storage in modified gas atmospheres

Red meats may be stored in modified gas atmospheres to delay bacterial spoilage and/or the development of surface discolouration. Atmospheres employed usually consist of carbon dioxide as the antibacterial component mixed with either air (when the purpose is solely to inhibit bacterial growth and the red colour will deteriorate), nitrogen (when the lean colour remains purple) or oxygen (when a red lean colour is required). Modified gas atmospheres have been used in ships' holds, transport vehicles and in gas-impermeable packages.

Carbon dioxide/air mixtures were the first modified gas atmospheres used for the storage of meat. Haines (1933b) demonstrated that 10% carbon dioxide in air lengthened the lag phase and doubled the generation time of *Pseudomonas* strains predominating on beef, while Newton *et al.* (1977) detected *Pseudomonas* and *Br. thermosphacta* as the predominant groups on lamb stored in 20% carbon dioxide in air.

Clark & Lentz (1973) demonstrated that meat spoilage is delayed in a gas atmosphere composed of carbon dioxide (20%) and oxygen (80%). Newton *et al.* (1977) reported the flora of lamb stored in this mixture to be predominantly *Br. thermosphacta*. Christopher *et al.* (1979a,b) reported that *Pseudomonas* strains were detectable in the dominant microbial flora on beef and pork stored in 20% carbon dixiode and 80% oxygen, but that lactic acid bacteria were predominant on both meats. These observations can be accounted for by the relative sensitivity of *Pseudomonas* strains to carbon dioxide and the insensitivity of *Br. thermosphacta* and lactic acid bacteria (Shaw & Nicol 1969; Roth & Clark 1975).

The rate of increase in total viable counts on meat stored in carbon dioxide (20%) and nitrogen (80%) is much reduced compared to that in air, thereby extending the storage life (Partmann *et al.* 1975). Christopher *et al.* (1979a, 1980) reported a predominance of lactobacilli on pork and beef stored in gas-impermeable plastic packages containing initially 20% carbon dioxide and 80% nitrogen. Newton *et al.* (1977) found that the ultimate flora on lamb stored in gas-impermeable plastic packages containing this same mixture depended on whether or not small quantities of oxygen diffusing into the packs were removed with an oxygen absorbent. When the absorbent was used the predominant groups were *Br. thermosphacta*, lactobacilli and *Enterobacteriaceae*; but if the oxygen was not removed the lactobacilli were replaced by *Pseudomonas* strains. This observation is difficult to explain in view of the known inhibitory effect of 20% carbon dioxide on *Pseudomonas* spp. but the authors suggest that it indicates an importance of oxygen limitation in suppressing the growth of these organisms on meat.

The microbiological consequences of storage in gas-impermeable plastic packages filled with single gases have also been studied. Huffman (1974) reported similar increases in total viable counts on beef stored in nitrogen and in air. Oxygen diffused slowly through the gas-impermeable plastic films used in nitrogen packing and permitted growth of *Pseudomonas* strains which became predominant (Enfors *et al.* 1979). The slight inhibition of pseudomonads by low concentrations of carbon dioxide which accumulated in the packs produced only a small extension in storage life (Enfors *et al.* 1979). There is also little microbiological benefit from packing meat in 100% O_2. Total viable counts increased at the same rate as in air (Huffman 1974), and although carbon dioxide accumulated in the packs *Pseudomonas* strains still formed a substantial part of the microflora after storage (Christopher *et al.* 1979*a*). Packaging in 100% CO_2 markedly reduced the rate of increase of total aerobic counts on meat (Huffman 1974; Enfors *et al.* 1979). *Pseudomonas* strains were inhibited and after storage the flora was completely dominated by lactic acid bacteria (Enfors *et al.* 1979; Erichsen & Molin 1981). Blickstad *et al.* (1981) have shown that storage in hyperbaric carbon dioxide (5 atm) inhibited these lactic acid bacteria and suggested this as a means of extending shelf life even further.

3. The Chemistry of Spoilage

A. Aerobic storage

(i) *Substrates for growth*

Prior to 1970, the chemical changes accompanying the growth of bacteria on meat during storage were poorly understood. The combined results of several authors (Saffle *et al.* 1961; Jay & Kontou 1964, 1967; Gardner 1965*b*; Gardner & Stewart 1966*a*,*b*) suggested that metabolism of glucose, lactic acid, certain amino acids, nucleotides, urea and sarcoplasmic proteins can all occur during storage. The incomplete, and in some instances contradictory, nature of the evidence made it impossible to draw general conclusions about the overall pattern of chemical changes to be expected. For example, it was not known whether each of the identified substrates was metabolized in all samples, nor whether they were used in a particular order. Because protein breakdown had only been detected at a very advanced stage of spoilage, initial microbial growth was thought to occur at the expense of the low molecular weight soluble compounds (Jay 1966).

The time course and extent of proteolysis during storage has been studied in more detail in the intervening years. In a series of publications from Michigan State University pure, single bacterial strains representing many of the major genera typically found on meat were inoculated into minced samples of pork and rabbit meat. Evidence of proteolysis was sought using a number of techniques including (a) the degree of extractability (i.e. solubility in buffers of standard ionic composition) of the sarcoplasmic, myofibrillar and connective tissue proteins (Borton *et al.* 1970*a*; Buckley *et al.* 1974); (b) the sedimentation pattern of myofibrillar proteins during sucrose density gradient centrifugation (Rampton *et al.* 1970); (c) changes in individual muscle proteins after separation by gel electrophoresis (Borton *et al.* 1970*b*; Hasegawa *et al.* 1970*a*,*b*); (d) the relative times to appearance of extracellular microbial proteinases, spoilage odours and protein breakdown in the meat tissues (Tarrant *et al.* 1971; Buckley *et al.* 1974); (e) changes in myofibrillar ultrastructure by electron microscopy (Dutson *et al.* 1971; Buckley *et al.* 1974). Although a number of examples of proteolytic attack on sarcoplasmic, myofibrillar and connective tissue components were revealed, the overwhelming majority were detected in grossly spoiled samples. *Pseudomonas fragi* was the most active organism tested, exhibiting activity against all three protein fractions (Tarrant *et al.* 1971). It also provided tentative evidence of pre-spoilage proteolysis because higher levels of proteolytic activity were extracted from inoculated pork samples after storage, but before overt signs of spoilage, than from initial or stored, uninoculated control samples. However, the increases were only significant with casein as the assay substrate and not with haemoglobin; and no direct evidence of breakdown of meat proteins was obtained. The authors suggested that protein breakdown might have been detected *in situ* if account had been taken of the surface nature of bacterial growth, and hence its chemical consequences, when sampling.

This possibility was examined by Dainty *et al.* (1975) who allowed intact pieces of beef to spoil either with a naturally attained flora or after inoculation with a reconstituted freeze-dried slime. Sarcoplasmic and myofibrillar proteins extracted from consecutively deeper layers of bacterially contaminated and sterile control samples were separated by gel electrophoresis and the stained gels compared. No evidence of pre-spoilage proteolysis was obtained but extensive breakdown in both protein fractions was apparent in the surface and adjacent layers of both the naturally contaminated and slime-inoculated samples after prolonged storage. Even so, deeper layers (1 cm from the surface) were unaffected, illustrating the surface nature of chemical spoilage. Proteolysis developed more rapidly, and was eventually more extensive, in the slime-inoculated samples while inoculation with single strains of the major types of pseudomonad in the slime inoculum produced different results to the slime itself. Hence it may be misleading to extrapolate findings with artificial inocula to natural situations as indicated in earlier work (Rampton *et al.* 1970).

Further evidence that significant proteolysis only occurs after spoilage has been provided by the immunological studies of Margitic & Jay (1970), in which cross-reactions between antibodies raised against naturally spoiled and fresh meat proteins were studied; the chemical and enzymological studies of Morrisey *et al.* (1980) with a non-pigmented *Pseudomonas* sp., *Ps. fluorescens*, *Aerobacter aerogenes* and a *Micrococcus* sp.; and the light microscopy studies by Gill & Penney (1977) of the mechanism of penetration of meat surfaces by various bacteria including *Salmonella typhimurium*, *Ps. fluorescens*, an *Enterobacter* sp., an *Acinetobacter* sp. and *Staphylococcus aureus*. Gill & Penney (1977) reported that only when maximum numbers were reached, a situation found post-spoilage, did the organisms begin to penetrate between the muscle fibres. It was assumed that they were able to do so because they had begun to produce an extracellular proteinase capable of degrading the connective tissue proteins. This interpretation has since been questioned on the basis of results obtained with *Ser. marcescens* whose penetration into samples of minced meat was not dependent upon its own collagenolytic activity but rather the water-holding capacity (WHC) of the meat (Sikes & Maxcy 1980).

Details of the mechanism of action of proteolytic enzyme preparations from *Ps. fragi* and *Ps. perolens* against porcine myofibrillar proteins have been revealed in the electron microscopy studies of Tarrant *et al.* (1973) and Buckley *et al.* (1974) respectively.

An effect of microbial growth on meat proteins less drastic than proteolytic breakdown has been inferred from measurements of the water-holding capacity of meat. Jay (1966) developed a rapid and simple filtration technique, which he called extract-release volume (ERV), to measure this phenomenon and suggested it be used as an objective measure of spoilage. No satisfactory explanation of the effect could be given at the time but it is now clear that pH changes are a major cause (Shelef 1974; Jay & Shelef 1976). The normal pH of meat is close to the isoelectric point of the major structural meat proteins, i.e. pH 5·5, and a relatively tight protein matrix exists as a result of maximum numbers of intermolecular salt linkages between oppositely charged groups on adjacent protein molecules (Hamm 1975). Consequently there is little space available for water and water-holding capacity is at a minimum. Increasing pH neutralizes some of the positive charges leading to an excess of negative charges, mutual charge repulsion and hence a loosening in the protein matrix, i.e. a greater capacity for water binding and lower ERV. Changes in pH during storage, even directly at the surface of meat, are normally small prior to spoilage being evident, but subsequently there is a marked increase (Gill 1976) and ERV mirrors these changes quite closely. Post spoilage, amino sugar components of the bacteria themselves have been shown to increase hydration capacity, and may therefore contribute to the changes. Binding of the amino sugars to the meat proteins with resulting increases in water-holding capacity may explain the greater susceptibility of the proteins to attack at this stage of spoilage (Jay & Shelef 1976).

It is now clear that microbial growth occurs initially at the expense of one or more of the low molecular weight compounds, as was demonstrated many years ago for fish (Beatty & Collins 1939). Analysis of surface layers of lamb *M. longissimus dorsi* inoculated with a fluorescent *Pseudomonas* spoilage strain showed glucose to be the first substrate utilized during aerobic storage (Gill 1976). A decrease in surface glucose concentration was first detected when cell numbers exceeded $10^7/cm^2$, but some 5 mm below the surface there was no reduction. When bacterial numbers had reached $10^8/cm^2$ glucose was no longer detectable at the surface and decreased concentrations were detectable up to 10 mm lower, i.e. a concentration gradient had developed. Shortly after, pH and ammonia concentrations began to increase at the surface and continued to do so until maximum cell numbers ($10^9/cm^2$) were reached. At this point changes in amino acid concentrations were first detected. The concentrations of some amino acids fell, indicating that they were being used as carbon and/or energy sources; that of others increased, indicating that proteolysis was also occurring. These observations were confirmed and extended by growing the organism in a liquid, beef extract medium when simultaneous utilization of lactic acid and amino acids occurred upon glucose depletion. An initial glucolytic phase was subsequently demonstrated for other common aerobic spoilage organisms including an *Enterobacter* sp., *Br. thermosphacta* and a non-fluorescent *Pseudomonas* sp. (Gill & Newton 1977). An *Acinetobacter* strain failed to utilize glucose and grew at the expense of amino acids and lactic acid.

On depletion of glucose the non-fluorescent pseudomonad metabolized lactic and amino acids like the fluorescent strain while *Br. thermosphacta* was reported to metabolize glutamic acid only. It is unlikely that glutamate was its major energy source because it does not possess a TCA cycle (Collins-Thompson *et al.* 1972), nor does glutamate support its growth in minimal medium (Grau 1979). Hence some other substrate, perhaps ribose or glycerol, must have been present. The *Enterobacter* strain metabolized glucose-6-phosphate after glucose and not until depletion of the former were lactic acid and amino acids metabolized. All of these patterns of substrate utilization are consistent with the well-established repressive and inhibitory effects of glucose. The fact that growth of each organism ceased before all available substrates had been depleted was attributed to oxygen limitation, since the continued accumulation of ammonia indicated that neither pH nor toxic metabolites had increased sufficiently to inhibit metabolism.

The very low concentrations, or even complete absence, of glucose and glucose-6-phosphate from DFD meat means that amino acids and lactic acid are the initial substrates for bacterial growth on this type of meat, which is probably the reason for its spoiling more rapidly than meat of normal pH. The pH of some samples of high pH beef devoid of glucose was reduced with citric acid and the glucose content of others was increased to levels found in normal pH meat (Newton & Gill 1978a). The time to spoilage and bacterial numbers at spoilage were then determined after inoculation with a *Pseudomonas* sp., beef of normal pH and glucose content being included as a control. Off odours were detected sooner, and at lower microbial numbers, in the samples devoid of glucose than in those containing it, irrespective of their pH. Bacterial growth rates were similar on all four samples although on the lower pH samples the lag phase was slightly longer. In a second experiment increases in ammonia concentrations were detected at lower bacterial numbers on samples of high pH mutton inoculated with the same pseudomonad than on similarly inoculated samples of normal pH mutton. The addition of glucose to DFD meat has therefore been suggested as a means of extending its aerobic shelf life and it has even been suggested by Shelef (1977) that the shelf life of normal pH meat can be extended by increasing glucose concentrations to *ca.* 5% (w/v).

(ii) *End products of growth*

The identity of the end products of bacterial growth on meat has, with a few exceptions, e.g. ammonia, been inferred from the subjective terms used to describe the odours developing during storage (e.g. Patterson & Gibbs 1977). Studies of the biochemical properties of bacteria growing in pure laboratory culture (Sutherland *et al.* 1975*a*) also helped. However, in a recent study of minced beef stored at temperatures between 5 and 20°C in oxygen concentrations ranging from 2 to 20% by volume, approximately 100 volatile compounds were extracted from samples analysed from freshness to rank spoilage (Stutz 1978). Hydrocarbons, alcohols, aldehydes, ketones, sulphur compounds, esters, aromatic compounds, amines and ammonia were all present in very low concentrations and concentrated on porous polymers, or by vacuum distillation, before analysis by combined gas liquid chromatography/mass spectrometry (g.l.c./m.s.). Both temperature and oxygen tension influenced the detailed composition of the volatiles but there were no obvious systematic changes.

In a more detailed study by the same author, 51 compounds were extracted from samples of minced beef stored for 5 days at 10°C in an atmosphere containing a constant 18% oxygen. Twenty-one were subsequently identified in the headspaces of meat samples stored under identical conditions after inoculation with pure cultures of bacteria representing the main genera previously isolated from naturally spoiling samples. Included were six hydrocarbons, two alcohols, two ketones, four sulphur compounds, two esters, three aromatic compounds, trimethylamine and ammonia. Those detected most consistently from sample to sample were methanol, acetone, methyl ethyl ketone, dimethylsulphide, ethyl acetate, toluene and xylene, and each normally constituted a major portion of the class of compounds to which it belongs. Ammonia and amines were not analysed in every sample and hydrogen sulphide could not be trapped by the techniques used, but it is probable that they should be included in the same category.

Of the bacteria tested, i.e. two fluorescent *Pseudomonas* spp., five non-fluorescent *Pseudomonas* spp., one *Moraxella* sp. and two *Acinetobacter* spp., only two of the non-fluorescent pseudomonads produced odours similar to the putrid odours of the naturally spoiling mince. From a comparison of their volatile profiles the author concluded that for a putrid odour to develop three requirements must be met: (a) total volatiles must exceed a minimum threshold value, (b) with the exception of the hydrocarbons, sulphur compounds must be major constituents of the volatiles, and (c) other classes of compounds must not be present in too large a quantity. For example, the volatile profile of the *Moraxella* sp. conformed to the first two requirements but contained large amounts of aromatic compounds and esters; an ester-like, rotten vegetable odour resulted.

From an analysis of the volatiles present at various times throughout storage at 5, 10 and 20°C only four compounds, acetone, methyl ethyl ketone, dimethyl sulphide and dimethyl disulphide, increased consistently and were therefore regarded by Stutz as a potential index of the microbial quality of the meat. Clearly more data are required before this suggestion can be assessed.

All the classes of compounds found by Stutz (1978) were detected in a study of the volatiles produced during growth of several species of bacteria on beef stored at an undefined temperature (Gibbs *et al.* 1979). Short chain fatty acids were produced by some organisms, e.g. *Br. thermosphacta*, while an *Aeromonas* sp., an *Alicaligenes* sp. and a *Moraxella*-like sp. produced two nitriles and two oximes. Three of these were identified by m.s. as isobutyronitrile, isobutyraldoxime *o*-methyl ether and methacrylnitrile, but their relevance to spoilage odours is not clear (Harper & Gibbs 1979). Volatiles produced during natural, mixed flora growth on normal and high pH beef stored at 15°C was studied by Patterson & Bolton (1981). Alcohols, ketones, esters, sulphur compounds, a hydrocarbon and acetic acid made up the 19 compounds detected, eight of which corresponded to compounds found by Stutz (1978). There were differences between meats of different pH but there were insufficient data to comment on their significance. End products resulting from the growth of *Br. thermosphacta* on intact slices of normal and high pH beef were studied by Dainty & Hibbard (1980). Acetoin, acetic acid, isobutyric and isovaleric acids were major components at both pH values and were thought to be responsible for the sickly sweet odours detected. High pH and low glucose concentration favoured fatty acid rather than acetoin formation during growth in laboratory media, thus explaining the reversed proportions of these compounds in the two kinds of meat. Patterson & Gibbs (1977) found acetoin to be a component of the buttery odours produced by several spoilage bacteria during growth on meat, including *Br. thermosphacta*.

Several studies of amine production during aerobic storage have been made. Increases in the concentrations of methylamine and dimethylamine plus trimethylamine were reported during storage of naturally contaminated pork at 5 and 20°C (Patterson & Edwards 1975). Although the changes were significant in the 20°C samples, the increased levels detected after prolonged storage at 5°C were still within the range defined for freshly slaughtered carcass meat in an earlier study (Patterson & Mottram 1974). A range of alkylamines, including the three mentioned above, was detected by Stutz (1978) in minced beef stored at 10°C for 5 days. Trimethylamine was most commonly detected and its formation seemed to be associated with the growth of fluorescent and non-fluorescent pseudomonads. Nakamura *et al.* (1979) found significantly increased levels of putrescine, spermidine and, to a lesser extent, cadaverine in minced pork stored at 4°C. Cadaverine levels increased much more during storage at 20°C. Large increases in concentrations of all three amines, together with smaller increases in histamine levels, were reported for samples of belly pork stored at 25°C (Lakritz *et al.* 1975). During storage at 5–8°C samples of both beef and pork inoculated with a mixture of three *Pseudomonas* spp., three *Enterobacter* spp., *Ser. liquefaciens* and *Klebsiella pneumoniae* showed greatly increased levels of putrescine and cadaverine (Slemr 1981). Even before evident spoilage the author reported finding 10-fold increases in the sum of the concentrations of the two amines and suggested their assay as an objective measure of quality. Pure culture experiments proved the pseudomonads to be the major source of putrescine while the Enterobacteriaceae produced most of the cadaverine.

(iii) *Lipid breakdown*

Oxidative (rancidity) and lipolytic changes in meat lipid components have been of interest for many years and chemical estimations of both processes, e.g. by thiobarbituric acid (TBA) and free fatty acid (FFA) values respectively have been suggested as useful, objective criteria of meat quality (Pearson 1968). Flavour defects attributable to one, or both, of these processes have been detected prior to off-odour development in lamb chops (Jeremiah *et al.* 1971) and in minced beef (Abo-Gnah 1978, cited by Branen 1978) but the role of bacteria in their formation has not been unequivocally established. Rancidity is usually attributed to autoxidation although many bacteria, including types commonly recovered from stored meat, produce and degrade oxidized derivatives of lipids (Alford *et al.* 1971). Many psychrotrophic bacteria produce lipases and growth of *Ps. fragi* on the surface of samples of beef stored at 1°C produced increased FFA values (Bala *et al.* 1977). The same organism's lipase was subsequently shown to catalyse similar changes (Bala *et al.* 1979).

Wheat-germ lipase has been shown to enhance the rate of myoglobin oxidation in minced beef stored at 0·6°C (Govindarajan *et al.* 1977). If microbial lipases have the same effect they may be involved in the deterioration of the colour of stored meat which is a frequent cause of loss of acceptability to the consumer (Bala *et al.* 1977, 1979).

Gill & Newton (1980) have shown that microbial growth can take place on fatty surfaces without any indication of lipid metabolism. A non-lipolytic strain of *Ps. fluorescens* metabolized glucose, amino acids and lactic acid when inoculated on a thin layer of agar placed on adipose tissue from lamb and incubated at 10°C. Only small concentrations of each substrate were present and amino acid metabolism was initiated at an early stage of spoilage. This resulted in off-odour production at relatively low cell numbers (*ca.* 10^6/cm^2) as in the case of lean DFD meat. No description of the odour was given but Berry *et al.* (1973) reported 'odours typical of putrefaction' during growth of a *Ps. fluorescens* strain on beef adipose tissue, while a *Flavobacterium* sp. produced fruity odours, i.e. similar terms to those used for off-odours of stored lean meat. In fact, Patterson & Bolton (1981) found a similar range of volatiles in fat and lean samples taken from normal (5·7) and high (6·3) pH beef carcasses and stored at 15°C, including sulphur compounds, alcohols, ketones, esters, a hydrocarbon and acetic acid. Experiments in which changes in lipid and non-lipid constituents are assayed in the same samples during storage are clearly required in order to clarify these apparent discrepancies.

B. Vacuum-packed storage

(i) *Substrates for growth*

Anaerobic growth of a *Lactobacillus* sp. previously isolated from vacuum-packed lamb, in either a beef juice medium pH 6·0 incubated at 30°C, or on the surface of sterile samples of lamb stored at 10°C, resulted in the depletion of glucose (Gill 1976). Lactic acid and ammonia concentrations, and the pH, remained unchanged in both systems. In the meat samples a reduction in surface glucose concentration was first detected when cell numbers reached 2×10^7/cm^2. Attainment of maximum cell numbers of 4×10^7/cm^2 coincided with the absence of detectable glucose at the surface. Despite the fact that arginine was subsequently utilized no further increase in numbers occurred because of the low energy yield from this substrate. Growth was thus limited by the rate of diffusion of glucose from underlying tissues. Substrate utilization by strains of three other bacterial genera typically found on vacuum-packed meat was studied in beef juice medium only (Newton & Gill 1978*b*; Gill & Newton 1979). *Brochothrix thermosphacta* grew at the sole expense of glucose while *Enterobacter liquefaciens* and

Alt. putrefaciens metabolized serine at the same time as glucose. On depletion of the latter, *E. liquefaciens* metabolized glucose-6-phosphate and lysine, arginine and threonine, but only the amino acids were used by *Alt. putrefaciens*. These data, together with the microbiological differences referred to earlier, help to explain the more rapid spoilage of vacuum-packaged DFD meat (Gill & Newton 1979). As under aerobic conditions, glucose deficiency results in earlier and more extensive utilization of amino acids which, with the growth of *Alt. putrefaciens*, results in the development of the H_2S-induced 'greening' and the offensive odours which typify DFD spoilage in vacuum packs. The shelf life can be extended to that of normal pH, vacuum-packed meat by acidification of the meat surface with citrate–lactate solutions (Gill & Newton 1979) but for commercial purposes the authors later recommended the use of 6 ml/kg of meat of 1·9 mol/l disodium citrate buffer pH 4·8 ± 0·3 mg/kg of glucose (Newton & Gill 1980). The citrate acts as an acidulant to prevent the growth of *Alt. putrefaciens* and, with glucose (if included), as an alternative source of carbon to the amino acids thus delaying off-odour production.

There is only one report of possible protein breakdown during vacuum packed storage, that of Sutherland *et al.* (1976). Using paper chromatography they detected new areas of ninhydrin positive material, in an area of the chromatogram believed to be occupied by peptides, in extracts of meat stored for 9 weeks at 0–2°C. The flora of the sample was dominated by lactic acid bacteria and *Br. thermosphacta*.

(ii) *End products of growth*

The off-odours detected on opening vacuum packs of normal pH meat stored for several weeks at chill temperatures are variously described as sour/acid/cheesy. They are thought to be associated, at least in part, with the accumulation of short chain fatty acids (Sutherland *et al.* 1976; Dainty *et al.* 1979). Both groups found acetic acid to be the major acid in beef stored for 9 weeks between 0–2°C, with smaller quantities of a number of straight and branched chain acids containing up to six carbon atoms. The main difference was the presence of significant amounts of *n*-butyric acid in the samples examined by Dainty *et al.* Most of the *n*-butyric acid, and part of the acetic acid, was later shown to be of non-microbial origin, while lactic acid bacteria, *Br. thermosphacta* and *Ser. liquefaciens* were found to be potential sources of the remaining acetic acid (Dainty 1981). Only *Br. thermosphacta* produced significant amounts of any other fatty acids, namely isobutyric and isovaleric acids. The non-microbial production of fatty acids could perhaps have resulted from inhibition of meat enzymes similar to that proposed by Shank *et al.* (1962) to explain an off-condition associated with fatty acid accumulation in beef in the absence of significant microbial growth.

Sutherland *et al.* (1976) reported the accumulation of unidentified amines in the samples they examined, while Dainty *et al.* (1979) identified methylamine, dimethylamine and trimethylamine. Trimethylamine, *n*-butylamine, methyl ethyl ketone, ethyl acetate, ethanol and ammonia were all found in low concentrations in minced beef stored anaerobically at 10°C for 10 days (Stutz 1978); the odour of the meat was described as bland and inoffensive. In none of the studies were the compounds shown to be of microbial origin and the amine concentrations reported by Dainty and co-workers were within the ranges reported for fresh pork (Patterson & Mottram 1974).

The offensive, putrid odours which develop during vacuum packed storage of DFD meat suggest that compounds other than, or in addition to, fatty acids are responsible. High concentrations of trimethylamine, together with hydrogen sulphide, were found in high pH (6·2–6·6) beef stored for 9 weeks at 1°C (Dainty *et al.* 1979). The samples also contained higher levels of acetic, *n*-butyric, isobutyric and isovaleric acids than the corresponding normal pH samples referred to above. Single strains of *Ser. liquefaciens*, *Alt. putrefaciens*, *Br. thermosphacta*, an *Aeromonas* sp. and a lactic acid bacterium each produced acetic acid during axenic growth on high pH beef, but only *Br. thermosphacta* produced the two branched chain acids (Dainty 1981). No microbial source of *n*-butyric acid was identified. The two Gram positive bacteria produced more fatty acids on high, than on low, pH meat and results from experiments in laboratory media showed this to be due to the combined effects of high pH and low glucose concentration.

4. Conclusions

To understand meat spoilage fully it is necessary to know (a) the types of bacteria which grow and their proportions within the flora, (b) the chemical changes produced during growth and the end products responsible for off-odours and off-flavours, and (c) the relative contributions of the various factors affecting the growth of spoilage bacteria.

Prior to 1970 most research on meat spoilage concentrated on the enumeration and identification of bacteria and this emphasis has continued in the last decade. There remain problems in the classification of the *Moraxella*-like strains and lactic acid bacteria which need to be solved if the microfloras of aerobically stored and vacuum packaged meats are to be described at a detailed level. There is also a need for more data on the bacterial changes occurring on carcasses and on fat during chilled storage. Otherwise we believe that there is sufficient information on the normal patterns of growth on meat during chilled aerobic or vacuum-packed storage and that more effort should be put into furthering our understanding of the chemistry of spoilage. There will remain a need for bacteriological studies in the assessment of new and untried systems of storage and in investigative work to explain previously unencountered spoilage problems. Explanation and interpretation of published research on the bacterial flora of packaged meat has been made difficult by the failure of many workers to document meat pH or gas composition above the meat. Any future studies on bacterial changes on meat should monitor and record all factors believed to be important in influencing them.

Substantial progress has been made since the review of Ingram & Dainty (1971) in identifying both the substrates for microbial growth and the resulting end products during aerobic and vacuum-packed storage. The preferential metabolism of glucose before amino acids, lactic acid and protein is in accord with earlier findings for fish and the widely established phenomenon of glucose repression. There would, therefore, appear to be little reason to doubt the relevance of the findings to natural, mixed flora spoilage even though the conclusions are based on data from pure culture inoculation experiments with a limited number of spoilage isolates. It should be borne in mind, however, that in the case of protein breakdown, discrepancies were apparent between the two types of spoilage situation. It would therefore be prudent to confirm that the pure culture findings are relevant to natural spoilage. Information on the metabolism of other substrates known to be present in significant concentrations but so far neglected, e.g. glycerol, glycogen, ribose (free, phosphorylated and in nucleoside form) and creatine, would also be of value. To help complete the picture more data are required on the nature and extent of lipid breakdown, together with an assessment of the relative importance of bacterial and autolytic processes.

From studies of end product formation, correlations between types of spoilage odour and the presence of particular classes of compounds are becoming evident. In addition certain specific compounds, which are produced consistently, and whose concentration is dependent on time of storage and/or bacterial numbers, are being proposed as possible spoilage indicators. At present the conclusions are based on too few investigations for their general validity to be assessed. More data are required, particularly with regard to the influence of relevant environmental factors, e.g. oxygen tension, pH and substrate availability on their formation. Possible differences between minced and non-minced samples should also be examined in view of the greater breakdown of lipid in the former.

Despite the reservations alluded to above, pure culture experiments are clearly essential if a complete understanding of spoilage processes is to be obtained. To date, choice of bacterial strains has been either arbitrary or based upon differences in properties of doubtful significance to spoilage, e.g. fluorescent pigment production in the case of the pseudomonads.

Recent progress in the classification of these organisms will now enable a more representative selection of strains to be used in future work, and hence provide a better understanding of their particular roles in spoilage. The studies will also reveal whether there is any correlation between these taxonomic groupings and the spoilage potential of the bacteria. Similar studies with other bacteria of interest, e.g. the lactic acid bacteria, are clearly dependent on future progress in their classification.

There is a considerable amount of published information on how factors such as temperature, pH, lactate concentrations, water activity, storage atmosphere, competition and substrate availability affect the development of the various groups of bacteria found on chilled meat (see the review of Gill & Newton 1978; also Gill & Newton 1977; Newton & Gill 1978b; Grau 1980). This can be used to provide explanations for most of the patterns of microbial changes observed under different storage conditions. In principle such data could be used to predict the extent of growth of all relevant groups of spoilage bacteria in defined conditions of storage. This, allied to data on the nature and extent of production of spoilage compounds by the different groups of organisms at various cell numbers and under different environmental conditions might enable accurate predictions of shelf life to be made using mathematical models. However, the effects of the factors have mostly been examined in insufficient details for this purpose and often without consideration of interactions. More systematic and forward looking studies are needed to satisfy this requirement.

5. References

ABO-GNAH, Y. S. 1978 *Use of antimicrobial substances from* Streptococcus diacetilactis *DRC1 and antioxidants to extend the keeping quality of refrigerated ground beef.* M.Sc. Thesis, Washington State University.

ALFORD, J. A., SMITH, J. L. & LILLY, H. D. 1971 Relationship of microbial activity to changes in lipids of foods. *Journal of Applied Bacteriology* **34**, 133–146.

AYRES, J. C. 1960 Temperature relationships and some other characteristics of the microbial flora developing on refrigerated beef. *Food Research* **25**, 1–18.

BALA, K., MARSHALL, R. T., STRINGER, W. C. & NAUMANN, H. D. 1977 Effect of *Pseudomonas fragi* on the colour of beef. *Journal of Food Science* **42**, 1176–1179.

BALA, K., MARSHALL, R. T., STRINGER, W. C. & NAUMANN, H. D. 1979 Stability of sterile beef and beef extract to protease and lipase from *Pseudomonas fragi*. *Journal of Food Science* **44**, 1294–1298.

BARLOW, J. & KITCHELL, A. G. 1966 A note on the spoilage of prepacked lamb chops by *Microbacterium thermosphactum*. *Journal of Applied Bacteriology* **29**, 185–188.

BEATTY, S. A. & COLLINS, V. K. 1939 Studies of fish spoilage. VI. The breakdown of carbohydrates, proteins and amino acids during spoilage of cod muscle press juice. *Journal of the Fisheries Research Board, Canada* **4**, 412–423.

BEEBE, S. D., VANDERZANT, C., HANNA, M. O., CARPENTER, Z. L. & SMITH, G. C. 1976 Effect of initial internal temperature and storage temperature on the microbial flora of vacuum packaged beef. *Journal of Milk and Food Technology* **39**, 600–605.

174   R. H. DAINTY *ET AL.*

BEM, Z., HECHELMANN, H. & LEISTNER, L. 1976 The bacteriology of DFD meat. *Fleischwirtschaft* **56**, 985–987.

BERRY, B. W., SMITH, G. C. & CARPENTER, Z. L. 1973 Growth of two genera of psychrotrophs on beef adipose tissue. *Journal of Food Science* **38**, 1074–1075.

BLICKSTAD, E., ENFORS, S.-O. & MOLIN, G. 1981 Effect of high concentrations of CO_2 on the microbial flora of pork stored at 4°C and 14°C. In *Psychrotrophic Micro-organisms in Spoilage and Pathogenicity* ed. Roberts, T. A., Hobbs, G., Christian, J. H. B. & Skovgaard, N. pp. 345–357. London & New York: Academic Press.

BORTON, R. J., BRATZLER, L. J. & PRICE, J. F. 1970a Effects of four species of bacteria on porcine muscle. 1. Protein solubility and emulsifying capacity. *Journal of Food Science* **35**, 779–782.

BORTON, R. J., BRATZLER, L. J. & PRICE, J. F. 1970b Effects of four species of bacteria on porcine muscle. 2. Electrophoretic patterns of extracts of salt soluble protein. *Journal of Food Science* **35**, 783–786.

BRANEN, A. L. 1978 Interaction of fat oxidation and microbial spoilage in muscle foods. In *Proceedings of the 31st Annual Reciprocal Meat Conference*, pp. 156–161.

BROWN, A. D. & WEIDEMANN, J. F. 1958 The taxonomy of the psychrophilic meat-spoilage bacteria: a reassessment. *Journal of Applied Bacteriology* **21**, 11–17.

BUCKLEY, D. J., GANN, G. L., PRICE, J. F. & SPINK, G. C. 1974 Proteolytic activity of *Pseudomonas perolens* and effects on porcine muscle. *Journal of Food Science* **39**, 825–828.

CAMPBELL, R. J., EGAN, A. F., GRAU, F. H. & SHAY, B. J. 1979 The growth of *Microbacterium thermosphactum* on beef. *Journal of Applied Bacteriology* **47**, 505–509.

CHRISTOPHER, F. M., SEIDEMAN, S. C., CARPENTER, Z. L., SMITH, G. C. & VANDERZANT, C. 1979a Microbiology of beef packaged in various gas atmospheres. *Journal of Food Protection* **42**, 240–244.

CHRISTOPHER, F. M., VANDERZANT, C., CARPENTER, Z. L. & SMITH, G. C. 1979b Microbiology of pork packaged in various gas atmospheres. *Journal of Food Protection* **42**, 323–327.

CHRISTOPHER, F. M., SMITH, G. C., DILL, C. W., CARPENTER, Z. L. & VANDERZANT, C. 1980 Effect of CO_2–N_2 atmospheres on the microbial flora of pork. *Journal of Food Protection* **43**, 268–271.

CLARK, D. S. & LENTZ, C. P. 1973 Use of mixtures of carbon dioxide and oxygen for extending the shelf-life of packaged fresh beef. *Canadian Institute of Food Science and Technology Journal* **6**, 194–196.

COLLINS-THOMPSON, D. L., SØRHAUG, T., WITTER, L. D. & ORDAL, Z. J. 1972 Taxonomic consideration of *Microbacterium lacticum*, *Microbacterium flavum*, and *Microbacterium thermosphactum*. *International Journal of Systematic Bacteriology* **22**, 65–72.

DAINTY, R. H. 1981 Volatile fatty acids detected in vacuum-packed beef during storage at chill temperatures. In *Proceedings of the 27th Meeting of European Meat Workers*, Vienna, pp. 688–690.

DAINTY, R. H. & HIBBARD, C. M. 1980 Aerobic metabolism of *Brochothrix thermosphacta* growing on meat surfaces and in laboratory media. *Journal of Applied Bacteriology* **48**, 387–396.

DAINTY, R. H., SHAW, B. G., DE BOER, K. A. & SCHEPS, E. S. J. 1975 Protein changes caused by bacterial growth on beef. *Journal of Applied Bacteriology* **39**, 73–81.

DAINTY, R. H., SHAW, B. G., HARDING, C. D. & MICHANIE, S. 1979 The spoilage of vacuum-packed beef by cold tolerant bacteria. In *Cold Tolerant Microbes in Spoilage and the Environment* ed. Russell, A. D. & Fuller, R. pp. 83–100. London & New York: Academic Press.

DAVIDSON, C. M., MOBBS, P. & STUBBS, J. M. 1968 Some morphological and physiological properties of *Microbacterium thermosphactum*. *Journal of Applied Bacteriology* **31**, 551–559.

DAVIDSON, C. M., DOWDELL, M. J. & BOARD, R. G. 1973 Properties of Gram negative aerobes isolated from meats. *Journal of Food Science* **38**, 303–305.

DUTSON, T. R., PEARSON, A. M., PRICE, J. F., SPINK, G. C. & TARRANT, P. J. V. 1971 Observations by electron microscopy on pig muscle inoculated and incubated with *Pseudomonas fragi*. *Applied Microbiology* **22**, 1152–1158.

EMPEY, W. A. & VICKERY, J. R. 1933 The use of carbon dioxide in the storage of chilled beef. *Journal of the Council of Scientific and Industrial Research, Australia* **6**, 233–243.

ENFORS, S. O., MOLIN, G. & TERNSTROM, A. 1979 Effect of packaging under carbon dioxide, nitrogen or air on the microbial flora of pork. *Journal of Applied Bacteriology* **47**, 197–208.

ERICHSEN, I. & MOLIN, G. 1981 The microbial flora of normal and high pH beef stored at 4°C in different gas environments. In *Psychrotrophic Micro-organisms in Spoilage and Pathogenicity* ed. Roberts, T. A., Hobbs, G., Christian, J. H. B. & Skovgaard, N. pp. 359–367. London & New York: Academic Press.

GARDNER, G. A. 1965a The aerobic flora of stored meat with particular reference to the use of selective media. *Journal of Applied Bacteriology* **28**, 252–264.

GARDNER, G. A. 1965b *Microbiological and biochemical changes in fresh meat during storage.* Ph.D. Thesis, Queens University of Belfast, N. Ireland.

GARDNER, G. A. 1966 A selective medium for the enumeration of *Microbacterium thermosphactum* in meat and meat products. *Journal of Applied Bacteriology* **29**, 455–460.

GARDNER, G. A. & STEWART, D. J. 1966a Changes in the free amino and other nitrogen compounds in stored beef muscle. *Journal of the Science of Food and Agriculture* **17**, 491–496.

GARDNER, G. A. & STEWART, D. J. 1966b The bacterial production of glutamic acid in stored comminuted beef. *Journal of Applied Bacteriology* **29**, 365–374.

GARDNER, G. A., CARSON, A. W. & PATTON, J. 1967 Bacteriology of pre-packed pork with reference to gas composition within the pack. *Journal of Applied Bacteriology* **30**, 321–333.

GIBBS, P. A., PATTERSON, J. T. & HARPER, D. B. 1979 Some characteristics of the spoilage of sterile beef by pure cultures of bacteria. *Journal of the Science of Food and Agriculture* **30**, 1109–1110.

GILL, C. O. 1976 Substrate limitation of bacterial growth at meat surfaces. *Journal of Applied Bacteriology* **41**, 401–410.

GILL, C. O. 1979 Intrinsic bacteria in meat—a review. *Journal of Applied Bacteriology* **47**, 367–378.

GILL, C. O. & NEWTON, K. G. 1977 The development of aerobic spoilage flora on meat stored at chill temperatures. *Journal of Applied Bacteriology* **43**, 189–195.

GILL, C. O. & NEWTON, K. G. 1978 The ecology of bacterial spoilage of fresh meat at chill temperatures. *Meat Science* **2**, 207–217.

GILL, C. O. & NEWTON, K. G. 1979 Spoilage of vacuum-packaged dark, firm, dry meat at chill temperatures. *Applied and Environmental Microbiology* **37**, 362–364.

GILL, C. O. & NEWTON, K. G. 1980 Development of bacterial spoilage at adipose tissue surfaces of fresh meat. *Applied and Environmental Microbiology* **39**, 1076–1077.

GILL, C. O. & PENNEY, N. 1977 Penetration of bacteria into meat. *Applied and Environmental Microbiology* **33**, 1284–1286.

GOVINDARAJAN, S., HULTIN, H. O. & KOTULA, A. W. 1977 Myoglobin oxidation in ground beef: mechanistic studies. *Journal of Food Science* **42**, 571–577, 582.

GRAU, F. H. 1979 Nutritional requirements of *Microbacterium thermosphactum*. *Applied and Environmental Microbiology* **38**, 818–820.

GRAU, F. H. 1980 Inhibition of the anaerobic growth of *Brochothrix thermosphacta* by lactic acid. *Applied and Environmental Microbiology* **40**, 433–436.

HAINES, R. B. 1933a The bacterial flora developing on stored lean meat, especially with regard to 'slimy' meat. *Journal of Hygiene* **33**, 175–182.

HAINES, R. B. 1933b The influence of carbon dioxide on the rate of multiplication of certain bacteria, as judged by viable counts. *Journal of the Society of the Chemical Industry* **52**, 13T.

HALLECK, F. E., BALL, O. E. & STIER, E. F. 1958 Factors affecting quality of prepackaged meat. IV. Microbiological studies. A. Cultural studies on bacterial flora of fresh meat; classification by genera. *Food Technology* **12**, 197–203.

HAMM, R. 1975 Water-holding capacity of meat. In *Meat* ed. Cole, B. J. A. & Lawrie, R. A. pp. 321–337. London: Butterworths.

HANNA, M. O., ZINK, D. L., CARPENTER, Z. L. & VANDERZANT, C. 1976 *Yersinia enterocolitica*-like organisms from vacuum-packaged beef and lamb. *Journal of Food Science* **41**, 1254–1256.

HARPER, D. B. & GIBBS, P. A. 1979 Identification of isobutyronitrile and isobutyraldoxime *O*-methyl ether as volatile microbial catabolites of valine. *Biochemistry Journal* **182**, 609–611.

HASEGAWA, T., PEARSON, A. M., PRICE, J. F. & LECHOWICH, R. V. 1970*a* Action of bacterial growth on the sarcoplasmic and urea soluble proteins from muscle. I. Effects of *Clostridium perfringens*, *Salmonella enteritidis*, *Achromobacter liquefaciens*, *Streptococcus faecalis* and *Kurthia zopfii*. *Applied Microbiology* **20**, 117–122.

HASEGAWA, T., PEARSON, A. M., RAMPTON, J. H. & LECHOWICH, R. V. 1970*b* Effect of microbial growth upon sarcoplasmic and urea-soluble proteins from muscle. *Journal of Food Science* **35**, 720–724.

HITCHENER, B. J., EGAN, A. F. & ROGERS, P. J. 1982 Characteristics of lactic acid bacteria isolated from vacuum-packaged beef. *Journal of Applied Bacteriology* **52**, 31–37.

HUFFMAN, D. L. 1974 Effect of gas atmospheres on microbial quality of pork. *Journal of Food Science* **39**, 723–725.

INGRAM, M. 1962 Microbiological principles in prepacking meats. *Journal of Applied Bacteriology* **25**, 259–281.

INGRAM, M. & DAINTY, R. H. 1971 Changes caused by microbes in spoilage of meats. *Journal of Applied Bacteriology* **34**, 21–39.

JAY, J. M. 1966 Influence of postmortem conditions on muscle microbiology. In *The Physiology and Biochemistry of Muscle as a Food* ed. Briskey, E. J., Cassens, R. G. & Trautman, J. C. pp. 387–402, Madison, Milwaukee: University of Wisconsin Press.

JAY, J. M. & KONTOU, K. S. 1964 Evaluation of the extract release volume phenomenon as a rapid test for detecting spoilage in beef. *Applied Microbiology* **12**, 378–383.

JAY, J. M. & KONTOU, K. S. 1967 Fate of free amino acids and nucleotides in spoiling beef. *Applied Microbiology* **15**, 759–764.

JAY, J. M. & SHELEF, L. A. 1976 Effect of micro-organisms on meat proteins at low temperatures. *Journal of Agriculture and Food Chemistry* **24**, 1113–1116.

JAYE, M., KITTAKA, R. S. & ORDAL, Z. J. 1962 The effect of temperature and packaging material on the storage life and bacterial flora of ground beef. *Food Technology* **10**, 95–98.

JEREMIAH, L. E., REAGAN, J. O., SMITH, G. C. & CARPENTER, Z. L. 1971 Ovine yield grades. I. Retail case-life. *Journal of Animal Science* **33**, 759–764.

KIRSCH, R. H., BERRY, F. E., BALDWIN, C. L. & FOSTER, E. M. 1952 The bacteriology of refrigerated ground beef. *Food Research* **17**, 495–503.

LAKRITZ, L., SPINELLI, A. M. & WASSERMAN, A. E. 1975 Determination of amines in fresh and processed pork. *Journal of Agricultural and Food Chemistry* **23**, 344–366.

LAUTROP, H. 1974 Genus *Moraxella*. In *Bergey's Manual of Determinative Bacteriology* 8th edn. ed. Buchanan, R. E. & Gibbons, N. E. pp. 433–436. Baltimore: Williams & Wilkins.

MARGITIC, S. & JAY, J. M. 1970 Antigenicity of salt-soluble beef muscle proteins held from freshness to spoilage at low temperatures. *Journal of Food Science* **35**, 252–255.

MCLEAN, R. A. & SULZBACHER, W. L. 1953 *Microbacterium thermosphactum*, spec. nov.: non-heat resistant bacterium from fresh pork sausage. *Journal of Bacteriology* **65**, 428–433.

MORRISSEY, P. A., BUCKLEY, D. J. & DALY, M. C. 1980 Effect of four species of bacteria on minced beef stored at 7°C. *Irish Journal of Food Science and Technology* **4**, 1–11.

NAKAMURA, M., WADA, Y., SAWAYA, H. & KAWABATA, T. 1979 Polyamine content in fresh and processed pork. *Journal of Food Science* **44**, 515–517, 523.

NEWTON, K. G. & GILL, C. O. 1978*a* Storage quality of dark, firm dry meat. *Applied and Environmental Microbiology* **36**, 375–376.

NEWTON, K. G. & GILL, C. O. 1978*b* The development of the anaerobic spoilage flora of meat stored at chill temperatures. *Journal of Applied Bacteriology* **44**, 91–95.

NEWTON, K. G. & GILL, C. O. 1980 Control of spoilage in vacuum packaged dark, firm, dry (DFD) meat. *Journal of Food Technology* **15**, 227–234.

NEWTON, K. G. & GILL, C. O. 1981 The microbiology of DFD fresh meats: a review. *Meat Science* **5**, 223–232.

NEWTON, K. G. & RIGG, W. J. 1979 The effect of film permeability on the storage life and microbiology of vacuum-packed meat. *Journal of Applied Bacteriology* **47**, 433–441.

NEWTON, K. G., HARRISON, J. C. L. & SMITH, K. M. 1977 The effect of storage in various gaseous atmospheres on the microflora of lamb chops held at −1°C. *Journal of Applied Bacteriology* **43**, 53–59.

NICOL, D. J., SHAW, M. K. & LEDWARD, D. A. 1970 Hydrogen sulphide production by bacteria and sulfmyoglobin formation in prepacked chilled beef. *Applied Microbiology* **19**, 937–939.

PARTMANN, W., BOMAR, M. T., HAJEK, M., BOHLING, H. & SCHLASZUS, H. 1975 Application of controlled atmospheres containing 20% CO_2 to the storage of beef. *Fleischwirtschaft* **55**, 1441–1442, 1445–1446, 1449–1451.

PATTERSON, J. T. 1970 Development of the bacteriological spoilage flora of lamb—results of some laboratory experiments using minced lamb and lamb chops. *Record of Agricultural Research in the Ministry of Agriculture Northern Ireland* **18**, 9–13.

PATTERSON, J. T. & BOLTON, G. 1981 Some odours produced by bacteria on high pH and normal beef. In *The Problems of Dark-cutting in Beef.* Commission of European Communities Seminar, Brussels, October 1980. *Current Topics in Veterinary Medicine and Animal Science* Vol. 10, ed. Hood, D. E. & Tarrant, P. V. The Hague: Martinus Nijhoff.

PATTERSON, J. T. & GIBBS, P. A. 1977 Incidence and spoilage potential of isolates from vacuum-packaged meat of high pH value. *Journal of Applied Bacteriology* **43**, 25–38.

PATTERSON, R. L. S. & EDWARDS, R. A. 1975 Volatile amine production in uncured pork during storage. *Journal of the Science of Food and Agriculture* **26**, 1371–1373.

PATTERSON, R. L. S. & MOTTRAM, D. S. 1974 The occurrence of volatile amines in uncured and cured pork meat and their possible role in nitrosamine formation in bacon. *Journal of the Science of Food and Agriculture* **25**, 1419–1425.

PEARSON, D. 1968 Assessment of meat freshness in quality control employing chemical techniques: a review. *Journal of the Science of Food and Agriculture* **19**, 357–363.

PIERSON, M. D., COLLINS-THOMPSON, D. L. & ORDAL, Z. J. 1970 Microbiological, sensory and pigment changes of aerobically and anaerobically packaged beef. *Food Technology* **24**, 1171–1175.

RAMPTON, J. H., PEARSON, A. M., PRICE, J. F., HASEGAWA, T. & LECHOWICH, R. V. 1970 Effect of microbial growth upon myofibrillar proteins. *Journal of Food Science* **35**, 510–513.

REY, C. R., KRAFT, A. A., TOPEL, D. G., PARRISH, F. C. & HOTCHKISS, D. K. 1976 Microbiology of pale, dark and normal pork. *Journal of Food Science* **41**, 111–116.

ROTH, L. A. & CLARK, D. S. 1972 Studies on the bacterial flora of vacuum-packaged fresh beef. *Canadian Journal of Microbiology* **18**, 1761–1766.

ROTH, L. A. & CLARK, D. S. 1975 Effect of lactobacilli and carbon dioxide on the growth of *Microbacterium thermosphactum* on fresh beef. *Canadian Journal of Microbiology* **21**, 629–632.

SAFFLE, R. L., MAY, K. N., HAMID, H. A. & IRBY, J. D. 1961 Comparing rapid methods of detecting spoilage in meat. *Food Technology* **15**, 465–467.

SEELYE, R. J. & YEARBURY, B. J. 1979 Isolation of *Yersinia enterocolitica*-resembling organisms and *Alteromonas putrefaciens* from vacuum-packed chilled beef cuts. *Journal of Applied Bacteriology* **46**, 493–499.

SEIDEMAN, S. C., VANDERZANT, C., SMITH, G. C., HANNA, M. O. & CARPENTER, Z. L. 1976 Effect of degree of vacuum and length of storage on the microflora of vacuum-packaged beef wholesale cuts. *Journal of Food Science* **41**, 738–742.

SHANK, J. L., SILLIKER, J. H. & GOESER, P. A. 1962 The development of a non-microbial off-condition in fresh meat. *Applied Microbiology* **10**, 240–246.

SHAW, B. G. & LATTY, J. L. 1981 The taxonomy of *Pseudomonadaceae* responsible for low-temperature aerobic spoilage of meats. In *Psychrotrophic Micro-organisms in Spoilage and Pathogenicity* ed. Roberts, T. A., Hobbs, G., Christian, J. H. B. & Skovgaard, N. pp. 259–268. London & New York: Academic Press.

SHAW, B. G. & LATTY, J. L. 1982 A numerical taxonomic study of *Pseudomonas* strains from spoiled meat. *Journal of Applied Bacteriology* **52**, 219–228.

SHAW, M. K. & NICOL, D. J. 1969 Effect of the gaseous environment on the growth on meat of some food poisoning and food spoilage organisms. In *Proceedings of the 15th European Meeting of Meat Research Workers, Helsinki*, pp. 226–232.

SHAW, B. G., HARDING, C. D. & TAYLOR, A. A. 1980 The microbiology and storage stability of vacuum packed lamb. *Journal of Food Technology* **15**, 397–405.

SHELEF, L. A. 1974 Hydration and pH of microbially spoiling beef. *Journal of Applied Bacteriology* **37**, 531–536.

SHELEF, L. A. 1977 Effect of glucose on the bacterial spoilage of beef. *Journal of Food Science* **42**, 1172–1175.

SHEWAN, J. M., HOBBS, G. & HODGKISS, W. 1960 A determinative scheme for the identification of certain genera of Gram-negative bacteria, with special reference to the *Pseudomonadaceae*. *Journal of Applied Bacteriology* **23**, 379–390.

SIKES, A. & MAXCY, R. B. 1980 Postmortem invasion of muscle food by a proteolytic bacterium. *Journal of Food Science* **45**, 293–296.

SLEMR, J. 1981 Biogene Amine als potentieller chemischer Qualitatsindikator fur Fleisch. *Fleischwirtschaft* **61**, 921–926.

SNEATH, P. H. A. & JONES, D. 1976 *Brochothrix*, a new genus tentatively placed in the family *Lactobacillaceae*. *International Journal of Systematic Bacteriology* **26**, 102–104.

STAINER, R. Y., PALLERONI, N. J. & DOUDEROFF, M. 1966 The aerobic pseudomonads: a taxonomic study. *Journal of General Microbiology* **43**, 159–271.

STRINGER, W. C., BILSKIE, H. E. & NAUMANN, H. D. 1969 Microbial profiles of fresh beef. *Food Technology* **23**, 97–102.

STUTZ, H. K. 1978 *The utilization of volatile compounds produced during microbial growth on ground beef to characterize spoilage*. Ph.D. Thesis, University of Massachusetts.

SUTHERLAND, J. P., PATTERSON, J. T., GIBBS, P. A. & MURRAY, J. G. 1975a Some metabolic and biochemical characteristics of representative microbial isolates from vacuum-packaged beef. *Journal of Applied Bacteriology* **39**, 239–249.

SUTHERLAND, J. P., PATTERSON, J. T. & MURRAY, J. G. 1975b Changes in the microbiology of vacuum packaged beef. *Journal of Applied Bacteriology* **39**, 227–237.

SUTHERLAND, J. P., GIBBS, P. A., PATTERSON, J. T. & MURRAY, J. G. 1976 Biochemical changes in vacuum packaged beef occurring during storage at 0–2°C. *Journal of Food Technology* **11**, 171–180.

TARRANT, P. J. V., PEARSON, A. M., PRICE, J. F. & LECHOWICH, R. W. 1971 Action of *Pseudomonas fragi* on the proteins of pig muscle. *Applied Microbiology* **22**, 224–228.

TARRANT, P. J. V., JENKINS, N., PEARSON, A. M. & DUTSON, T. R. 1973 Proteolytic enzyme preparation from *Pseudomonas fragi*: its action on pig muscle. *Applied Microbiology* **25**, 996–1005.

TAYLOR, A. A. & SHAW, B. G. 1977 The effect of meat pH and package permeability on putrefaction and greening in vacuum packed beef. *Journal of Food Technology* **12**, 515–521.

THORNLEY, M. J. 1967 A taxonomic study of *Acinetobacter* and related genera. *Journal of General Microbiology* **49**, 211–257.

WAUTERS, G. 1981 The pathogenic significance of the different Yersinia groups in man. In *Psychrotrophic Micro-organisms in Spoilage and Pathogenicity* ed. Roberts, T. A., Hobbs, G., Christian, J. H. B. & Skovgaard, N. pp. 401–403. London & New York: Academic Press.

WOLIN, E. F., EVANS, J. B. & NIVEN, C. F. JR 1957 The microbiology of fresh and irradiated beef. *Food Research* **22**, 682–686.

Microbial Spoilage of Cured Meats

G. A. GARDNER

Ulster Curers' Association, Belfast, UK

Contents

1. Introduction

MICROBIAL spoilage may be defined as the state reached when the normal characteristics of a meat are so changed by the growth and metabolic activities of micro-organisms as to render it unfit for human consumption. Such a state may be perceived by one or more of the senses of sight, smell, taste and touch.

A cured meat can be defined as one to which NaCl has been added and in which the native meat pigment myoglobin is, as a result of reaction with nitric oxide, mainly in the nitroso-form. The precursor of NO is the other essential curing salt, $NaNO_2$ or KNO_2, which may be added directly or result from nitrate reduction (e.g. Bard & Townsend 1971). These salts, together

FOOD MICROBIOLOGY
ISBN 0 12 589670 0

with other curing adjuncts, are usually added to the meat in a solution (brine).

In the British Isles the most important meat species used for curing is pork and, to a lesser extent, beef (corned beef) and poultry. In this paper the various types of microbial spoilage will be illustrated and discussed primarily in relation to cured pig meat. It is not the intention of the author to discuss all forms of cured meat spoilage, but to highlight and discuss in detail the most common forms found, and those which are currently of economic importance to the food industry in the British Isles. The spoilage of a wide range of European comminuted meats and other products, which by the definition given above may be classified as cured meats, has not been included.

2. Raw Cured Meats

Commercially the most important product in this class is Wiltshire style bacon, which is pork cured as a side (the trimmed half of a carcass) by a process of brine injection, tanking, draining and maturation (Gibbons 1953; Gardner 1973a, 1982). Most cured sides are marketed without further processing ('green'), but some are partially dried ('pale dried') or fully dried and smoked.

In the last decade or so many technological changes have taken place in the production of bacon, all of which contribute to some extent to changes in the microbiology (stability) of the product. Public demand has indicated a marked preference for much milder (less salt) cures and legislation has reduced the permitted levels of sodium nitrate and sodium nitrite in the meat. Many manufacturers no longer use nitrate in the injection brines used for bacon curing. Mechanical injection of brine into pork sides is now widely practised, but where direct recirculation is employed, the bacteriological condition of the brine is markedly inferior to those used formerly (Gardner & Patton 1978). Lower salt contents lead to decreased stability of surface tissues, hence the need for refrigeration and a quick turnover of bacon sales. In an economic sense this is reflected in lower weight losses during the process, which has obvious commercial acceptance.

To increase the stability of the surface of Wiltshire bacon, a dry salting process has been developed both in the UK and in Denmark, whereby bacon sides are passed through a cloud of dry salt immediately on removal from the curing tank and matured on pallets. This process markedly extends the shelf life of the side without significantly contributing to the saltiness of the bacon (Gardner, unpublished data).

Microbial spoilage of bacon may be recognized by one or more characteristics: surface slime, off-odours and flavours, or visible discolourations.

A. Surface slime

The most common form of microbial spoilage of Wiltshire bacon sides or cuts (shoulder, middle and gammon) is slime formation. In sides this is first detected in the rib-cage, flank, under the foreleg, on the rind (skin) or at the head of the femur in the gammon. Initially this form of spoilage can be recognized as a greasiness or stickiness of the surface tissues and at a more advanced stage the slime can be seen. Depending on the stage of development, slimes may be accompanied by relatively non-obnoxious off-odours and/or discolourations.

The slime is composed of coalesced microbial colonies growing on the bacon surface, but may be enhanced by organisms which produce large amounts of extracellular material, e.g. dextran-producing *Leuconostoc* spp. on the surface of Paris hams (Fournaud & Raibaud 1964).

As a butchered Wiltshire bacon side is largely covered by rind, membranes and fat, with only a small proportion of exposed muscle, such slimes in the initial stages of spoilage can be physically removed by washing. Washing such bacon with curing brine has long been practised without detriment to the product. The spoilage is often limited to the surface, but if it has penetrated the musculature, the bacon is beyond recovery. Washed bacon, because of the high potential to 're-slime', is normally sprinkled with dry salt, dried and smoked.

In an attempt to relate subjective assessment of slime formation (i.e. visible growth) to bacterial numbers Gibbons (1940) showed that the pleural membrane was never slimy when the counts were $5 \times 10^6/cm^2$ or less, but were always slimy when the counts exceeded $9{\cdot}3 \times 10^7/cm^2$. Between these figures the membrane may or may not exhibit slime formation. For adequate stability Gibbons (1940) stated that sides with initial counts of $<10^5$ organisms/cm^2 would be satisfactory under their conditions of marketing (20–25 days). Where such long keeping quality is not important, a more practical standard may be $10^6/cm^2$, i.e. to give a keeping quality of 7–14 days. Gardner (1982) has shown that 93% of Wiltshire bacons would meet this criterion and 80% would meet that of Gibbons (1940).

Non-singed rind samples have the highest initial count, followed by singed rind and finally 'meat' (internal carcass) surfaces (Gardner & Patton 1969). High initial counts on unsinged rinds probably explain the observation that the rind under the foreleg on most bacon sides is one of the first areas to exhibit slime formation.

The species of bacteria found on a Wiltshire bacon side are very varied (Gardner & Patton 1969). They include Micrococcaceae, *Acinetobacter*, *Vibrio* and many other species including the lactic acid bacteria, *Brochothrix thermosphacta*, yeasts, *Alcaligenes*, flavobacteria, *Aeromonas*, *Pseudomonas–Achromobacter* and Enterobacteriaceae. Hence the 'initial' flora on a bacon side may be diverse and will include species from the pork from which the bacon originated together with others acquired from the brining process.

The composition of the microflora on matured bacons also depends on the site, i.e. rind, lean meat or fat. Thus bacon matured for 14 days at 5°C (counts between 10^6–10^7/cm^2 on 4% NaCl agar) had a flora in which cocci outnumbered rods on the rind and rods greatly outnumbered cocci on the meat surface (Tofte Jespersen & Riemann 1958). Thus slime can result from the growth of different species at different sites on a Wiltshire bacon side. Bacon sides are usually stored in stacks on pallets, where there are at least three ecological situations; exposed rind, exposed meat and rind/meat interface. Thus in any discussion on the nature of slime formation it is crucial to have information about the sampling. As the necessary detail is not available in most published work, it is not usually possible, accurately, to relate slime composition and sample site.

Members of the genus *Micrococcus* have long been known to predominate on Wiltshire bacon (e.g. Garrard & Lochhead 1939; Ingram 1960) and they can cause slime formation (e.g. Kitchell 1958). They belong mostly to *Micrococcus* subgroup 5 (Baird-Parker 1962) or by the classification of Shaw *et al.* (1951) to *Staphylococcus* subgroup 3; data recently summarized by Gardner (1982). These bacteria have been isolated from the skin of the pig, pork carcasses and tank curing brine, so their origins are well known (e.g. Kitchell 1958). They can grow in salt concentrations found on bacons and at refrigeration temperatures. Their greater importance on rind microfloras may be a result of higher salt levels (*ca.* 10% NaCl) and much higher pH levels (*ca.* 7·0) than those found on the meat surfaces.

During the spoilage of bacon lean nitrate is converted to nitrite, and when all the nitrate has been converted, the accumulated nitrite is then reduced (e.g. Gardner 1971). Whereas the majority of bacon-spoilage micrococci strains can reduce nitrate (Kitchell 1962), only *ca.* 20% can reduce nitrite (e.g. Gardner 1971).

Kitchell (1962) has also reviewed the proteolytic and lipolytic activity of micrococci and it is sufficient for the purpose of this paper to recognize these potential spoilage reactions.

Gram negative bacteria of the genus *Vibrio* can also be found in spoiled bacon. Recent studies (Gardner 1980–81) have shown that there are three distinct groups of halophilic vibrios: Group 1, *V. costicola*; Group 2, *V. costicola* subsp. *liquefaciens* and Group 3, unidentified vibrios. The Group 3 type was most frequently found in spoiled (slimy) Wiltshire bacon sides, though in some cases Groups 1 and 2 were also found. Originally *V. costicola* (Smith 1938) was found to cause rib taint in bacons (see Section 1.D) and was isolated from brines and bacons but not from pork. Similar findings were made by Gardner (1980–81) for *V. costicola* and its subsp. *liquefaciens*, but the Group 3 unidentified vibrios were also found in the lairage and slaughter/butchery (no-salt) areas of the bacon factory. Wiltshire bacons are likely to become contaminated during production by the immersion brines, most of which contain *Vibrio* spp. (Gardner 1973*b*, 1980–81, 1982).

The psychrotrophic, obligately halophilic nature of these bacteria (Smith 1938) is ideally suited to growth on aerobically stored Wiltshire bacon. All species can reduce nitrate, many reduce nitrite, and they undoubtedly contribute to the breakdown of these curing salts during bacon spoilage (Gardner 1971). Only *V. costicola* subsp. *liquefaciens* was proteolytic towards casein and gelatin, but most strains of all groups could produce H_2S from cystine (Gardner 1980–81), but were non-lipolytic (Smith 1938).

As with the micrococci, vibrios have a strong spoilage potential in a metabolic sense, but there is a scarcity of knowledge in this area. Nevertheless, *Vibrio* and *Micrococcus* will account for most cases of surface sliming in green Wiltshire bacon.

B. Surface discolourations

Discolouration of the surface of raw cured meats may be caused by a wide range of pigmented bacteria (Jensen 1954). Moreover, the surface pigments of the meat may be altered prior to the recognition of a slime and are usually associated with off-odour.

Formerly, sufficient nitrate was absorbed when sprinkled on sides of bacon before curing and, on reduction by microbial activity, could yield 0·5–1·0% nitrite in the bacon. Such amounts cause 'nitrite burn', easily recognized as a bright green discolouration of the cured meat pigment.

Sulphmyoglobin, a green pigment, can be found when the flora has a large proportion of H_2S-producing micro-organisms (Jensen 1954). This can readily be seen in sulphide spoilage in internal taints (e.g. pocket taint) or on spoiled, vacuum-packed, cured meats, when exposed to air (see Section 3.B).

Normally the surface colouration of bacon is caused by a mixture of nitrosomyoglobin, which is bright red, and the oxidized compounds nitrosometmyoglobin or metmyoglobin, which are brown in colour. Bacon heavily contaminated by aerobic micro-organisms usually appears a bright fiery red colour, probably due to the highly reducing conditions on the surface, as the bacteria compete successfully with the pigment for oxygen.

C. Surface moulds

In dried and smoked bacons bacterial slimes are very rare, due to the relatively low a_w of the surface tissues. Instead, mould growth occurs. Late summer and early autumn is termed the 'mould season' for smoked bellies of bacon (Jensen 1954). Many species of mould can grow on the bacons; some are psychrophilic but most are mesophilic, and they all have a high resistance to salt and the residual aldehydes from smoke. They include species of *Aspergillus, Alternaria, Monilia, Oidium, Fusarium, Mucor, Rhizopus, Botrytis* and *Penicillium* (Jensen 1954). Apart from the mycelial growth on the bacon, many species are highly pigmented, e.g. *Aspergillus candidus*, which produces small reddish patches on the bacon. Such mould growths in the early stages are relatively easy to remove.

Another product with low moisture levels on the surface is the American 'country cured' ham. These are dry-cured hams, which still have both rind and bones and have a keeping quality of over 6 months. Advances in the technology of producing this product have been published (Anon. 1980). Mould growth on this ham, particularly on the meat, occurs frequently and is regarded by some people as an index of a well-matured product. The mould may contribute to the flavour of the ham, but generally such visible growth is recognized as an undesirable feature. Leistner & Ayres (1968) found that the moulds most frequently isolated from 'country cured' hams belong to the genera *Aspergillus, Penicillium* and *Cladosporium* (Table 1).

TABLE 1

Genera of moulds isolated from 40 country-cured hams

Genus	Percentage of hams where moulds were found
Aspergillus	90
Penicillium	82·5
Cladosporium	30
Rhizopus	12·5
Alternaria	12·5
Scopulariopsis	7·5
Paecilomyces	7·5
Oospora	7·5
Epicoccum	7·5
Mortierella	2·5
Syncephalastrum	2·5
Fusarium	2·5

Data from Leistner & Ayres (1968).

During the early stages of ripening the mould flora was dominated by penicillia, but in older hams aspergilli were predominant, probably due to their greater tolerance of low a_w.

Penicillium and *Aspergillus* can be found in the gut contents of slaughter animals (Klare 1970) and in natural spices (Hadlok 1969). They are probably ubiquitous in meat processing plants, so that one must assume all products will be contaminated. Mould growth can be retarded by spraying potassium sorbate on the hams (Baldock *et al.* 1979).

Although Jensen (1954) stated that none of the moulds produced 'substances deleterious to health', more recent studies have shown that toxigenic species of *Aspergillus* and *Penicillium* could be isolated from cured meats (e.g. Wu *et al.* 1974), although there was no evidence that the affected meats contained toxic metabolites. Leistner & Eckardt (1979) examined 761 penicillia from meat products and by chemical and biological methods showed that *ca.* 78% of isolates were toxigenic. They listed 16 mycotoxins which could be produced by the penicillia in synthetic media, fermented sausages and raw hams. Thus it can be concluded that apart from being a visible defect on a meat product, mould growth may be of public health significance.

D. Internal taints

(i) *'Pocket' taint*

In a side of Wiltshire bacon the scapula is removed leaving a cavity in the shoulder known as the 'pocket'. Either before or after brine injection a quantity of NaCl (170 g) is placed with a salt stuffer into the pocket to prevent microbial taints developing. With the advent of machine brine injection some manufacturers omit this salting during the winter months, but at times of high ambient temperature it is re-introduced. After the tanking process prior to maturation it has been the practice for many years to put a tablet of salt (28 or 56 g) into the pocket, again to minimize the risk of taint. These measures indicate that the pocket is regarded as a 'taint risk' area.

Bacterial slimes of similar composition to those on the surface can develop in the pocket. Provided the membrane which surrounds the scapula has not been broken or damaged during the removal of the bone, such slime can be removed physically, leaving the adjacent meat free of taint. However, if, as happens in many cases, the membrane is damaged and meat exposed, the taint is more serious and the meat is spoiled.

'Pocket' taints may be caused by many different bacteria, but the most frequent genera found include *Vibrio*, *Micrococcus*, *Alcaligenes* and, on occasions, *Proteus* (Gardner & Patterson 1975). This form of spoilage is influenced largely by the standard of butchery in scapula removal, the correct application of salt to the cavity and storage temperature. It must also be realized that the pH of the musculature surrounding the pocket is probably the highest to be found in the carcass (>6·2) and is thus more liable to rapid spoilage.

(ii) *Rib taint*

A taint in the ribs of bacon manufactured in Australia, and associated with the summer months, has been described by Smith (1938). When the ribs of affected sides were removed and broken for examination, a strong taint was evident. The causative organism was identified as *V. costicolus*, which could be isolated from the curing brines but never from the meat or bones before curing. Maintenance of high salt levels (>25%) in the tank brines and the use of nitrate to alter the redox potential were recognized as practical methods to minimize spoilage. These bacteria, together with as yet unidentified strains of *Vibrio*, have more recently been recognized in other cured meat spoilages (Gardner 1980–81).

(iii) *Bone taint*

Bone taint is recognized as a souring or putrefactive form of spoilage in the deep meat and bone marrows of cured meats. Hams are the most important product to which this applies, although taints in the foreleg of the shoulder cuts also occur.

A comprehensive study of bone taints in American hams was published by Jensen & Hess (1941*a*). They detailed six or more different forms of ham souring and considered the practical reasons for their occurrence. The most important bacteria causing these taints included *Achromobacter, Bacillus, Pseudomonas, Proteus, Serratia, Clostridium* (particularly the psychrophile *Cl. putrefaciens*), micrococci and streptobacilli. At the time of their study hams were either dry cured without prior injection (60–80 day cure) or they were pumped with brine via the vascular system (20 day cure). Consequently their findings may not be directly applicable to modern ham production.

In similar material, the southern country-style hams, Mundt & Kitchen (1951) identified the causative organisms of stifle joint taint as belonging to the genus *Clostridium*. The species found include *Cl. bifermentans, Cl. mucosum, Cl. sporogenes, Cl. parabifermentans, Cl. septicum, Cl. paraputrificum* and *Cl. putrefaciens*. Also, in hams prepared with little or no brine injection, Ingram (1952) found that taints were caused by a salt- and acid-sensitive microflora, whose optimum growth temperature was 37°C, and would be classified as clostridia and faecal streptococci.

Another cured meat product which is not pumped with brine is the Italian type (e.g. Parma) ham, which may be matured for up to 16 months. These hams are prepared by applying salt (NaCl) to the ham which is stored in salt for 1 month, washed, dried and finally ripened in controlled conditions for 6–12 months.

The internal flora of normal hams has an average count of $1 \cdot 16 \times 16^6$/g and is composed mainly of micrococci and lactobacilli. Spoiled hams may have counts $>10^8$/g and the microflora is more heterogeneous, including in addition to micrococci and lactobacilli, species of *Arthrobacter*, corynebacteria, *Bacillus*, *Sarcina* and *Aerococcus* (Cantoni *et al.* 1969). These workers also studied in detail the chemical changes in hams during maturation, not all of which were associated with the microflora. However, the production of highly objectionable sulphides by the microflora was shown (Table 2).

TABLE 2

Production of H_2S and CH_3SH by the microflora of spoiled Parma hams

Organism	No. of strains	No. of strains able to produce	
		H_2S only	H_2S and CH_3SH
Micrococcus	21	10	0
Arthrobacter/corynebacteria	5	0	3
Lactobacillus	10	6	3
Pediococci	8	3	0
Yeasts	5	0	3

Data from Cantoni *et al.* (1969).

Proteus spp., particularly *Pr. rettgeri*, is the major cause of internal taints in unpumped hams (Zeller & Renz 1967). The organism can grow at temperatures about 8°C and in salt (NaCl) concentrations up to 8%. The spoilage was associated with non-chilled meat, but was also found in chilled meat with a low salt content, such as might have resulted from incorrect curing.

In pumped hams Ingram (1952) recognized that internal taints were caused by a salt-tolerant flora, whose optimum growth temperature was 25°C and which belonged mainly to the micrococci group. Both *Proteus* (Gardner & Patterson 1975) and halophilic species of *Vibrio* (Gardner 1980–81) have been isolated from femur taints of Wiltshire bacon.

In summary, the microflora one might expect to find in tainted hams will be influenced largely by the manufacturing procedure of the cure. Even within one type of ham a wide and varied microflora may be found.

Hams cured without brine injection will have a flora which has been described as intrinsic (Ingram 1952), which is mainly mesophilic and salt-sensitive, and is associated with the animal and the pork. Such bacteria and their mode of access to the deep meat and bone have been reviewed (Gill 1979). The growth of these deep-seated bacteria will be influenced largely by temperature regime, levels of curing salts, particularly NaCl, and the pH of the surrounding meat (Gardner 1982).

By far the most important type of ham production involves brine injection either by hand or by machine. The bacteriological count of such brines can be high and may include potential spoilage species (Gardner & Patton 1978). This extrinsic flora (Ingram 1952), which includes salt-tolerant and psychrophilic species, gains access to the 'deep' situations in the pork leg, and under the same conditions affecting the intrinsic flora, growth and spoilage may occur.

3. Vacuum-packed Raw Cured Meats

Wiltshire cured bacon sides or cured cuts (e.g. shoulder, backs, streaks) may be further processed in 'green' or smoked form for marketing in the form of vacuum-packed slices. Technologically this type of consumer pack has two main advantages. Firstly, the bright red pigment of the bacon (nitrosomyoglobin) and the fat are protected from deteriorative oxidative changes, and secondly the shelf life of the product is extended from *ca.* 1 to 3 weeks or more.

The microbiology of vacuum-packed bacon has received a great deal of attention (e.g. Ingram 1960; Cavett 1962; Kitchell & Ingram 1963; Spencer 1967). In a vacuum pack the spoilage of the bacon cannot be readily detected. Even a grossly spoiled product in the pack can appear quite normal, and it may only be on opening that off-odours are detected; these can vary considerably.

A. Souring

Cavett (1962) examined the spoilage of both low- and high-salt bacons under non-refrigeration storage temperatures (20 and 30°C): the spoilage odours and associated microfloras are given in Table 3. Detailed bacteriological (Cavett 1962) and chemical (Tonge *et al.* 1964) data concerning these spoilages have been published. The effects of salt and high storage temperatures are self-evident, Micrococcaceae being dominant in these situations over the lactic acid bacteria.

However, vacuum-packed bacons stored at temperatures from 5 to 21°C

TABLE 3

Spoilage characteristics and predominant microflora of bacons stored at
20 or 30°C for 22 days

Storage temperature (°C)		Low-salt bacon	High-salt bacon
20	Spoilage odour	Scented—sour	Cheesy
	Dominant groups	Micrococci, Gp.D streptococci, lactobacilli, pediococci, leuconostocs	Micrococci, leuconostocs
30	Spoilage odour	Putrid	Cheesy
	Dominant groups	Coagulase-negative staphylococci, Gp.D streptococci, lactobacilli, pediococci, leuconostocs	Coagulase-negative staphylococci, leuconostocs

Data from Cavett (1962).

spoil with a sour/sweet, rancid, cheesy off-odour as a result of the proliferation of lactobacilli, particularly atypical streptobacteria (Kitchell & Ingram 1963; Spencer 1967). They also dominate the flora of vacuum-packed baconburgers, a comminuted, raw, cured meat product (Gardner 1968). The isolation and identity of these atypical streptobacteria have been the subject of comprehensive reviews (Kitchell & Shaw 1975; Reuter 1975). It is evident (Kitchell & Ingram 1967; Spencer 1967) that these bacteria are ubiquitous in a bacon factory, in that they can be isolated from the faeces and straw in the lairage, the pork carcass, the curing brines and the finished bacon ready for packaging.

Smoking of bacon prior to vacuum packaging had a greater effect than vacuum packaging alone in reducing the proportion of Gram negative rods and micrococci in the flora after storage for 7 days at 15°C (Kitchell & Ingram 1966). Lactobacilli are resistant to the antimicrobial effects of smoke constituents (Handford & Gibbs 1964; Kitchell & Ingram 1966).

The nitrite content of the bacon influences the rate of spoilage by lactobacilli (Taylor & Shaw 1975); the lower the nitrite content, the more rapid is the spoilage. The inclusion of nitrate in bacon delays spoilage of high pH muscles (e.g. shoulder or collar bacon) (Shaw & Harding 1978).

Lactobacilli will probably be found in all packaged bacon, particularly in the meat, the characteristics of which (e.g. pH and salt content) together with environmental influences (e.g. temperature) will determine their rate of growth and hence the shelf life of the bacon.

B. Sulphide odour

Putrefactive spoilage of vacuum-packed bacon can occur particularly during the summer months, when high ambient temperatures prevail, and is characterized by sulphide odours resulting from the growth and activity of Gram negative bacteria in the genera *Vibrio*, *Proteus* and *Enterobacter–Hafnia* (Gardner 1979). *Proteus morganii*, *Pr. vulgaris* and *Pr. mirabilis* are commonly found in the spoilage flora of vacuum-packed bacons held at 22°C (Gardner & Patterson 1975).

A related spoilage condition, 'cabbage odour' of vacuum-packed bacon, has also been described (Gardner & Patterson 1975). This was due to the production of methane thiol from methionine in the meat by a strain of *Pr. inconstans*. This organism, *in vitro*, grew at 10°C with 6% NaCl at pH 5·0 and with 8% NaCl at pH 7·0, conditions which would be found in most bacons. With *in vivo* studies this spoilage was more prevalent in bacons with a pH >6·0 and a salt content of <4% NaCl. The prevalence of such high pH bacon and the modern tendency for milder (low salt) bacons coupled with inadequate, or lack of, refrigeration will increase the incidence of this form of spoilage.

The conditions which favour the three main groups of bacteria responsible for spoilage of vacuum-packed bacon are summarized in Table 4.

4. Vacuum-packed Sliced Cooked Cured Meats

In the last 20 years there has been a marked increase in the marketing of sliced, cooked, cured meats in vacuum packs. The wide range of meats in this category include cooked ham, corned beef, emulsion-type sausages and luncheon meats. All have been cured, cooked and then sliced for vacuum packaging. The salt contents are usually in the range 2–4%, pH values are normally >6·0, and residual nitrite levels vary from 10–200 μg/g but are mostly <100 μg/g.

TABLE 4

Effect of environmental factors on the occurrence of the main spoilage bacteria in vacuum-packed bacon

| Factor | Level | Group of micro-organisms | | |
		Micrococcaceae	Lactobacilli	Gram negative bacteria
Storage temperature	High	+	−	+
	Low	−	+	−
Level of NaCl in the bacon	High	+	−	−
	Low	−	+	+
pH of the bacon	High	+	−	+
	Low	−	+	−
Use of smoke		−	+	−

+, most likely to be found; −, may be found, but as a minor fraction of the microflora.

TABLE 5

Principal forms of spoilage of vacuum-packed, sliced cooked cured meats

Spoilage type	Causative bacteria	Reference
A. Sweet/sour odour	Lactobacilli, streptococci, leuconostocs	Mol *et al.* (1971)
B. Cheesy/pungent odour	*Brochothrix thermosphacta*	Egan *et al.* (1980)
C. Sulphide odour	*Vibrio*, Enterobacteriaceae	Gardner & Patton (1971)
D. Discolouration of pigment (greening)	H_2O_2-producing Lactobacilliaceae	Shipp (1964)

Bacterial spoilage of these products will be influenced largely by the nature of the meat, together with environmental factors and the degree and composition of the initial microflora. The most common spoilages can be classified into four main types (Table 5).

A. Sweet/sour odour

The bacteria isolated most frequently from this type of spoiled product belong to the group 'atypical' streptobacteria. The range of pH values and salt and nitrite levels of the products examined did not retard the growth of these organisms (Mol *et al.* 1971) which are similar to those lactobacilli which can spoil vacuum-packed, raw, cured meats. As these organisms are not heat resistant, the cooked meats are contaminated at the slicing and packaging operations, and because the organisms are ubiquitous, such contamination is inevitable. Thus, shelf life will be largely determined by the initial microbial load.

It is interesting to note that vacuum-packed, sliced, cooked ham can have lactobacilli counts $>10^8$/g and yet be deemed organoleptically sound (Qvist & Mukherji 1979).

B. Cheesy/pungent odour

In some instances a cheesy, pungent odour is found in spoiled, vacuum-packed, sliced, cooked meats and is organoleptically more objectionable than the 'normal souring'. The odour is due to the growth and activity of *Brochothrix thermosphacta* (e.g. Gardner & Patton 1971; Qvist & Mukherji 1979; Egan *et al.* 1980). The importance of this organism in the spoilage of meats has recently been reviewed (Gardner 1981).

Brochothrix thermosphacta grows more rapidly than lactobacilli in these products (Egan *et al.* 1980) and causes spoilage with lower numbers, *ca.* 10^7/g (Qvist & Mukherji 1979). Roth & Clark (1975) found that lactobacilli could inhibit the growth of *Br. thermosphacta* in vacuum-packed beef and attributed this to an antibiotic effect rather than to pH or lactate. However, Grau (1980) has shown that lactate does inhibit *Br. thermosphacta* under anaerobic conditions. Meat with pH $\leq 5 \cdot 7$ contains sufficient lactate to inhibit anaerobic growth. Qvist & Mukherji (1979) found no inhibitory effect of lactobacilli on *Brochothrix* in vacuum-packed, sliced ham and the latter organism determined shelf life. Collins-Thompson & Rodriguez Lopez (1980) showed that *Br. thermosphacta* increased in number during the early phases of storage of vacuum-packed, sliced, bologna sausage, but later, lactobacilli predominated. The lactobacilli markedly inhibited the growth of *Br. thermosphacta*.

However, *Br. thermosphacta* is a psychrophile and can tolerate the salt concentrations found in these products, and although less resistant to nitrite than lactobacilli (Brownlie 1966), it can grow well under anaerobic conditions in the presence of 50 μg/g of nitrite (Collins-Thompson & Rodriguez Lopez 1980).

The reasons for the dominance of lactobacilli in some vacuum-packed, sliced, cooked, cured meats are still not fully understood.

C. Sulphide odour

Occasionally spoilage characterized by strong sulphide (H_2S) odours is encountered. As discussed earlier (Section 3.B), the flora is dominated by *Vibrio* and Enterobacteriaceae. Vibrios have the potential to produce H_2S (Gardner 1980–81) and are commonly found in stored, vacuum-packed, cooked ham. They can inhibit the growth of leuconostocs, *Lactobacillus viridescens* and *L. casei-plantarum in vitro*, but it is still not known whether they do so in the packaged meat (Gardner & Patton 1971). Sulphide spoilage is usually associated with unrefrigerated storage, as is the case with vacuum-packed, raw meats.

D. Discolouration of pigment ('greening')

This form of spoilage is easily recognized. On removal from the pack, or when the pack is opened (i.e. oxygen becomes available), the pink, cooked, cured meat pigment, nitrosohaemochrome, may rapidly (0·5–2 h) fade to a grey, green, or in extreme cases, a yellow or white colour. The most common form is the grey–green discolouration. This change is brought about by bacteria which grow on the surface of the slices within the pack. On exposure to O_2, their metabolism changes and they produce H_2O_2, which reacts with the meat pigment, resulting in the 'greening'. These bacteria are all catalase-negative members of the Lactobacillaceae, both *Lactobacillus* and *Streptococcus*. The heterofermentative species, *L. viridescens* (e.g. Sharpe 1962) and the commonly found atypical streptobacteria are responsible. Mol *et al.* (1971) found 66% of strains, and Kitchell & Shaw (1975) reported 92% of their strains, could cause the 'greening' reaction on meat. In addition, pediococci may be involved (Reuter 1975). Streptococci, particularly those resembling *Streptococcus faecalis* or *Strep. faecium*, may also cause 'greening' in ham. In fact, any catalase-negative bacterium capable of growing on the cooked, cured meat without a vacuum pack is a potential 'greening' organism. Isolation of such strains can be made by the method of Shipp (1964) with subsequent confirmation by inoculation onto cooked ham and observation, after incubation, of the discolouration.

It is highly improbable, but still possible, that 'greening' bacteria would survive the process of cooking (e.g. Gardner 1967), so post-cooking contamination is the commonest route of their gaining access to the meat.

Finally it must be realized that these forms of spoilage are four arbitrary groups. The likelihood of one or more of these manifestations being found in one instance is high (e.g. souring odour followed by 'greening' of the meat). Nor is the list of 'causative' species complete. Many other bacteria are known to be associated with the above defects, but those listed are by far the most common as reported in the literature and in the experience of the author.

5. Semi-preserved Cured Meats

This class of product includes hams, bacons and sausages packed in an impervious container (can or flexible film) and cooked to an internal temperature of between 65 and 75°C, which kills most vegetative micro-organisms and gives the product a keeping quality of 6 months below 5°C (Type 1 product of Leistner 1979). Spores of bacilli and clostridia can survive this heating process and the systems which largely prevent their subsequent outgrowth have been reviewed (Duncan 1970; Lechowich et al. 1978). For a comprehensive discussion of the microbiology of food pasteurization the reader is referred to Ingram's treatise (1971).

As with other foods in hermetically sealed containers, leakage spoilage can occur and grossly under-processed material can be produced. In these situations microfloras composed of non-heat resistant bacteria (e.g. Enterobacteriaceae, flavobacteria, *Alcaligenes* and *Pseudomonas*) may be found (Ingram & Hobbs 1954).

However, the types of spoilage to be discussed are those which relate to non-leakage situations, such as marginal under-processing and/or factors such as inadequate levels of curing salts and inappropriate storage temperature.

Using canned ham as a typical semi-preserved cured meat, a number of broad spoilage types (Table 6) may be found. In many instances more than one form of spoilage can be found in one can.

TABLE 6

Common spoilages of semi-preserved cured meats

Type	Spoilage characteristic	Causative bacteria	Reference
A	Gas production	*Bacillus* and *Clostridium*	Ingram & Hobbs (1954)
B	Souring	Lactobacillaceae	
C	Gelatin breakdown	*Strep. faecalis* var. *liquefaciens*	Gardner & Patton (1975)
D	Discolourations	Lactobacillaceae	Sharpe (1962)
E	Starch breakdown	*Bacillus* sp.	Mitrica & Granum (1979)

A. Gas production

Canned hams, luncheon meats, etc. can be spoiled by gas production by species of *Bacillus* and *Clostridium* and on rare occasions by members of the heterofermentative Lactobacillaceae.

Gas production accompanied by offensive, putrefactive odours is mainly caused by species of *Clostridium*. They can survive the heat process of semi-preserved cured meats and such spoilages are usually associated with high-temperature storage (>15°C) and low salt levels (<5% NaCl) (Ingram & Hobbs 1954). The number of clostridial spores in cured meats is low (Roberts & Smart 1976) and hence this form of spoilage is not common.

Gas production without obnoxious odours may be found in canned, comminuted cured meats (Jensen & Hess 1941*b*), canned hams (Ingram & Hobbs 1954) and canned bacon (Eddy & Ingram 1956) and is caused by species of the genus *Bacillus*: the gas is produced from carbohydrates and/or nitrate. Jensen (1954) and Eddy & Ingram (1956) found difficulty in classifying their strains to species level. Ingram & Hobbs (1954) identified the gas-producing species of *B. coagulans*, *B. mesentericus*, *B. subtilis* and *B. pumilis*, and noted that these isolates had a greater resistance to salt than is commonly associated with these species. *Bacillus* spp. may also be found in various cooked sausages and the most common species were *B. subtilis*, *B. pumilis*, *B. licheniformis* and *B. cereus* (Berkel & Hadlok 1976).

Like the clostridia, bacilli will survive the heat process, and low salt levels and high storage temperatures are conducive to growth and gas production. Enterococci found in these products commonly produce a metabolite which acts antagonistically on the outgrowth of *Clostridium*, *Bacillus* and *Lactobacillus* (Kafel & Ayres 1969) and may contribute to the preservation of canned ham.

Also in this category, gas production without putrefaction may be caused by heterofermentative lactobacilli and leuconostocs. Some of these species are sufficiently heat-resistant to survive the heat process and subsequently to grow in the product.

B. Souring

Probably the most common form of bacterial spoilage of semi-preserved cured meats is the recognition, on opening the pack, of a sour, cheesy odour, coupled on occasions with slight surface slime formation. This is caused by many members of the Lactobacillaceae, but the most common belong to the enterococcus group and the homofermentative and heterofermentative lactobacilli (Sharpe 1962). Their presence in these meats is indicative of marginal under-processing often coupled with a high degree of contamination of the raw material. The heat process may be sufficient to give the product a cooked appearance, but is insufficient to kill the spoilage bacteria. Their salt-tolerance coupled with their ability to grow even at refrigeration temperature in the anaerobic conditions of the pack are the major factors which permit their growth. Ensuring that all cans are adequately and properly heat-treated will prevent this spoilage.

C. Gelatin breakdown

Gelatinase-producing bacteria (e.g. *Strep. faecalis* subsp. *liquefaciens*) can grow in under-processed cured meats and liquefy the gelatin on the surface of the canned ham (Buttiaux 1953). If only the centre of the ham is under-processed, the gelatinase activity of these bacteria will be manifest in the breakdown of the muscle structure, a condition known as 'soft core' (Coretti & Enders 1964; Gardner & Patton 1975). This spoilage may or may not be associated with other recognizable defects (odour/colour).

D. Discolourations

'Greening' of canned hams similar to that discussed earlier (Section 4.D) is primarily due to under-processing. The causative H_2O_2-producing organisms of the Lactobacillaceae survive the process and multiply in the pack, which, when opened and exposed to air, permits the rapid production of H_2O_2 which oxidizes the meat pigment. This form of spoilage is relatively rare. Most cases of 'greening' on canned ham have resulted from contamination by these bacteria *after* removal from the can.

'Centre' or 'core greening' may also be found which, as with 'soft core' (Section 5.C), is the result of survival and growth of spoilage species inside under-processed meats.

Bacterial 'greening' of large, cooked sausages (e.g. frankfurters or bologna) has long been recognized (e.g. Jensen 1954). On a freshly cut, cooked, cured sausage the green discolouration may be one of three forms: (1) green rim; (2) green ring, or (3) green zone (Bartels *et al.* 1961). Types 1 and 3 are identical to the surface and centre 'greening' of ham discussed above. Type 2 is similar to Type 1, in that it can be recognized on cutting as a ring some 2–4 mm below the surface. These spoilages are associated with the processing of highly contaminated materials (e.g. re-use of old production material) prior to heat processing, as well as inadequate thermal treatment.

Lactobacillus viridescens (Niven & Evans 1957) is a common species found in both forms of 'greening' in hams and sausages, but other lactobacilli, the *Streptococcus* spp. *faecalis* and *faecium*, and unclassified enterococci, can often be found (Gardner, G. A., unpublished data). All these species, and undoubtedly others, are ubiquitous in a meat processing plant and may even grow in the raw material. Thus strict hygiene measures coupled with the proper heat treatment will eliminate these types of spoilage. *Streptococcus faecium* is regarded as the most heat-resistant organism in this group. This feature has been extensively reviewed and studied by Houben (1980). The addition of nisin to the product prior to the thermal process has been shown to protect frankfurters from bacterial discolourations (Stankiewicz-Berger 1969).

Finally, greyish green spots on slices of cooked, Paris ham have been shown to be due to the lack of bacterial activity during the cure (Fournaud & Raibaud 1964). In France nitrite was not a permitted additive until the mid-sixties and the manufacturers depended on the bacterial reduction of nitrate (permitted) in the curing brine as a source of nitrite to produce the cured pigment, nitrosomyoglobin. Lack of such activity will lead to inadequate nitrite production, which can result in a mis-cure. The colour differences may be regarded as uncured ham (i.e. pork), which is grey after cooking, within the cured ham which is pink.

E. Starch breakdown

During storage the packing liquid surrounding semi-preserved canned sausages may become cloudy, and the sausages themselves may become slimy on the surface or soft in consistency (Mitrica & Granum 1979). Of 13 strains of amylase-producing organisms isolated, nine were *Bacillus subtilis*, three were *B. amyloliquefaciens* and one was *B. macerans*. Their sources were vegetable products such as starch and onion powder. The spores of these strains survived the pasteurizing heat process, germinated and grew in the product, producing enough amylase to cause a breakdown of the texture.

6. Conclusions

It is now apparent that the nature of microbial spoilage in cured meats is influenced primarily by the nature of the product coupled with environmental factors (Table 7). Whilst the processor has control over many of these factors, the inherent variation in the raw material and the even distribution of curing salts in any product leads to each product having, in microbiological terms, a wide range of ecological situations.

The nature of microbial spoilage will be influenced by many different combinations of these factors and the forms of spoilage discussed in this paper refer only to those most frequently encountered. Because technologi-

TABLE 7

Factors influencing the form of microbial spoilage of cured meats

Product characteristic	Environmental factors
Nature of tissue (fat, lean, etc.)	Atmosphere of storage (O_2, CO_2)
pH of tissue	Storage temperature/time
Moisture content	Heat process
Levels of curing salts (NaCl, nitrite, nitrate)	
Smoke components	
Polyphosphates, sugars and other curing adjuncts	

cal developments in the meat curing industry are continuous, much of the early published work on cured meat microbiology is now out of date. The use of machine injection systems has created a new set of conditions, in that potential spoilage bacteria can be introduced directly into the meat. Modern smoking systems can give the product the desired characteristic flavour without at the same time reducing the moisture content. Shorter tanking times, coupled with the application of dry salt to the surface of Wiltshire bacon sides and cuts, have also made a significant contribution to the keeping quality of the product, by replacing the rapidly growing putrefactive Gram negative bacteria by Gram positive cocci. With modern equipment there is a greater degree of control of process temperatures and an awareness in production units of the importance of adhering to a strict production code of practice. Routine monitoring of curing brines (Gardner 1973c) and bacons (Gardner 1975) is only one example of the application of microbiology in the quality control of cured meat production. Such quality control is now more widely practised in the cured meat industry, with the consequence that major outbreaks of microbial spoilages are becoming less frequent, if not a rarity.

Finally one must ask the question: what is a spoilage micro-organism? Micrococci and lactobacilli are widely used as starter cultures in the manufacture of European dry sausages; vibrios are used for ham production (Petäjä *et al.* 1972), yet in different circumstances, as discussed, these bacteria can be regarded as major spoilage species. A spoilage organism can be likened to a weed, in that it is the wrong species in the situation at the wrong time.

7. References

ANON. 1980 Country-cured meats: the latest technology in an ancient art. *Meat Processing* **19** (11), 10, 12, 16, 24.

BAIRD-PARKER, A. C. 1962 The occurrence and enumeration, according to a new classification, of micrococci and staphylococci in bacon and on human and pig skin. *Journal of Applied Bacteriology* 25, 352–361.

BALDOCK, J. D., FRANK, P. R., GRAHAM, P. P. & IVEY, F. J. 1979 Potassium sorbate as a fungistatic agent in country ham processing. *Journal of Food Protection* 42, 780–783.

BARD, J. & TOWNSEND, W. E. 1971 Meat curing. In *The Science of Meat and Meat Products* 2nd edn, ed. Price, J. F. & Schweigert, B. S. pp. 452–470. San Francisco: W. H. Freeman & Co.

BARTELS, H., CORETTI, K. & SCHADECK, M. 1961 Grünverfärbungen bei Brühwürsten und ihre Verhütung. *Die Fleischwirtschaft* 13, 991–995.

BERKEL, H. & HADLOK, R. 1976 Differenzierung und Speziesverteilung von aus Brüh- und Kochwürsten stammenden Mikroorganismen der Gattung *Bacillus. Die Fleischwirtschaft* 56, 387–392.

BROWNLIE, L. E. 1966 Effect of some environmental factors on psychrophilic microbacteria. *Journal of Applied Bacteriology* 29, 447–454.

BUTTIAUX, R. 1953 The bacteriological examination of canned ham. Parts 1 and 2. *Food Manufacture* 28, 112, 135.

CANTONI, C., BIANCHI, M. A., RENON, P. & D'AUBERT, S. 1969 Ricerche sulla putrefazione del prosciutto crudo. *Archivo Veterinario Italiano* 20, 354–370.

CAVETT, J. J. 1962 The microbiology of vacuum packed sliced bacon. *Journal of Applied Bacteriology* 25, 282–289.

COLLINS-THOMPSON, D. L. & RODRIGUEZ LOPEZ, G. 1980 Influence of sodium nitrite, temperature and lactic acid bacteria on the growth of *Brochothrix thermosphacta* under anaerobic conditions. *Canadian Journal of Microbiology* 26, 1416–1421.

CORETTI, K. & ENDERS, P. 1964 Enterokokken als Ursache von Kernerweichungen bei Dosenfleischwaren. *Die Fleischwirtschaft* 16, 304–308.

DUNCAN, C. L. 1970 Arrest of growth from spores in semi-preserved foods. *Journal of Applied Bacteriology* 33, 60–73.

EDDY, B. P. & INGRAM, M. 1956 A salt-tolerant denitrifying *Bacillus* strain which 'blows' canned bacon. *Journal of Applied Bacteriology* 19, 62–70.

EGAN, A. F., FORD, A. L. & SHAY, B. J. 1980 A comparison of *Microbacterium thermosphactum* and lactobacilli as spoilage organisms of vacuum-packed sliced luncheon meats. *Journal of Food Science* 45, 1745–1748.

FOURNAUD, J. & RAIBAUD, P. 1964 Bacteriological causes of certain defects observed during the manufacture of Paris hams. *Proceedings of 10th European Meeting of Meat Research Workers, Roskilde*. Paper G9.

GARDNER, G. A. 1967 Discolouration in cooked ham. *Process Biochemistry* 2, 49–52.

GARDNER, G. A. 1968 Effects of pasteurization or added sulphite on the microbiology of stored vacuum packed baconburgers. *Journal of Applied Bacteriology* 31, 462–478.

GARDNER, G. A. 1971 Microbiological and chemical changes in lean Wiltshire bacon during aerobic storage. *Journal of Applied Bacteriology* 34, 645–654.

GARDNER, G. A. 1973a The microbiology of Wiltshire style curing. *Proceedings of the Institute of Food Science and Technology, UK* 6, 130–136.

GARDNER, G. A. 1973b The occurrence of *Vibrio* in the microflora of Wiltshire cured bacon sides. *Proceedings of 19th European Meeting of Meat Research Workers, Paris*, Vol. 3, pp. 1071–1080.

GARDNER, G. A. 1973c Routine microbiological examination of Wiltshire bacon curing brines. In *Sampling–Microbiological Monitoring of Environments* ed. Board, R. G. & Lovelock, D. pp. 21–27. London & New York: Academic Press.

GARDNER, G. A. 1975 The enumeration of micro-organisms on Wiltshire bacon sides: a scheme for use in quality control. *Journal of Food Technology* 10, 181–189.

GARDNER, G. A. 1979 Sulphide spoilage in vacuum-packed bacon. *Journal of Applied Bacteriology* 47, iii–iv.

GARDNER, G. A. 1980–81 Identification and ecology of salt-requiring *Vibrio* associated with cured meats. *Meat Science* 5, 71–81.

GARDNER, G. A. 1981 *Brochothrix thermosphacta* (*Microbacterium thermosphactum*) in the spoilage of meats: a review. In *Psychrotrophic Micro-organisms in Spoilage and Pathogen-*

icity, ed. Roberts, T. A., Hobbs, G., Christian, J. H. B. & Skovgaard, N. pp. 139–173. London & New York: Academic Press.

GARDNER, G. A. 1982 Microbiology of processing: bacon and ham. In *Meat Microbiology*, ed. Brown, M. pp. 129–178. London: Applied Science Publishers.

GARDNER, G. A. & PATTERSON, R. L. S. 1975 A *Proteus inconstans* which produces 'cabbage odour' in the spoilage of vacuum-packed sliced bacon. *Journal of Applied Bacteriology* **39**, 263–271.

GARDNER, G. A. & PATTON, J. 1969 Variations in the composition of the flora on a Wiltshire cured bacon side. *Journal of Food Technology* **4**, 125–131.

GARDNER, G. A. & PATTON, J. 1971 Bacteriocinogenic activity of ham microflora. *Proceedings of 17th European Meeting of Meat Research Workers, Bristol*, pp. 247–251.

GARDNER, G. A. & PATTON, J. 1975 A note on the heat resistance of a *Streptococcus faecalis* isolated from a 'soft-core' in canned ham. *Proceedings of 21st European Meeting of Meat Research Workers, Berne* pp. 52–54.

GARDNER, G. A. & PATTON, J. 1978 The bacteriology of bacon curing brines from multi-needle injection machines *24. Europäisches Fleischforscherkongress, Kulmbach*, Vol. 3, Paper K6.

GARRARD, E. H. & LOCHHEAD, A. G. 1939 A study of bacteria contaminating sides of Wiltshire bacon with special consideration of their behaviour in concentrated salt solutions. *Canadian Journal of Research* **D17**, 45–58.

GIBBONS, N. E. 1940 Canadian Wiltshire bacon. VI. Quantitative bacteriological studies on product. *Canadian Journal of Research* **D18**, 202–210.

GIBBONS, N. E. 1953 Wiltshire bacon. *Advances in Food Research* **4**, 1–35.

GILL, C. O. 1979 Intrinsic bacteria in meat. *Journal of Applied Bacteriology* **47**, 367–378.

GRAU, F. H. 1980 Inhibition of the anaerobic growth of *Brochothrix thermosphacta* by lactic acid. *Applied and Environmental Microbiology* **40**, 433–436.

HADLOK, R. 1969 Schimmelpilzkontamination von Fleischerzeugnissen durch naturbelassene Gewürze. *Die Fleischwirtschaft* **49**, 1601–1609.

HANDFORD, P. M. & GIBBS, B. M. 1964 Antibacterial effects of smoke constituents on bacteria isolated from bacon. In *Microbial Inhibitors in Food*, ed. Molin, N. pp. 333–346. Uppsala: Almqvist & Wiksell.

HOUBEN, J. H. 1980 *Thermoresistentie van* Streptococcus faecium *in gepaisteuriserde ham*. Utrecht: Drukkerij Elinkwijk BV.

INGRAM, M. 1952 Internal bacterial taints ('bone taint' or 'souring') of cured pork legs. *Journal of Hygiene, Cambridge* **50**, 165–181.

INGRAM, M. 1960 Bacterial multiplication in packed Wiltshire bacon. *Journal of Applied Bacteriology* **23**, 205–215.

INGRAM, M. 1971 *The Microbiology of Food Pasteurization*. Report No. 292 pp. A1–A44. Göteborg: Svenska Institutet for Konserveringsforskning.

INGRAM, M. & HOBBS, B. C. 1954 The bacteriology of 'pasteurized' canned hams. *Royal Sanitary Institute Journal* **74**, 1151–1163.

JENSEN, L. B. 1954 *Microbiology of Meats* 3rd edn. Champaign, Illinois: Garrard.

JENSEN, L. B. & HESS, W. R. 1941*a* A study of ham souring. *Food Research* **6**, 273–326.

JENSEN, L. B. & HESS, W. R. 1941*b* Fermentation in meat products by the genus *Bacillus*. *Food Research* **6**, 75–83.

KAFEL, S. & AYRES, J. C. 1969 The antagonism of enterococci on other bacteria in canned hams. *Journal of Applied Bacteriology* **32**, 217–232.

KITCHELL, A. G. 1958 The micrococci of pork and bacon and of bacon brines. In *The Microbiology of Fish and Meat Curing Brines*, ed. Eddy, B. P. pp. 191–196. London : HMSO.

KITCHELL, A. G. 1962 Micrococci and coagulase negative staphylococci in cured meats and meat products. *Journal of Applied Bacteriology* **25**, 416–431.

KITCHELL, A. G. & INGRAM, M. 1963 Vacuum packed sliced Wiltshire bacon. *Food Processing and Packaging* **32**, 3–9.

KITCHELL, A. G. & INGRAM, M. 1966 The selective effect of salt and smoking on the bacterial flora of bacon, with special reference to vacuum packing. *Abstracts of 2nd International Congress of Food Science and Technology, Warsaw*, pp. 149–150.

KITCHELL, A. G. & INGRAM, G. C. 1967 A survey of a factory for the sources of lactobacilli characteristic of vacuum packed Wiltshire cured bacon. *Proceedings of 13th Meeting of European Meat Research Workers, Rotterdam*, Paper B2.

KITCHELL, A. G. & SHAW, B. G. 1975 Lactic acid bacteria in fresh and cured meat. In *Lactic Acid Bacteria in Beverages and Food* ed. Carr, J. G., Cutting, C. V. & Whiting, G. C. pp. 209–220. London & New York: Academic Press.

KLARE, H. J. 1970 Die Bedeuting des Darminhaltes von Schlachttieren als Ursache für die Kontamination von Fleisch und Fleischerzeugnissen mit Schimmelpilzen. *Die Fleischwirtschaft* **50**, 1507–1510.

LECHOWICH, R. V., BROWN, W. L., DEIBEL, R. H. & SOMMERS, I. I. 1978 The role of nitrite in the production of canned cured meat products. *Food Technology* **32** (5), 45–58.

LEISTNER, L. 1979 Mikrobiologische Einteilung von Fleischkonserven. *Die Fleischwirtschaft* **59**, 1452–1455.

LEISTNER, L. & AYRES, J. C. 1968 Molds and meats. *Die Fleischwirtschaft* **48**, 62–65.

LEISTNER, L. & ECKARDT, C. 1979 Vorkommen toxinogener Penicillien bei Fleischerzeugnissen. *Die Fleischwirtschaft* **59**, 1892–1896.

MITRICA, L. & GRANUM, P. E. 1979 The amylase producing microflora of semi-preserved canned sausages. Identification of the bacteria and characterization of their amylases. *Zeitschrift für Lebensmittel Untersuchung und -Forschung* **169**, 4–8.

MOL, J. H. H., HIETBRINK, J. E. A., MOLLEN, H. W. M. & VAN TINTEREN, J. 1971 Observations on the microflora of vacuum packed sliced cooked meat products. *Journal of Applied Bacteriology* **34**, 377–397.

MUNDT, J. O. & KITCHEN, H. M. 1951 Taint in Southern country-style hams. *Food Research* **16**, 233–238.

NIVEN, C. F. & EVANS, J. B. 1957 *Lactobacillus viridescens* nov. spec., a heterofermentative species that produces a green discolouration of cured meat products. *Journal of Bacteriology* **73**, 758–759.

PETÄJÄ, E., LAINE, J. J. & NIINIVAARA, F. P. 1972 Starterkulturen bei der Pökelung von Fleisch. *Dei Fleischwirtschaft* **52**, 839–842, 845.

QVIST, S. & MUKHERJI, S. 1979 *Microbacterium thermosphactum*—forekomst og betydning. *Dansk. Vet. Tidsskrift* **62**, 963–971.

REUTER, G. 1975 Classification problems, ecology and some biochemical activities of lactobacilli of meat products. In *Lactic Acid Bacteria in Beverages and Food*, ed. Carr, J. G., Cutting, C. V. & Whiting, G. C. pp. 221–229. London & New York: Academic Press.

ROBERTS, T. A. & SMART, J. L. 1976 The occurrence and growth of *Clostridium* spp. in vacuum-packed bacon with particular reference to *Cl. perfringens* (*welchii*) and *Cl. botulinum*. *Journal of Food Technology* **11**, 229–244.

ROTH, L. A. & CLARK, D. S. 1975 Effect of lactobacilli and carbon dioxide on the growth of *Microbacterium thermosphactum* on fresh beef. *Canadian Journal of Microbiology* **21**, 629–632.

SHARPE, M. E. 1962 Lactobacilli in meat products. *Food Manufacture* **37**, 582–589.

SHAW, B. G. & HARDING, C. D. 1978 The effect of nitrate and nitrite on the microbial flora of Wiltshire bacon after maturation and vacuum-packed storage. *Journal of Applied Bacteriology* **45**, 39–47.

SHAW, C., STITT, J. F. & COWAN, S. T. 1951 Staphylococci and their classification. *Journal of General Microbiology* **5**, 1010–1023.

SHIPP, H. L. 1964 *The Green Discoloration of Cooked Cured Meats of Bacterial Origin*. Technical Circular No. 266. Leatherhead: BFMIRA.

SMITH, F. B. 1938 An investigation of a taint in rib bones of bacon: The determination of halophilic vibrios (n. spp.). *Proceedings of the Royal Society of Queensland* **49**, 29–53.

SPENCER, R. 1967 *A Study of the Factors Affecting the Quality and Shelf Life of Vacuum Packaged Bacon, and of the Behaviour of Wiltshire Cured Bacon Packed and Stored Under Controlled Conditions*. Research Report No. 136. Leatherhead: BFMIRA.

STANKIEWICZ-BERGER, H. 1969 Effect of nisin on the lactobacilli that cause greening of cured meat products. *Acta microbiologica polonica Ser. B* **1** (18), 117–120.

TAYLOR, A. A. & SHAW, B. G. 1975 Wiltshire curing with and without nitrate. I. Vacuum packed sliced back bacon. *Journal of Food Technology* **10**, 157–167.

TOFTE JESPERSEN, N. J. & RIEMANN, H. 1958 The numbers of salt-tolerant bacteria in curing brine and on bacon. In *The Microbiology of Fish and Meat Curing Brines* ed. Eddy, B. P. pp. 177–182. London: HMSO.

TONGE, R. J., BAIRD-PARKER, A. C. & CAVETT, J. J. 1964 Chemical and microbiological changes during storage of vacuum packed sliced bacon. *Journal of Applied Bacteriology* **27**, 252–264.

WU, M. T., AYRES, J. C. & KOEHLER, P. E. 1974 Toxigenic aspergilli and penicillia isolated from aged cured meats. *Applied Microbiology* **28**, 1094–1096.

ZELLER, M. & RENZ, H. 1967 Untersuchungen über die Fäulnis von Rohschinken. *Die Fleischwirtschaft* **47**, 825–828.

Effect of Packaging and Gaseous Environment on the Microbiology and Shelf Life of Processed Poultry Products

G. C. MEAD

Agricultural Research Council Food Research Institute,
Norwich, Norfolk, UK

Contents

1. Introduction

ALTHOUGH most poultry meat is sold in the form of whole, oven-ready carcasses, there is now an increasing demand for cut-up portions and a variety of other further-processed products, both raw and cooked.

Raw poultry meat is a perishable commodity of relatively high pH (*ca.* 5·7–6·7) which readily supports the growth of micro-organisms when stored under chill conditions. Its shelf life will depend upon the combined effects of certain intrinsic and extrinsic factors, including the numbers and types of psychrotrophic spoilage organisms present initially, the storage temperature, muscle pH and type (red or white), as well as the kind of packaging material used and gaseous environment of the product.

During refrigerated storage there is a marked change in the relative proportions of the different psychrotrophs present, and the organisms which usually predominate on the spoiled product are pigmented and non-pigmented strains of *Pseudomonas* spp. with somewhat smaller numbers of *Acinetobacter/Moraxella* spp. These organisms grow mainly deep in feather holes and on cut muscle surfaces, and 'off' odours, which appear to originate largely from non-protein nitrogenous compounds (Jay & Shelef 1978), are detectable when total viable counts at 1 or 20°C reach *ca.* 10^8 organisms/cm^2. Treatments aimed at extending shelf life usually do so by restricting the

FOOD MICROBIOLOGY
ISBN 0 12 589670 0

growth of pseudomonads and permitting the development of a slower-growing microflora which often produces less objectionable spoilage odours and may include any of the following: *Aeromonas* spp., *Alteromonas putrefaciens*, *Brochothrix thermosphacta*, cold-tolerant coliforms, the atypical lactobacilli of Thornley & Sharpe (1959) and yeasts.

Apart from the uneviscerated (New York Dressed) bird which is sold in butchers' shops, the majority of chilled poultry products are sold wrapped, the immediate objectives being to prevent the surface drying out and to avoid microbial cross-contamination. With the packaged, oven-ready product which may have been either air- or water-chilled, conditions are usually such that microbial growth is not limited by low a_w. In a recent EEC study (Anon. 1978) it was found that the surface a_w of water-chilled chickens was 0·996 whilst that of air-chilled carcasses was only 0·970; after packaging, however, the latter figure increased to 0·990 during storage at 2°C for 6 h. Taking data for packaged carcasses from four different countries, water-chilled birds had a mean shelf life at 2°C of 8·6 days which was comparable with the mean of 8·8 days for those chilled in air. Barnes & Impey (1975) showed that the shelf life of unwrapped, air-chilled carcasses stored at 4°C varied considerably due, presumably, to the extent to which the cut muscle had dried out. Wrapping a parallel batch in polyethylene reduced the variability but also the mean shelf life.

Apart from the ability to retain moisture, the most important property of a packaging film in relation to shelf life is its permeability to O_2 and CO_2. All packaging films will modify to some extent the gaseous environment of a raw flesh food stored under chill conditions because O_2 is consumed by residual tissue respiration and by the metabolic activities of contaminating micro-organisms, while CO_2 is produced. Hence, in the case of the more impermeable films, not only does the O_2 concentration within the pack diminish because atmospheric O_2 is largely excluded but the CO_2 concentration tends to increase. Although pseudomonads and certain other psychrotrophic spoilage organisms are obligate aerobes, their growth is not inhibited until very low O_2 tensions are achieved. For example, Shaw & Nicol (1969) showed that the growth of a pseudomonad on meat was not affected until the O_2 concentration was reduced below 0·8%, a level unlikely to be achieved in commercial vacuum packs (Ingram 1962).

Of greater significance in practice is the accumulation of CO_2. It was shown by Coyne (1933) and Haines (1933) that 10–20% CO_2 delayed the growth of pseudomonads and certain other spoilage organisms if the temperature remained below 4°C, ideally close to 0°C. However, even *ca.* 3% CO_2 was found by Gardner & Carson (1967) to inhibit the growth of a pseudomonad on porcine muscle incubated at 2°C. In contrast, the same CO_2 concentration was stimulatory for some of the lactic acid bacteria which were known to grow in vacuum packs. The possible mechanism of the antibacterial effect of CO_2 has been considered by Gill & Tan (1980) and Wolfe (1980) and will not be discussed here.

2. Control of Gaseous Environment

Whilst vacuum-packaging may be used to encourage the development of an atmosphere around the product which delays microbial spoilage, other systems in use involve an initial addition of *ca.* 20% CO_2 prior to storage. In such cases the gaseous environment should be immediately inhibitory to the usual spoilage organisms on the chilled product but the composition of the atmosphere is unlikely to remain constant throughout the storage period because of absorptive and biochemical processes associated with the meat and possible changes due to slow permeation of gases through the container. These systems are more correctly described as 'modified atmosphere' (MA) storage to distinguish them from 'controlled atmosphere' (CA) systems used mainly for commodities other than poultry where the required concentrations of gases are actively maintained throughout the storage period. Modified atmosphere storage is adequate for extending the shelf life of poultry meat and on economic grounds is considered preferable (Wolfe 1980). With poultry products, MA storage usually involves either individually packaged items or bulk packs of varying size in an O_2-impermeable plastic film.

Another method of altering the product environment in order to extend shelf life is that of hypobaric storage which involves precise control of temperature, humidity and pressure with a closely regulated rate of air exchange and is thus a type of CA system. Within the storage chamber the atmospheric pressure is reduced, thus lowering proportionately the partial pressure of oxygen. For meat storage purposes, the O_2 concentration is decreased below 0·2% to limit the growth of aerobic spoilage bacteria but the humidity is kept high to prevent moisture loss and shrinkage of the product. The commercial system described by Mermelstein (1979) gives a recommended storage life of 28 days at *ca.* $-1°C$ for chicken carcasses and portions.

In relation to bulk storage of poultry, little use appears to have been made of the type of CA system first applied in the 1930s for shipments of red meat from Australia and New Zealand to the British Isles. This system involves transporting the meat in gas-tight holds where an atmosphere containing *ca.* 10% CO_2 is maintained.

The most widely studied methods of extending the shelf life of chill-stored poultry products are vacuum-packaging and the use of gas-flush packs containing CO_2. The microbiological implications of these types of pack will be discussed below.

A. Vacuum packaging and the use of oxygen-impermeable film

In a study of chicken carcasses stored at 1°C, Shrimpton & Barnes (1960) showed that the use of an evacuated and heat-shrunk O_2-impermeable film (vinylidene chloride–vinyl chloride co-polymer) extended shelf life on the basis of off-odour development from 11 to 16 days when compared with carcasses in O_2-permeable film. Not only was microbial growth slower in the vacuum packs but a different microflora predominated at spoilage, with pseudomonads being replaced by an organism later identified as *Alteromonas putrefaciens* (Barnes & Melton 1971). It was shown that the O_2 concentration in the vacuum packs decreased during storage whilst the levels of CO_2 increased to 9–10% (Table 1). This is still one of the few studies of poultry meat in which changes in the gaseous environment within the packs have been monitored.

A similar reduction in the rate of microbial growth was observed by Debevere & Voets (1973) for chicken carcasses stored at 4°C and packed in a shrinkable polyvinyl chloride film with a low permeability to O_2. However, there was little effect when the birds were wrapped in a stretchable polyvinyl chloride film with a high O_2-permeability although both films inhibited

TABLE 1

Influence of packaging material on the gaseous environment of chicken carcasses stored at 1 °C in relation to microbial growth

| Storage period (days) | Concentration (% v/v) of | | | | Total count at 20°C* (Log_{10} organisms/cm²) | |
| | O_2 | | CO_2 | | | |
	A	B	A	B	A	B
0	20·8	15·1	0·2	4·4	5·2	5·2
8	20·2	6·5	0·5	6·2	6·7	5·7
12	17·1	2·8	2·1	9·1	8·4†	6·8
14	20·3	3·9	1·0	9·3	Nt	6·9
16	Nt	6·7	Nt	10·0	Nt	7·8†

A, polyethylene; B, vinylidene chloride–vinyl chloride co-polymer.
Nt, not tested.
*Mean of three carcasses in each case.
†Definite off-odour detected.
Data of Shrimpton & Barnes (1960).

microbial ammonia production when compared with carcasses wrapped in a polyethylene film or with unwrapped controls. In this study shelf life was evaluated only in microbiological terms.

Clearly, the permeability of the film used to wrap the product is critical in relation to shelf life and would appear to explain why some studies have failed to obtain any useful delay in spoilage despite the use of vacuum packs. For example, Taylor *et al.* (1966) used a heat-shrinkable type of polyethylene (presumably O_2-permeable) and stored chicken carcasses at 1–2°C in packs which were either (i) sealed without being evacuated, (ii) evacuated, (iii) evacuated and heat-shrunk at 100°C for 2 s or (iv) evacuated, heat-shrunk and coated with a paraffin-base wax. Only the last treatment gave any extension of shelf life, i.e. *ca.* 2 days, when judged on the basis of off-odour development.

A type of polyethylene pack was also used by Arafa & Chen (1975) for chicken breast and thigh portions stored at 2–4°C in either (i) unsealed bags, (ii) bags sealed without being evacuated or (iii) vacuum-packs. Although the film was said to be of low O_2-permeability, neither treatment (ii) nor (iii) significantly extended shelf life and all packs produced off-odours within about 12 days. However, the majority of micro-organisms isolated from the vacuum packs were *Enterobacter* spp. with only a small proportion of *Pseudomonas* spp. while the position was reversed in the case of the unsealed bags.

Apart from the lack of data on changes in the gaseous environment within the pack in most studies concerning poultry meat, there is a need for more information on the effect of storage temperature on the O_2- and CO_2-permeability of the different packaging materials used (Patterson 1980). Usually permeability tests are made at a temperature which bears no relation to that at which the packaged product will be stored and Patterson (1980) reported that permeability to O_2 and CO_2 for two overwrap 'cling' films was reduced by about two-thirds at 4°C in comparison with values obtained at 20°C. According to Eustace (1981), O_2 transmission rates at 3·5°C for films of the O_2-impermeable polyvinylidene chloride type were only 5–15% of those at 25°C. The transmission rate of nylon-based films increased under conditions of high humidity such as would be found in a meat pack, although the effect was more marked at 25°C than at 3·5°C. Humidity had a negligible effect on the O_2 transmission rate of films containing polyvinylidene chloride. Shrinking of heat-shrinkable films reduced the transmission rate approximately in proportion to the reduction in surface area.

The inhibitory effect of the CO_2 accumulating within a vacuum-pack on the growth of pseudomonads and other aerobic spoilage bacteria is likely to be reduced when packs are kept at 4°C or above because the organisms are more sensitive to CO_2 at temperatures close to 0°C (Haines 1933). In our laboratory, vacuum packed, eviscerated turkeys wrapped in a heat-shrunk, O_2-impermeable film (Lovac: Horsens Plastics A/S, Horsens, Denmark) and similar carcasses wrapped in O_2-permeable polyethylene were stored at different temperatures. At 4°C, off-odours were evident within 7 days in the polyethylene-wrapped birds and the vacuum packs extended shelf life by only about 1 day. At spoilage, counts of psychrotrophs were only slightly lower for the vacuum-packed birds. However, at 1°C off-odours were not detected until 16 days in the case of the vacuum packs, compared with 13 days for carcasses wrapped in polyethylene. Although the growth of psychrotrophs was slower at 1°C for both types of pack, the rate of microbial growth declined much more markedly towards the end of the storage period in the case of the vacuum packs and bacterial levels remained at least 10-fold lower than those in the O_2-permeable packs. The off-odour which developed in the vacuum-packed birds was sweetish and less offensive than that detected in the controls.

The advantage in combining vacuum-packaging of poultry in an O_2-impermeable film with storage below 0°C was demonstrated by Barnes *et al.* (1979), who compared the shelf life of oven-ready ducks stored at 2 or −1°C when wrapped in a low density O_2-impermeable polyethylene film or vacuum packed in a heat-shrunk, O_2-impermeable 'Barrier Bag' (W. R. Grace & Co., London, UK).

At 2°C, off-odours were produced in about 10 days in carcasses wrapped in the O_2-permeable film compared with *ca.* 19 days at −1°C. The use of vacuum packs extended shelf life by more than 50% at either temperature. At spoilage, the predominant organisms were atypical lactobacilli (Thornley & Sharpe 1959) and enterobacteria, whilst counts of *Pseudomonas* spp. were 100- to 1000-fold lower.

The same types of film were used by Jones *et al.* (1982) in comparing the shelf life of turkey breast and leg portions. During storage at 1°C in either O_2-permeable film or vacuum packs, it was found that vacuum-packaging delayed off-odour development from 16 to 25 days in the case of breast fillets (pH 5·9–6·0) and from 14 to 20 days for drumsticks (pH 6·1–6·3). Pseudomonads predominated in the spoilage microflora of both breast and leg portions in the O_2-permeable packs but in vacuum packs the predominant organisms included two groups of atypical lactobacilli, one of which was more common on the breast portions, the other on leg. Surprisingly, deleterious flavour changes tended to precede the development of off-odours in vacuum packs of both types of muscle and the results indicate that in the future more attention should be given to flavour evaluation in studies concerned with extending the shelf life of poultry meat.

There is evidence that vacuum-packaging can be combined successfully with other treatments aimed at extending shelf life. For example, with turkey breast portions, Taylor *et al.* (1965) showed that immersion in a 6% solution of a tripolyphosphate (Kena) for 6 h reduced the rate of microbial growth during chill storage but counts were lower when the polyphosphate treatment was combined with packaging in a heat-shrunk polyvinyl film than one of cellophane.

B. Gas-flush packs

Although gas-packaging is yet to be used extensively for poultry meat, it has been applied commercially on a limited scale, especially in the USA. One such system (Timmons 1976) involves bulk packs containing *ca.* 30 kg of product. Each pack is first evacuated and then back-flushed with CO_2 and heat-sealed, a treatment said to give an 18–21 day shelf life at −2 to +1°C. In view of the build-up of metabolic CO_2 in a vacuum pack to a level which effectively suppresses the usual spoilage microflora, it is reasonable to question whether there is any advantage in adding this gas during packaging.

The most comprehensive study of the use of CO_2-enriched atmospheres for extending the shelf life of poultry meat was carried out in the USA by Ogilvy & Ayres (1951). In storing various cut-up chicken portions at 0–10°C, two different experimental systems were used. First, portions were transferred to gas-tight jars through which was passed either a gas mixture containing a given concentration of CO_2 in air or air alone. Within the range 0–25% CO_2, the ratio of shelf life in CO_2 to that in air was a linear function of CO_2 concentration (Fig. 1), despite the data being obtained from meat stored at 4·4°C which is above the optimum for CO_2 inhibition of spoilage bacteria (Haines 1933). However, it was confirmed that the added CO_2 was more effective in extending shelf life at lower storage temperatures and that the presence of the gas affected both the lag phase and doubling time of the micro-organisms present. In relation to meat quality, the maximum usable CO_2 concentration was 25% since above this the meat became discoloured and even at 15% a loss of 'bloom' was sometimes noted.

An interesting aspect of the CO_2 effect, shown by Ogilvy & Ayres (1951), is that transferring CO_2-stored poultry meat to normal atmospheric conditions of cold storage still gives an extended shelf life which tends to be intermediate between the keeping time in air alone and that in the CO_2 atmosphere. A similar after-effect was reported for poultry meat by Bailey *et al.* (1979) and by Silliker & Wolfe (1980) and could be of value to industry in relation to retail distribution and display. It has not been established whether the effect is due to a lag phase in microbial growth induced by a change of gaseous environment or merely to the slow diffusion of CO_2 from the tissues and hence a 'residue' effect.

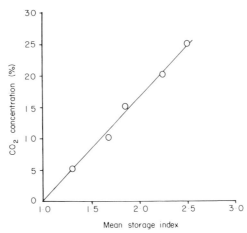

Fig. 1. Relationship between CO_2 concentration and the ratio of shelf life in CO_2 to that in air for commercially prepared chicken thighs stored at 4·4°C (determined from microbial growth curves) (data of Ogilvy & Ayres 1951). Reprinted from *Food Technology*, 1951, **5**, 97–102, with permission from the authors and publisher. Copyright © by the Institute of Food Technologists.

In the second type of storage system used by Ogilvy & Ayres (1951), chicken portions were sealed in a CO_2 atmosphere which could not be regulated throughout the holding period (MA storage). Here, solution of CO_2 in the tissues resulted in a reduced level in the atmosphere surrounding the product and eventually complete utilization of the O_2 present was observed. Some extension of shelf life was reported but it tended to be less than that obtained with the CA method where both the desired CO_2 concentration and aerobic conditions could be maintained. At the end of the MA storage period the predominant microflora comprised mainly two groups of bacteria: one resembling lactobacilli, the other being unidentified, Gram negative rods.

This MA storage system is akin to the gas-flush pack referred to previously. Although there is evidence that CO_2 packaging can be more effective in extending shelf life than the vacuum pack, most studies in which an appreciable shelf life extension has been demonstrated have involved CO_2 concentrations above the 25% limit suggested by the work of Ogilvy & Ayres (1951). For example, Gardner *et al.* (1977) found that chicken carcasses packaged in sealed bags containing 50 or 80% CO_2 were still acceptable after 23 days at 1·5°C whereas in the absence of CO_2 spoilage was evident after 14–20 days. With bulk-packed chicken carcasses stored at 1°C Sander & Soo (1978) showed that the use of an O_2-impermeable film and evacuation of packs followed by addition of $7·22 \times 10^{-4}$ m^3 CO_2/kg body weight gave a shelf life of 27 days compared with only 15 days for vacuum packs without added CO_2. In relation to chicken breast portions wrapped in a low-permeability film and held at $1.5°C$, Gibbs & Patterson (1977) found that 100% CO_2 markedly reduced the growth rate of micro-organisms whereas the growth rate in 20% CO_2 was only slightly less than that in vacuum packs.

In the studies described above, no adverse effects of high CO_2 concentrations on meat quality were reported. On the other hand, Wabeck *et al.* (1968) used a CA storage system similar to that of Ogilvy & Ayres (1951) and found that chicken carcasses stored at $1°C$ in air containing 10 or 20% CO_2 developed off-odours several days earlier than those held in air alone. Thompson (1971) also reported the early development of an off-odour, described as 'sweetish', when CO_2 snow was added to carrier boxes held at $0.5°C$. Apparently, the conditions under which CO_2-enriched atmospheres can be used to extend the shelf life of chill-stored poultry have yet to be clearly defined. In particular, little or no work has been done in relation to poultry species other than chicken.

3. Influence of Packaging on Manufactured Products

Although numerous manufactured poultry products have been developed, including rolls, roasts, burgers and sausages, in most cases little information is available on either keeping quality or the influence on shelf life of particular packaging materials.

In order to preserve both colour and organoleptic properties, cured meat products are normally vacuum-packed in a film of low O_2-permeability. The effect of vacuum-packaging in such a film on microbial growth and shelf life of an experimental turkey ham product was studied by Mead & Adams in this laboratory. The product was turkey thigh meat wrapped in pork fat and contained NaCl, polyphosphate and sodium nitrite. After smoking at $50°C$ and cooking to an internal temperature of $70°C$, the 'ham' was cooled to $2°C$. A second type of product received the same heat treatment but was unsmoked and contained no NaCl, polyphosphate or sodium nitrite. Both products were sliced and either tray-wrapped in an O_2-permeable film or vacuum-packed in a laminated, impermeable film prior to storage at $1°C$.

Changes in total counts at $1°C$ occurring during storage of the products are shown in Table 2. In the uncured version wrapped in O_2-permeable film, bacterial counts of $10^7–10^8/g$ occurred in only 5 days but growth was considerably slower in the vacuum packs. Smoking and curing reduced initial counts by 100- to 1000-fold and vacuum packs were held for 27 days before counts at $1°C$ reached $10^7/g$. This was at least 1 week longer than the same product wrapped in O_2-permeable film.

TABLE 2

Changes in total counts at 1°C during storage of further-processed turkey products at the same temperature

Product	Initial pH value	Type of pack	Total bacterial count* (mean $\log_{10}$ organisms/g) after storage period of (days)				
			5	12	20	27	35
Cooked, uncured	6·5	O_2-permeable	7·8	8·7	Nt	Nt	Nt
		O_2-impermeable	4·9	6·4	7·7	8·1	Nt
Cooked, cured	6·0	O_2-permeable	3·6	5·3	8·0	8·4	Nt
		O_2-impermeable	Nt	5·3	6·7	7·6	7·9

Nt, not tested.
*Mean of 3 samples in each test.

At spoilage, the predominant microflora of the cured, vacuum-packed product was mainly *Brochothrix thermosphacta* and at least two different groups of atypical lactobacilli (Thornley & Sharpe 1959). Some of the latter organisms, when isolated from raw turkey portions, were shown by Jones *et al*. (1982) to be capable of surviving at 60°C for 30 min, indicating their likely persistence in heat-treated products.

Sensory evaluation of the turkey ham by a trained taste-panel showed that deleterious flavour changes were detectable after 35 days at 1°C which suggests a considerably shorter shelf life than the period of *ca*. 84 days reported by Mulder (1974) for vacuum-packed, smoked chicken stored at the same temperature.

4. Microbiological Safety Considerations

The question of whether the packaging of flesh foods in O_2- and CO_2-impermeable film increases the risk of food poisoning due to multiplication of clostridia has long been a matter for debate (Cavett 1968; Silliker & Wolfe 1980). However, studies on poultry and other meats suggest that the low-oxygen conditions provided by vacuum packs do not promote the growth of food poisoning clostridia when the product is held at temperatures well above the normal range for chilled meats (0–5°C).

Clostridium perfringens is a common contaminant of poultry meat and a frequent cause of food poisoning. Roberts (1972) showed that a non-haemolytic strain grew readily even in air on partly cooked chicken carcasses when they were inoculated prior to cooking with a spore suspension and subsequently held at room temperature. Irrespective of whether the organism was injected into deep muscle or merely sprayed into the abdominal cavity, numbers frequently exceeded 1×10^6 organisms/g of meat after overnight storage.

Growth of food poisoning clostridia is prevented by storing poultry products under chill conditions while, in the case of raw poultry meat, temperature abuse leads to very rapid spoilage, due to other organisms, even when the meat is packed in a heat-shrunk O_2-impermeable film (Barnes & Shrimpton 1968), thus providing a clear indication that the meat is unfit for consumption. However, in heat-treated products, particularly 'cook-in-the-bag' items, anaerobic spore formers are frequently among the few types of surviving bacteria and will grow readily with little competition from other organisms, but without always producing obvious signs of spoilage, if the products are not held under appropriate conditions of cold storage. As far as *Cl. botulinum* is concerned, the organism is very rarely found on raw poultry meat but it could be introduced into manufactured products with certain ingredients including spices and other meats, as indicated by the work of Roberts & Smart (1976) on pork and bacon.

An organism which grows well at chill temperatures and is known to be a cause of food-borne enteritis is *Yersinia enterocolitica*. Although other cold-tolerant coliform bacteria such as *Serratia liquefaciens* often multiply extensively on vacuum-packed poultry held under chill conditions (Barnes 1976), and *Y. enterocolitica* has been isolated from chicken carcasses (Leistner *et al.* 1975), there is no evidence that any problem of consumer health has arisen. Organisms resembling *Y. enterocolitica* reached high levels during the chill storage of vacuum-packed beef and lamb (Hanna *et al.* 1976).

The effect of CO_2-enriched atmospheres on the behaviour of food poisoning organisms in poultry products has received little attention. Using a laboratory medium incubated at 43°C, Parekh & Solberg (1970) found no significant difference between the effect of CO_2 and N_2 atmospheres on the growth of eight strains of *Cl. perfringens* although two of the strains showed a slight delay in initiating growth under CO_2. In a similar experiment carried out in the present author's laboratory using a medium at pH 6·2 and incubated at 20°C, the lag phase for a strain of *Cl. perfringens* was 31 h under N_2 and 68 h under CO_2. This work suggests that clostridial growth could at least be delayed by the presence of CO_2 within the pack under marginal growth-temperature conditions, and needs to be extended to commercial poultry products. In the case of the experimental turkey ham referred to previously, neither sulphite-reducing clostridia nor *Staphylococcus aureus* multiplied during storage for 3 days at 25°C or 12 days at 10°C.

5. Conclusions

Sufficient data are available to show that vacuum-packaging of poultry carcasses, cut-up portions and certain other further-processed products can extend shelf life provided that the product is held under chill conditions, especially below 0°C. Thus, the effectiveness of vacuum-packaging is temperature-dependent, a fact which is not always appreciated by producers and retailers; it is also influenced by the nature of the packaging material and its permeability to O_2 and CO_2 over the temperature range of commercial chill storage. Where little or no extension of shelf life has been demonstrated experimentally with chill-stored vacuum-packs, the suitability of the packaging material is open to question and in most studies, as Patterson (1980) has indicated to prove the point, data are lacking because the changes in gaseous environment occurring within the packs have not been determined.

Atmospheres enriched with CO_2 have also been advocated for extending the shelf life of processed poultry products but the precise composition of the atmospheres necessary to achieve the desired result have yet to be clarified in the published literature. Partly owing to the lack of readily available information, CO_2 storage of poultry has not been widely used commercially.

Most studies on possible methods of extending the shelf life of poultry meat have been on chicken with little attention to other poultry species. Obviously, it cannot be assumed that the keeping quality and shelf life characteristics of turkeys and ducks are the same in all respects as those of chickens, particularly in relation to flavour changes under different storage conditions which do not always correlate with off-odour development. Equally, much more information is needed on the behaviour of food-poisoning organisms in poultry products of all kinds, especially with regard to the use of different packaging materials and manipulation of the gaseous environment.

6. References

ANON. 1978 Microbiology and Shelf-life of Chilled Poultry Carcasses. *Commission of the European Communities, Information on Agriculture* No. 61. London: HMSO.

ARAFA, A. S. & CHEN, T. C. 1975 Effect of vacuum packaging on micro-organisms on cut-up chickens and in chicken products. *Journal of Food Science* **40**, 50–52.

BAILEY, J. S., REAGAN, J. O., CARPENTER, J. A. & SCHULER, G. A. 1979 Microbiological condition of broilers as influenced by vacuum and carbon dioxide in bulk shipping packs. *Journal of Food Science* **44**, 134–137.

BARNES, E. M. 1976 Microbiological problems of poultry at refrigerator temperatures—a review. *Journal of the Science of Food and Agriculture* **27**, 777–782.

BARNES, E. M. & IMPEY, C. S. 1975 The shelf-life of uneviscerated and eviscerated chicken carcasses stored at 10°C and 4°C. *British Poultry Science* **16**, 319–326.

BARNES, E. M. & MELTON, W. 1971 Extracellular enzymic activity of poultry spoilage bacteria. *Journal of Applied Bacteriology* **34**, 599–609.

BARNES, E. M. & SHRIMPTON, D. H. 1968 The effect of processing and marketing procedures on the bacteriological condition and shelf life of eviscerated turkeys. *British Poultry Science* **9**, 243–251.

BARNES, E. M., IMPEY, C. S. & GRIFFITHS, N. M. 1979 The spoilage flora and shelf-life of duck carcasses stored at 2 or −1°C in oxygen permeable or oxygen impermeable film. *British Poultry Science* **20**, 491–500.

CAVETT, J. J. 1968 The effects of newer forms of packaging on the microbiology and storage life of meats, poultry and fish. *Progress in Industrial Microbiology* **7**, 77–123.

COYNE, F. P. 1933 The effect of carbon dioxide on bacterial growth. *Proceedings of the Royal Society, Series B* **113**, 196–217.

DEBEVERE, J. M. & VOETS, J. P. 1973 Influence of packaging materials on quality of fresh poultry. *British Poultry Science* **14**, 17–22.

EUSTACE, I. J. 1981 Some factors affecting oxygen transmission rates of plastic films for vacuum packaging of meat. *Journal of Food Technology* **16**, 73–80.

GARDNER, G. A. & CARSON, A. W. 1967 Relationship between carbon dioxide production and growth of pure strains of bacteria on porcine muscle. *Journal of Applied Bacteriology* **30**, 500–510.

GARDNER, F. A., DENTON, J. H. & HATLEY, S. E. 1977 Effects of carbon dioxide environments on the shelf-life of broiler carcasses. *Poultry Science* **56**, 1715–1716.

GIBBS, P. A. & PATTERSON, J. T. 1977 Some preliminary findings on the microbiology of vacuum- and gas-packaged chicken portions. In *The Quality of Poultry Meat* ed. Scholtyssek, S. pp. 198–200. Stuttgart: Universität Hohenheim.

GILL, C. O. & TAN, K. H. 1980 Effect of carbon dioxide on growth of meat spoilage bacteria. *Applied and Environmental Microbiology* **39**, 317–319.

HAINES, R. B. 1933 The influence of carbon dioxide on the rate of multiplication of certain bacteria as judged by viable counts. *Journal of the Society of Chemical Industry* **52**, 13T–17T.

HANNA, M. O., ZINK, D. L., CARPENTER, Z. L. & VANDERZANT, C. 1976 *Yersinia enterocolitica*-like organisms from vacuum-packaged beef and lamb. *Journal of Food Science* **41**, 1254–1256.

INGRAM, M. 1962 Microbiological principles in prepacking meats. *Journal of Applied Bacteriology* **25**, 259–281.

JAY, J. M. & SHELEF, L. A. 1978 Microbial modifications in raw and processed meats and poultry at low temperatures. *Food Technology* **32**, 186–187.

JONES, J. M., MEAD, G. C., GRIFFITHS, N. M. & ADAMS, B. W. 1982 Influence of packaging on microbiological, chemical and sensory changes in chill-stored turkey portions. *British Poultry Science* **23**, 25–40.

LEISTNER, L., HECHELMAN, H., KASHIWAZAKI, M. & ALBERTZ, R. 1975 Nachweis von *Yersinia enterocolitica* in Faeces und Fleisch von Schweinen, Rindern und Geflügel. *Die Fleischwirtschaft* **55**,1599–1602.

MERMELSTEIN, N. H. 1979 Hypobaric transport and storage of fresh meats and produce earns 1979 Food Technology Industrial Achievement Award. *Food Technology* **33**, 32–40.

MULDER, R. W. A. W. 1974 Microbiological shelf-life of vacuum packed smoked chickens. *Archiv für Lebensmittelhygiene* **25**, 252–254.

OGILVY, W. S. & AYRES, J. C. 1951 Post-mortem changes in stored meats. II. The effect of atmospheres containing carbon dioxide in prolonging the storage life of cut-up chicken. *Food Technology* **5**, 97–102.

PAREKH, K. G. & SOLBERG, M. 1970 Comparative growth of *Clostridium perfringens* in carbon dioxide and nitrogen atmospheres. *Journal of Food Science* **35**, 156–159.

PATTERSON, J. T. 1980 Factors affecting the shelf-life and spoilage flora of chilled, eviscerated poultry. In *Meat Quality in Poultry and Game Birds* ed. Mead, G. C. & Freeman, B. M. pp. 227–237. Edinburgh: British Poultry Science Ltd.

ROBERTS, D. 1972 Observations on procedures for thawing and spit-roasting frozen dressed chickens, and post-cooking care and storage: with particular reference to food-poisoning bacteria. *Journal of Hygiene*, **70**, 565–588.

ROBERTS, T. A. & SMART, J. L. 1976 The occurrence and growth of *Clostridium* spp. in vacuum-packed bacon with particular reference to *Cl. perfringens* (*welchii*) and *Cl. botulinum. Journal of Food Technology* **11**, 229–244.

SANDER, E. H. & SOO, H. 1978 Increasing shelf life by carbon dioxide treatment and low temperature storage of bulk pack fresh chickens packaged in nylon/surlyn film. *Journal of Food Science* **43**, 1519–1523, 1527.

SHAW, M. K. & NICOL, D. J. 1969 Effect of the gaseous environment on the growth on meat of some food poisoning and food spoilage organisms. *Proceedings of the 15th European Meeting of Meat Research Workers, Helsinki*, pp. 226–232.

SHRIMPTON, D. H. & BARNES, E. M. 1960 A comparison of oxygen permeable and impermeable wrapping materials for the storage of chilled, eviscerated poultry. *Chemistry and Industry* pp. 1492–1493.

SILLIKER, J. H. & WOLFE, S. K. 1980 Microbiological safety considerations in controlled-atmosphere storage of meats. *Food Technology* **34**, 59–63.

TAYLOR, M. H., SMITH, L. T. & MITCHELL, J. D. 1965 The effect of packaging materials and a tripolyphosphate (Kena) on the shelf life of turkey steaks. *Poultry Science* **44**, 297–298.

TAYLOR, M. H., HELBACKA, N. V. & KOTULA, A. W. 1966 Evacuated packaging of fresh broiler chickens. *Poultry Science* **45**, 1207–1211.

THOMPSON, J. E. 1971. Dry ice in various shipping boxes for chilled poultry: effect on microbiological and organoleptic quality. *Journal of Food Science* **36**, 74–77.

THORNLEY, M. J. & SHARPE, M. E. 1959 Micro-organisms from chicken meat related to both lactobacilli and aerobic spore formers. *Journal of Applied Bacteriology* **22**, 368–376.

TIMMONS, D. 1976 'Dryer fryers'—is CVP the ultimate bulk pack? *Broiler Business* December pp. 10–17.

WABECK, C. J., PARMELEE, C. E. & STADELMAN, W. J. 1968 Carbon dioxide preservation of fresh poultry. *Poultry Science* **47**, 468–474.

WOLFE, S. K. 1980 Use of CO- and CO_2-enriched atmospheres for meats, fish and produce. *Food Technology* **34**, 55–63.

Microbial Spoilage of Fish

G. HOBBS

Torry Research Station, Aberdeen, UK

Contents

1. Introduction

Spoilage of flesh foods such as fish and shellfish results from changes brought about by chemical reactions such as oxidation, reactions due to the fish's own enzymes and the metabolic activities of micro-organisms.

There is a wide variety of species of fish and shellfish living in diverse habitats. After catching, as a result of handling and processing, a number of products can be made from each of these. Fish and shellfish products therefore have a range of different storage and spoilage characteristics. Nevertheless similarities exist among members of large groups of fish, for instance amongst the teleosts, freshly caught 'non-fatty' fish such as cod and haddock have a water content of *ca*. 80%, a protein content of *ca*. 18%, a carbohydrate content of 1% and a lipid content of <1%. On the other hand, 'fatty' fish such as herring or mackerel have about the same levels of protein and carbohydrate as these 'non-fatty' fish but the water and lipid contents vary enormously. The sum of water and lipid remains approximately the same, about 80%, the lipid content, however, can be as low as 1% or less and as high as 25% in herring or even 30% in mackerel. These differences are associated with the seasonal spawning cycle of the fish.

Besides the main classes of chemical compounds, there are other minor components of fish such as minerals, vitamins and an important group generally referred to as extractives (i.e. extractable into water or aqueous solutions). These are primarily free amino acids, sugars and various nitrogenous bases, many of which contribute directly to the characteristic odours and flavours of the particular fish.

FOOD MICROBIOLOGY
ISBN 0 12 589670 0

In live healthy fish the muscle tissue or flesh is sterile. Micro-organisms are present on the outer surfaces and in the intestinal tract. Since they are cold-blooded animals this microflora is susceptible to environmental changes and in general reflects that of the environment and feed.

Immediately *post mortem* a whole series of tissue enzyme reactions begin the process which leads eventually to spoilage. At some stage during this process the action of micro-organisms, primarily bacteria, becomes significant. There are several reviews in the literature relating to the bacteriology of fish spoilage; the most recent of these are by Shewan (1977), Liston (1980) and Hobbs & Hodgkiss (1982). It is clear from the literature that most of the objectionable changes occurring during the storage of fish result from bacterial activity. One notable exception is the rancid flavour by fat oxidation, particularly important in fish with a high fat content. In the case of 'non-fatty' fish, pieces of sterile muscle can be stored at chill temperatures (0–5°C) for several weeks without the development of objectionable sensory changes. Non-sterile fish under the same conditions becomes unpalatable after 10–15 days.

In eviscerated whole cod stored in melting ice, autolytic enzyme reactions predominate for 4–6 days after which the products of bacterial activity become increasingly evident. Undesirable odours and flavours begin to appear, mostly resulting from bacterial metabolism of the extractives. As spoilage proceeds there is a gradual invasion of the flesh by bacteria from the outer surfaces. This was demonstrated histologically by Shewan & Murray (1979) who concluded that it was a slow process and that the development of objectionable odours and flavours resulted mainly from bacterial activity in the surface slime and in the integuments of the muscle. Breakdown of the muscle structure only occurs after spoilage has proceeded well beyond the point of rejection.

The *post mortem* changes leading to spoilage thus depend on many factors: the chemical composition of the fish, its microbial flora and subsequent handling, processing and storage. The chemical composition and microbial flora vary considerably between species, different fishing grounds and season.

2. The Microbial Flora of Freshly Caught Fish

The numbers and types of bacteria isolated from fish will depend on the medium and incubation conditions employed, which in most studies have been selected to isolate that part of the flora active under the particular storage conditions being investigated. For marine fish from colder waters, which are stored in melting ice after catching, the resident flora on the live

TABLE 1

Total viable counts (20°C) from marine fish and their environment

Source	Count
Fish	
Surface of skin	10^3–10^5/cm^2
Gills	10^3–10^4/g
Intestines	10^2–10^9/g
Sea-water	
Surface (5 fathoms)	10^2–10^3/ml
Deep (40 fathoms)	10^1–10^2/ml
Sea sediment	10^6–10^7/g
Ice	
Freshly made	10^2/ml
After 5 days in ship's bunker	10^5–10^7/ml

fish is psychrophilic in an environment with *ca.* 3% salt. During storage the flora developing will be selected by the freshwater environment.

In practice total viable counts on fish in media made with sea-water or fresh water are seldom very different provided there is some salt present in the latter (usually 0·5%). Incubation temperature, between 0 and 20°C, has no detectable effect on total numbers, but at 37°C the viable counts are only about one-tenth of those in the lower temperature range.

Table 1 gives total viable counts (at 20°C) obtained from marine fish and their environment. As might be expected the variations are large partly because the sampling problems are formidable. In the literature there are examples of data which show a seasonal variation in the numbers of bacteria on fish (see reviews quoted earlier). The higher numbers may be associated with plankton blooms in spring and late summer. In other experiments these variations are not obvious. In the face of the sampling problems associated with such large populations or masses of water, it is doubtful if it would be practicable to examine enough samples to establish unequivocally such seasonal variations. An example of the kind of results that can be obtained is presented in Fig. 1. These experiments were carried out at Torry Research Station during spoilage studies on mackerel caught off the coast of south-west England. Each value is an average of duplicate counts on each of three fish. Analysis of variance was carried out on the logarithms of the count in each set and based on the resulting variances 'between duplicates' and 'between sample', 95% confidence limits being established for each mean.

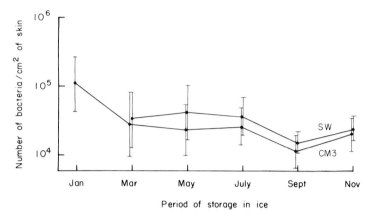

Fig. 1. Seasonal variations in viable counts on freshly caught mackerel skin on sea-water (SW) and fresh-water (CM3) nutrient agar.

The small differences obtained using sea-water and freshwater media when isolating bacteria from freshly caught marine fish can be seen in these results. Another example of the range of variation in total counts with a given species of fish is shown in Fig. 2. These data are taken from Cann *et al.* (1971) and show the distribution of counts at 20 and 37°C from freshly caught scampi (*Nephrops norvegicus*). The relationship of counts at the two temperatures is also obvious in this data.

Many studies have attempted to characterize the flora of fish and shellfish (see reviews quoted above). Again, the media and incubation conditions will influence the outcome of such studies and most have been carried out using either a sea-water or fresh water based nutrient agar incubated at 20°C. If the interest is in the flora active on the skin of live fish then sea-water media are appropriate; if, however, the interest centres on what will grow during subsequent storage in ice, then freshwater media should be used.

Whilst there are considerable variations in the details of the flora reported, the same relatively few genera predominate. Shewan (1977) listed nine out of 17 cases where the microflora comprised more than 80% Gram negative rods, two where it was 65–67% and one where it was over 40%. These examples were all from colder waters; fish taken from warmer or tropical waters more often showed a predominance of Gram positive bacteria.

Examples of the flora found on a number of species of fish and shellfish are given in Table 2. Many more can be found in the reviews by Shewan (1971, 1977), Horsley (1977) and Liston (1980).

Characterization of the bacteria isolated in such studies has presented many problems. In much of the earlier work there were difficulties not only as a result of the lack of determinative criteria and suitable laboratory tests,

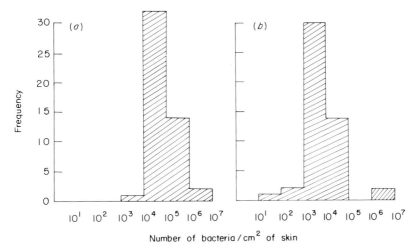

Fig. 2. Distribution of viable counts on scampi (*Nephrops norvegicus*) at first landing. (*a*) Count based on cultures incubated at 20°C; (*b*) count based on cultures incubated at 37°C. Frequency, number of samples from a total of 49. (From Cann *et al.* 1971.)

but also because of the confused state of the nomenclature and taxonomy of the organisms at that time (Ingram & Shewan 1960). At best, strains could be placed into relatively large groups or genera (Shewan *et al.* 1960*a*). Since then newer methods and tests for identification and changing views on nomenclature and taxonomy have resulted in changes in the classification of many of the bacteria concerned (Gibson *et al.* 1977). Thus organisms which were earlier assigned to the genus *Achromobacter* were a mixed collection.

TABLE 2

Bacterial flora of the skin of various fish

| | Bacterial flora (% of total numbers) on skin of | | | |
Genus or type of bacteria	Cod	Herring	Skate	Lemon sole
Pseudomonas/Alteromonas	47·5	32·2	52·5	60·3
Moraxella/Acinetobacter	36·6	26·4	18·8	22·3
Flavobacterium/Cytophaga	3·9	17·6	9·4	5·0
Vibrio	4·7	2·6	12·5	8·7
Coryneform bacteria	5·1	9·7	1·5	0·9
Micrococcus	0·9	2·6	3·4	1·4
Unidentified	1·3	8·9	1·9	1·4

This genus as defined in the sixth edition of Bergey's Manual (Breed *et al.* 1948) included (1) non-motile rods which are now called either *Moraxella* (Baumann *et al.* 1968*a*) or *Acinetobacter* (Baumann *et al.* 1968*b*), depending on their oxidase activity; (2) motile peritrichous rods which would now be assigned to the genus *Alcaligenes* (Hendrie *et al.* 1974; Buchanan & Gibbons 1974) and (3) motile, non-fluorescent rods with polar flagella now identified as *Pseudomonas* or *Alteromonas* strains. It is important to take account of these changes when referring to earlier work.

The bacterial flora of freshly caught fish thus consists in the main of organisms belonging to the Gram negative genera *Pseudomonas*, *Alteromonas*, *Moraxella*, *Acinetobacter*, *Vibrio*, *Flavobacterium* and *Cytophaga*, and to the Gram positive coryneform group and genus *Micrococcus*. The proportion of these genera vary considerably, not least because the flora reflects that of the environment which is variable. In temperate and sub-arctic waters, however, the Gram negative genera *Pseudomonas*, *Alteromonas*, *Moraxella* and *Acinetobacter* generally predominate in the skin flora of marine fish. There are relatively few data on freshwater fish. Available information (Herborg & Villadsen 1975; Cantoni *et al.* 1976; Wyatt 1978) shows that the flora of freshwater fish is comprised of the same relatively few genera as that of marine fish and again it reflects that of the environment.

3. Changes in Chill-stored Fish

Almost all the work on spoilage of fish has been carried out with fish stored at chill temperatures, usually 0°C in melting ice. Bacteriological analyses have, for the most part, been carried out at around 20°C because most of the bacteria which grow at 0°C have an optimum growth temperature of 18–20°C. Thus there is a considerable saving of incubation time at the higher temperature without the loss of significant data. Whilst this is largely true there are elements of the flora which grow rapidly at 18–20°C but do not grow below *ca.* 5°C. Interpretation of, for instance, a series of flora analyses determined at 20°C, on fish stored at 0°C, must therefore take this into account. The Gram positive flora in general presents this problem; most coryneforms and micrococci grow much slower than the Gram negative flora at 0°C or not at all. This is not always obvious when, as usually occurs, the components of the flora are presented as a percentage of the total numbers; it becomes more obvious if the actual numbers are quoted. Examples of this can be seen in the data presented by Shewan *et al.* (1960*b*). Table 3 is taken from this work and shows the sort of changes that occur during storage in melting ice. Similar patterns of flora changes have been described in a variety of species of fish and shellfish (see Hobbs & Hodgkiss 1982). Work of this

TABLE 3

Changes in the flora of the skin of cod stored in melting ice

| | Counts after storage in melting ice for (days) | | | | | | | |
| | 0 | | 5 | | 10 | | 15 | |
	P	N	P	N	P	N	P	N
Pseudomonas/Alteromonas	26	24·49	33	162·49	84	11,088	82	45,805
Moraxella/Acinetobacter	33	31·09	26	128·18	7	924	13	7262
Flavobacterium/Cytophaga	0	0	8	39·44	0	0	0	0
Coryneform bacteria	25	23·55	12	59·16	8	1056	3	1676
Micrococcus	14	13·19	21	103·53	0	0	0	0
Unidentified	2	1·88	0	0	1	132	2	1117
Total count		94·20		493		13,200		55,860

P, percentage of total flora; N, number $\times 10^3/cm^2$ of skin.
Data from Shewan *et al.* (1960*b*).

kind has established that the total number of bacteria growing at 0–20°C increases systematically during storage with ice and that the Gram negative genera *Pseudomonas, Alteromonas, Moraxella* and *Acinetobacter* become increasingly predominant so that at the end of the shelf life (in terms of edibility) they comprise most of the flora. This has established the numerical significance of these genera in the spoilage process.

It might appear from these considerations that a total viable count at, say, 20°C could be used as a measure of spoilage. There are, however, some severe limitations to this. Firstly, as already stated, the initial count after catching can be quite variable. Secondly two apparently similar storage systems can give rather different results. Figure 3 shows data obtained at Torry Research Station from some recent work on mackerel spoilage. There were no significant seasonal variations and these curves were produced from the pooled data obtained at six different times of the year. Each point was therefore obtained from 18 samples each counted in duplicate. Confidence limits were established as for Fig. 1.

In both cases the fish were stored at 0°C but in ice the fish skin would be in a highly aerobic environment whereas in the chilled sea-water it would be micro-aerophilic, since a stream of nitrogen was used to keep the fish, ice and water mixed. Sensory assessment showed that the mackerel stored in ice developed off-flavours 1 or 2 days before that in chilled sea-water but with no differences between them in the early stages (Smith *et al.* 1980). The bacterial counts indicate a greater difference in quality during the first half of the storage period. Incidently, the trimethylamine levels (see later) were higher in the chilled sea-water fish, indicating a poorer quality.

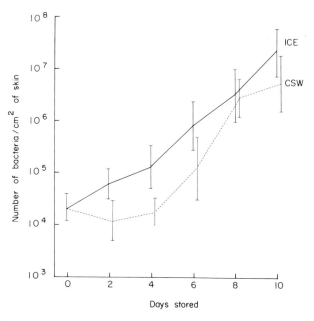

Fig. 3. Changes in viable counts during storage of mackerel in ice and chilled sea-water.

Obviously in the situation described the different environmental conditions could well select a different flora. It has also been known for some time
that not all the bacteria on fish will produce spoilage odours and flavours. For
example, Shewan & Murray (1979) reviewed a number of experiments
where sterile fish muscle had been inoculated with pure cultures of bacteria.
Moraxella and *Acinetobacter* strains did not produce typical spoilage odours
and flavours whereas *Pseudomonas* and *Alteromonas* strains did.

Many attempts have been made to identify and enumerate the 'active
spoilers' rather than the total numbers of bacteria (see Herbert *et al.* 1971).
Also the accumulation of various chemical compounds in fish has been used
for many years as an index of spoilage (Fields *et al.* 1968; Connell & Shewan
1980). Many of these will be the products of bacterial growth. As already
stated, the characteristic flavours and odours of fresh fish are contributed
mainly by the extractive fractions, and this is also true of most spoilage
odours and flavours. Examples of substances produced by bacteria from the
extractives are trimethylamine from trimethylamine oxide; hydrogen sulphide, dimethyl sulphide and methyl mercaptan from sulphur-containing
amino acids; various amines and ammonia from amino acids; lower fatty
acids from sugars such as glucose and ribose; various carbonyl compounds

from lipids and indole, skatole, tyrosine, putrescine and cadaverine from proteins (Connell & Shewan 1980).

Enumeration of those bacteria able to bring about these reactions would be some measure of the 'active spoilers'. Though a number of attempts have been made to do this, none has resulted in a practical usable method, though some progress has been made in identifying which components of the bacterial flora produce the spoilage changes.

Enumeration of bacteria producing hydrogen sulphide has recently been re-investigated by Levin (1968) who concluded that *Alteromonas putrefaciens* (*Pseudomonas putrefaciens*) was an important fish spoilage organism. Using gas-liquid chromatography (g.l.c.) and mass spectrometry (m.s.) Chai *et al*. (1968) also demonstrated that phenylethyl alcohol and phenol detected in haddock fillets were metabolic products of *Moraxella*-like strains. Studies on iced cod (Herbert *et al*. 1975) showed that, together with hydrogen sulphide, methyl mercaptan and dimethyl sulphide were responsible for the 'sulphide-like' odours associated with spoilage. Herbert & Shewan (1976) then showed that all 13 strains of *Pseudomonas* and *Alteromonas* tested were active in the production of hydrogen sulphide and methyl mercaptan, and six of these strains also produced dimethyl sulphide. Volatile sulphur compounds did not arise from autolytic reactions.

Using sterile fish muscle homogenates Miller *et al*. (1972, 1973*a*,*b*,*c*) demonstrated a wide range of metabolic products using pure cultures of various spoilage bacteria. They were able to relate some of these to specific off-odours, for instance *Ps. fragi* produces 'fruity' off-odours which could be attributed to the synergistic flavour interactions of ethyl esters of fatty acids.

Trimethylamine has long been associated with bacterial spoilage of fish and indeed has been used for many years as an index of quality (Connell & Shewan 1980). Trimethylamine oxide occurs in appreciable quantities in most marine fish and is reduced directly to trimethylamine by bacteria. Laycock & Regier (1971) showed that in the fish they examined *A. putrefaciens* was mainly responsible for this reaction. This was borne out by the work of van Spreekens (1977) though the work of Gibson *et al*. (1977) shows that other *Alteromonas* species are also involved.

One reason for the association of trimethylamine with fish spoilage is apparent from the work of Easter *et al*. (1982) who showed that many spoilage bacteria, including *A. putrefaciens*, are able to utilize trimethylamine oxide as a terminal hydrogen acceptor when oxygen becomes depleted. Thus such non-fermentative bacteria are able to grow rapidly under the micro-aerophilic or anaerobic conditions that exist in spoiling fish tissues. The high level of trimethylamine oxide in marine fish was therefore offered as one reason why they spoil more rapidly than other animal flesh foods, including freshwater fish.

It is well known that the rates of spoilage of different kinds of fish vary partly because of substantial differences in initial chemical composition. Another factor which might have wider implications was described by Murray & Fletcher (1976). They showed that the skin of plaice, but not that of cod, possesses a powerful lysozyme, which possibly explains why plaice keeps longer than cod under similar storage conditions.

There is an increasing amount of evidence that fish caught in tropical waters have a longer shelf life in ice than similar species caught in temperate or cold water (Cann 1977; Shewan 1977). The usual explanation offered is that the tropical fish do not carry a predominantly psychrotophic bacterial flora when caught whereas fish from colder waters do. In both cases the spoilage flora that eventually emerges during ice-storage is the same.

The spoilage processes described can be modified by handling and processing. Indeed the purpose of processing is often deliberately to prevent or delay spoilage.

Heat sterilization, freezing and cold storage virtually eliminate microbial activity though enzymic activity still continues at cold storage temperatures. Freezing and cold storage also preserves bacteria as well as the fish, thus after thawing, spoilage will proceed as in unfrozen fish (Hobbs & Hodgkiss 1982).

Other processes which extend shelf life, but do not totally prevent microbial growth, rely on killing a proportion of the flora. Examples are cooking and heat or irradiation pasteurization, reduction of water activity by salt curing and drying, for instance, and reduction of pH by fermentation. All these processes are more effective against Gram negative, rather than Gram positive, bacteria. Thus the greatest reduction of numbers is amongst those bacteria responsible for the spoilage of chilled foods. These processes combined with subsequent, proper chill storage can therefore result in large extensions of shelf life.

Handling and processing, especially if carried out under unhygienic conditions, can, of course, introduce bacteria, including those able to cause food poisoning. Different spoilage bacteria, or merely increased numbers of the same spoilage bacteria, can contaminate the fish; an extreme example is the introduction of strictly halophilic bacteria to heavily salt-cured fish. The bacteria spoiling this kind of product originate entirely from the salt.

4. Conclusions

There is clear evidence that when fish and shellfish are stored at chill temperatures a Gram negative bacterial flora develops. The numerically most important genera present are *Pseudomonas*, *Alteromonas*, *Moraxella*, and *Acinetobacter*. The undesirable spoilage odours and flavours which develop during storage are a result of bacterial activity and the evidence indicates that strains of the *Pseudomonas* and *Alteromonas* genera are largely responsible.

Attempts to use total numbers of bacteria as an index of spoilage or freshness have had limited success. In a particular production situation where the raw material and storage conditions are known, total counts can indicate where faulty handling or processing has occurred. They are of limited value for indicating the shelf life or degree of spoilage of end products. Selective methods of enumerating the spoilage bacteria have not yet been developed to the point of being generally useful, though there is now a growing amount of evidence to indicate which bacteria are involved.

Estimation of the level of bacterial metabolic products has been more fruitful. Trimethylamine and total volatile base determinations have both been used with some success but the correlation of such tests with sensory evaluation is not always entirely satisfactory. The bacteriological, chemical and sensory data from many experiments do not agree when comparing storage in ice with that in chilled sea-water.

Most of the research into fish spoilage has so far been carried out on fish stored at or around 0°C. In commercial practice temperature fluctuations are common, thus enabling other components of the bacterial flora to grow, resulting in a different sequence of sensory changes. Adequate objective tests for spoilage applied to end products of unknown history will only be achieved with a fuller understanding of the spoilage changes occurring under different storage conditions. This will also present opportunities to propose novel ways of preventing spoilage and making better use of existing methods.

Statistical work associated with the data in Figs 1 and 3 was performed by G. Smith of the Torry Research Station.

5. References

BAUMANN, P., DOUDOROFF, M. & STANIER, R. Y. 1968*a*. A study of the *Moraxella* group. I. Genus *Moraxella* and the *Neisseria catarrhalis* group. *Journal of Bacteriology* **95**, 58–73.

BAUMANN, P., DOUDOROFF, M. & STANIER, R. Y. 1968*b* A study of the *Moraxella* group. II. Oxidase-negative species (genus *Acinetobacter*). *Journal of Bacteriology* **95**, 1520–1541.

BREED, R. S., MURRAY, E. G. D. & HITCHENS, A. P. (eds) 1948 *Bergey's Manual of Determinative Bacteriology* 6th edn. Baltimore: Williams & Wilkins Co.

BUCHANAN, R. E. & GIBBONS, N. E. (eds) 1974 *Bergey's Manual of Determinative Bacteriology* 8th edn. Baltimore: Williams & Wilkins Co.

CANN, D. C., HOBBS, G., WILSON, B. B. & HORSLEY, R. W. 1971 The bacteriology of 'scampi' (*Nephrops norvegicus*). II. Detailed investigation of the bacterial flora of freshly caught samples. *Journal of Food Technology* **6**, 153–161.

228   G. HOBBS

CANN, D. C. 1977 Bacteriology of shellfish with reference to International Trade. In *Proceedings of the Conference on the Handling, Processing and Marketing of Tropical Fish*, pp. 377–394. London: Tropical Products Institute.

CANTONI, C., CATTANEO, P. & AUBERT, S. D. 1976 Bacteriology and evaluation of freshness of freshwater fish. *Industrie Alimentari* **15**, 105–111.

CHAI, T., CHEN, C., ROSEN, A. & LEVIN, R. E. 1968 Detection and incidence of specific species of spoilage bacteria on fish. II. Relative incidence of *Pseudomonas putrefaciens* and fluorescent pseudomonads on haddock fillets. *Applied Microbiology* **16**, 1738–1741.

CONNELL, J. J. & SHEWAN, J. M. 1980 Sensory and non-sensory assessment of fish. In *Advances in Fish Science and Technology* ed. Connell, J. J. pp. 56–65. Farnham, Surrey, England: Fishing News Books Ltd.

EASTER, M. C., GIBSON, D. M. & WARD, F. B. 1982 A conductance method for the assay and study of bacterial trimethylamine oxide reduction. *Journal of Applied Bacteriology* **52**, 357–365.

FIELDS, M. L., RICHMOND, B. S. & BALDWIN, R. E. 1968 Food quality as determined by metabolic by-products of microorganisms, *Advances in Food Research* **16**, 161–229.

GIBSON, D. M., HENDRIE, M. S., HOUSTON, N. C. & HOBBS, G. 1977 The identification of some Gram negative heterotrophic aquatic bacteria. In *Aquatic Microbiology* ed. Skinner, F. A. & Shewan, J. M. pp. 135–159. London & New York: Academic Press.

HENDRIE, M. S., HOLDING, A. J. & SHEWAN, J. M. 1974 Emended descriptions of the genus *Alcaligenes* and of *Alcaligenes faecalis* and proposal that the generic name *Achromobacter* be rejected. Status of the named species of *Alcaligenes* and *Achromobacter*. *International Journal of Systematic Bacteriology* **24**, 534–550.

HERBERT, R. A., HENDRIE, M. S., GIBSON, D. M. & SHEWAN, J. M. 1971 Bacteria active in the spoilage of certain sea foods. *Journal of Applied Bacteriology* **34**, 41–50.

HERBERT, R. A., ELLIS, J. R. & SHEWAN, J. M. 1975 Isolation and identification of the volatile sulphides produced during chill storage of North Sea cod. *Journal of the Science of Food and Agriculture* **26**, 1187–1194.

HERBERT, R. A. & SHEWAN, J. M. 1976 Roles played by bacterial and autolytic enzymes in the production of volatile sulphides in spoiling North Sea cod (*Gadus morrhua*). *Journal of the Science of Food and Agriculture* **27**, 89–94.

HERBORG, L. & VILLADSEN, A. 1975 Bacterial infection/invasion in fish flesh. *Journal of Food Technology* **10**, 507–513.

HOBBS, G. & HODGKISS, W. 1982 The bacteriology of fish handling and processing. In *Developments in Food Microbiology I.* ed. Davies, R. Barking, Essex, England: Applied Science Publishers Ltd.

HORSLEY, R. W. 1977 A review of the bacterial flora of teleosts and elasmobranchs including methods for its analysis. *Journal of Fish Biology* **10**, 529–553.

INGRAM, M. & SHEWAN, J. M. 1960 Introductory reflections on the *Pseudomonas–Achromobacter* group. *Journal of Applied Bacteriology* **23**, 373–378.

LAYCOCK, R. A. & REGIER, L. W. 1971 Trimethylamine-producing bacteria on haddock (*Melanogrammus aeglefinus*) fillets during refrigerated storage. *Journal of the Fisheries Research Board of Canada* **28**, 305–309.

LEVIN, R. E. 1968 Detection and incidence of specific species of spoilage bacteria on fish. I. Methodology. *Applied Microbiology* **16**, 1734–1737.

LISTON, J. 1980 Microbiology in fishery science. In *Advances in Fish Science and Technology* ed. Connell, J. J. pp. 138–157. Farnham, Surrey, England: Fishing News Books Ltd.

MILLER, A., SCANLAN, R. A., LEE, J. S. & LIBBEY, L. M. 1972 Quantitative and selective gas chromatographic analysis of dimethyl- and trimethylamine in fish. *Journal of Agricultural and Food Chemistry* **20**, 709–711.

MILLER, A., SCANLAN, R. A., LEE, J. S. & LIBBEY, L. M. 1973a Volatile compounds produced in sterile fish muscle (*Sebastes melanops*) by *Pseudomonas perolens*. *Applied Microbiology* **25**, 257–261.

MILLER, A., SCANLAN, R. A., LEE, J. S. & LIBBEY, L. M. 1973b Identification of the volatile compounds produced in sterile fish muscle (*Sebastes melanops*) by *Pseudomonas fragi*. *Applied Microbiology* **25**, 952–955.

MILLER, A., SCANLAN, R. A., LEE, J. S. & LIBBEY, L. M. 1973c Volatile compounds produced in sterile fish muscle (*Sebastes melanops*) by *Pseudomonas putrefaciens*, *Pseudomonas fluorescens* and an *Achromobacter* species. *Applied Microbiology* **26**, 18–21.

MURRAY, C. K. & FLETCHER, T. C. 1976. The immunohistochemical localization of lysozyme in plaice (*Pleuronectes platessa* L) tissues. *Journal of Fish Biology* **9**, 329–334.

SHEWAN, J. M. 1971 The microbiology of fish and fishery products—a progress report. *Journal of Applied Bacteriology* **34**, 299–315.

SHEWAN, J. M. 1977 The bacteriology of fresh and spoiling fish and the biochemical changes induced by bacterial action. In *Proceedings of the Conference on the Handling, Processing and Marketing of Tropical Fish*. pp. 51–66. London: Tropical Products Institute.

SHEWAN, J. M. & MURRAY, C. K. 1979 The microbial spoilage of fish with special reference to the role of psychrophiles. In *Cold Tolerant Microbes in Spoilage and the Environment* ed. Russell, A. D. & Fuller, R. pp. 117–136. Society for Applied Bacteriology Technical Series No. 13. London & New York: Academic Press.

SHEWAN, J. M., HOBBS, G. & HODGKISS, W. 1960a A determinative scheme for the identification of certain Gram negative bacteria, with special reference to the Pseudomonadaceae. *Journal of Applied Bacteriology* **23**, 379–390.

SHEWAN, J. M., HOBBS, G. & HODGKISS, W. 1960b The *Pseudomonas* and *Achromobacter* groups of bacteria in the spoilage of marine white fish. *Journal of Applied Bacteriology* **23**, 463–468.

SMITH, J. G. M., HARDY, R. & YOUNG, K. W. 1980 A seasonal study of the storage characteristics of mackerel stored at chill and ambient temperatures. In *Advances in Fish Science and Technology* ed. Connell, J. J. pp. 372–378. Farnham, Surrey, England: Fishing News Books Ltd.

VAN SPREEKENS, K. J. A. 1977 Characterization of some fish and shrimp spoiling bacteria. *Antonie van Leeuwenhoek* **43**, 283–303.

WYATT, L. E. 1978 Microbiological profiles of fresh water catfish. *Dissertation Abstracts International, B* **38**, 5206.

The Microbial Ecology of Prepared Raw Vegetables

P. C. KOEK, Y. DE WITTE AND J. DE MAAKER

Sprenger Instituut, Wageningen, The Netherlands

Contents

1. Introduction

Cleaned and cut raw vegetables are used in great quantities in the kitchens of hospitals, hotels, old people's homes and army units, and also in private households. Working people, especially, often buy such prepared vegetables. In view of the known spoilage and health hazards associated with food processing, the Microbiological Section of the Sprenger Instituut, Wageningen, is conducting applied research on the storage and processing of horticultural produce, especially fruits and vegetables. Some investigations on prepared cut, sliced and diced vegetables have already been completed.

During 1978 investigations on the microbial loading of vegetables were started using sliced endive bought from shops in and around Wageningen. Some of the samples were stored at four different temperatures (1, 8, 14 and 20°C) for 24 or 48 h and others were investigated microbiologically at once. These experiments were set up in summer with endive crops grown in the open field, and in winter with endive grown in greenhouses or imported from Italy or France.

In all cases high counts of mesophilic aerobes ($ca.$ $10^7/g$), coli-aerogenes group ($ca.$ $10^7/g$) and non-fermentative Gram negative rods ($ca.$ $10^6/g$) were found: the storage temperatures in this experiment were clearly too high.

FOOD MICROBIOLOGY
ISBN 0 12 589670 0

There was a correlation between the numbers of *Escherichia coli*, and the storage time and temperature. In this experiment there were too many different factors beyond our control, such as the origin of the product, mostly unknown, and the use of different machines for slicing the endive. The following year a line study was made at two important plants for preparing vegetables. Their machines and methods differed completely so the results could not usefully be compared. Despite these difficulties our impression was that preparation, cooling and the latest 'sell-by' date should be better standardized.

2. Good Manufacturing Practice is Recommended

In the USA and Canada the Food and Drug Administration (FDA) has, in co-operation with the food industries, developed the concept of 'hazard analysis and critical control points'. Recommended codes of practices are being developed in the United Nations Food Standard Program, and the term good manufacturing practice (GMP) can also be used for the same set of principles.

One of the main points of GMP is that non-potable water should not be brought into contact with the food being processed. Another important principle is that the machines should reach the specified hygiene requirements (Anema 1978). It is desirable, therefore, that the designers of machines for slicing or dicing vegetables should co-operate with micro-biologists. GMP concerns the whole industrial process, not simply the machines (Mossel 1977): hygiene of the staff, the plant, packing materials and stores are also important (Shewan 1976).

3. The Quality of Vegetables to be Prepared

In preparing raw, sliced or diced vegetables the primary material quality must be properly controlled; only the best quality produce should be bought for processing.

However good the quality the produce will already have been contaminated naturally with various bacteria and fungi (Mossel & Westerdijk 1949; Tamminga *et al.* 1978). Leafy vegetables are not only susceptible to bacterial and mould diseases, but they can also be infected by secondary bacteria such as *Erwinia* and *Pseudomonas* species (Webb & Mundt 1978). For instance, Green *et al.* (1974) found soil to be a reservoir of *Pseudomonas aeruginosa* and discovered that this bacterium has the capacity to colonize the upper parts of plants. Splashing water can carry these bacteria to aerial parts of plants where they enter water-soaked areas.

Bulb vegetables, such as onions, and root vegetables, such as carrots, are more likely to be contaminated by fungi and bacteria than are leafy vegetables. Root vegetables are usually cleaned with water to improve their appearance at auction. It is important, however, to use good quality water or the GMP rules will be infringed. In the Netherlands it is now forbidden by law to use ditch-water for this purpose (Tamminga *et al*. 1978).

4. Experiments with Four Vegetables

A. Machines used to prepare the vegetables

Different slicing and dicing machines were used in the microbial ecological study of prepared vegetables. For slicing endive and leek an in-plant machine unit, consisting of a slicing machine, a washing machine, a container and a spin-drier was used (Fig. 1).

The bottom of each plant was first cut off and the outer leaves removed manually. The product was then transported by a conveyor belt to an S-shaped knife which sliced it. The sliced product was then passed to the washing machine containing 2000 litres of tap-water in which it was agitated by metal blades which moved forwards in the washing basin. At the end of the basin the blades returned above the water to the starting point where

Fig. 1. The machine unit for prepared raw vegetables. From right to left: the slicing machine, the washing machine, the container and the spin-drier.

they were lowered into the water again, thereby making a circulating movement. In the meantime, recirculated water was sprayed on top of the floating sliced product. At the end of the basin the sliced product was transferred to a conveyor belt made of small, stainless-steel chains and washed simultaneously from above with tap-water. At the end of the conveyor belt the slices fell into a stainless-steel container. The sliced product was then put in a spin-drier to remove surplus water. Finally, the product was collected in polyethylene bags which were then put in cleaned plastic boxes.

Carrots were scraped mechanically with carborundum, then diced and cut in a special machine, using different cutting discs. Diced carrots were collected in a plastic box. A similar machine was used for slicing leeks. At the plant used for the experimental work the plastic boxes were brushed and cleaned by a special machine, a method which was very effective and hygienic and which should be introduced into other plants where vegetables are prepared. For commercial use the product is normally packed in new, clean polyethylene bags; large quantities are intended for use in hospitals, restaurants, etc.

B. Materials and methods

The experiments on endive, leek, carrots and cauliflower were conducted according to GMP rules which meant that all parts of the machines had to be cleaned before use. In most cases this was satisfactory, but the construction of the washing machine prevented thorough cleaning.

Contamination of prepared vegetables by micro-organisms was monitored according to the following criteria: (1) colony counts of the Enterobacteriaceae, as a tentative taxonomic grouping of psychrotrophic and thermotrophic micro-organisms; (2) colony counts of the mesophilic and psychrotrophic groups of Gram negative and oxidase-positive rods, such as members of the Pseudomonadaceae; (3) numbers of Lancefield group D streptococci, yeasts and moulds; (4) numbers of lactobacilli (as these also occur naturally on vegetables they were enumerated separately).

(i) Sampling

Ten grams were taken at random from each sample of vegetable, mixed with 90 g of tryptone–soya broth and macerated in a Colworth Stomacher for 2·5 min.

(ii) Monitoring micro-organisms in samples

As there is no standardized method for the microbial monitoring of foods, particularly prepared raw vegetables, we used the system of Mossel *et al.*

(1977) to isolate Gram negative bacteria, using tryptone–soya agar containing 2 mg/l of crystal violet following repair of the cells by pre-incubation in tryptone–soya broth. The strains isolated by ourselves, using this method, were identified by the staff of Professor Mossel's laboratory at Utrecht. Most isolates were pseudomonads.

For isolating the Enterobacteriaceae a dilution of 1 : 10 was poured on a solid repair medium of tryptone–soya agar and incubated for 5 h, then overlaid with violet-red–bile–glucose agar (Mossel 1978). Strains isolated from the colonies were then identified to genus level.

Two methods were used to monitor *E. coli* in the samples. In the first method a dilution of the sample was poured into tryptone–soya broth, as a liquid repair medium, and incubated for 3 h. Tryptone–soya broth to which a double strength solution of brilliant green-ox bile had been added, was then added to the culture and incubation continued.

In the second method the dilution was poured on tryptone–soya agar, as a solid repair medium, and after 5 h of resuscitation was overlaid with McConkey agar.

For isolating Lancefield group D streptococci the same methods were used for resuscitation of the organisms, with tryptone–soya broth and tryptone–soya agar as liquid and solid repair media, respectively. Afterwards, kanamycin–aesculin–azide medium was used to establish their presence or absence and to isolate from the colonies.

Oxytetracycline–gentamicin yeast agar was used to isolate yeasts and fungi. Yeasts were present in fewer cases than expected.

Lactobacilli were resuscitated on a solid repair medium of tryptone–soya agar for 5 h. Rogosa agar was then overlaid on the solid medium and the plates placed in GasPak jars for anaerobic incubation (Mossel & Tamminga 1980).

C. Results

The results of the work on sliced endive and leek treated with standard and controlled industrial methods are shown in Fig. 2. For each vegetable the results of microbiological analysis for the standard industrial method are shown on the left of the figure (*a*), and the results, after applying the GMP rules as stringently as possible, are shown on the right (*b*). Each column represents the $\log_{10}$/g of the total numbers of micro-organisms indicated. Lancefield group D streptococci and *E. coli* are not shown in the histograms. *Escherichia coli* was identified only once, in the sliced endive prepared in the standard way, but even then they were not numerous.

There was a difference between the numbers of bacteria on the sliced endive prepared in the standard way and that prepared by the GMP-controlled method. More bacteria of the Enterobacteriacea and Pseudomonadaceae

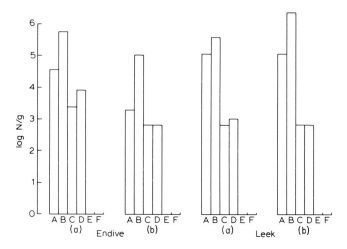

Fig. 2. Survey of micro-organisms in sliced endive and leek after processing by the standard and the GMP-controlled methods. For each product the results after the standard industrial method (*a*) are compared with the results after applying the GMP rules (*b*). A. Enterobacteriaceae; B. Pseudomonadaceae; C, yeasts; D, lactobacilli; E. Lancefield group D streptococci; F, *Escherichia coli.*

were isolated from the former product than from the latter. A completely different result was obtained with sliced leek. Compared with the sliced endive the total numbers of Enterobacteriaceae and Pseudomonadaceae were higher, and unexpectedly, the controlled, sliced leek had exceptionally high numbers of pseudomonads. The total number of Enterobacteriaceae was the same in both cases, and the numbers of yeasts and lactobacilli were very low.

In Fig. 3 the results for the four sliced or diced vegetables prepared by the GMP-controlled method are compared. The highest counts of Enterobacteriaceae were given by carrots, the next highest by leeks and the lowest by sliced endive. One reason for this could be that carrots and leeks grow in the soil and are in consequence more prone to contamination with soil bacteria than are endive and cauliflower. Neither Lancefield group D streptococci nor *E. coli* were isolated from any product. Cauliflower was an exception in this investigation; no appreciable numbers of micro-organisms in any of the groups tested for were found.

Cauliflower, which is related botanically to cabbage, merits special attention. Tamminga *et al.* (1978) and Yildiz & Westhoff (1981) reported that cauliflower produced substances which partially inhibited the growth of bacteria in media. Pederson & Fisher (1944) had already mentioned this phenomenon with the juice of cabbage and other vegetables, and Clapp *et*

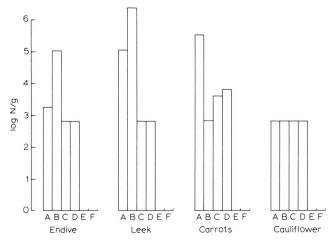

Fig. 3. Distribution of micro-organisms in sliced endive, leek, diced carrots and cauliflower after the GMP-controlled industrial method of preparation. A. Enterobacteriaceae; B. Pseudomonadaceae; C, yeasts; D. Lactobacilli; E. Lancefield group D streptococci; F, *Escherichia coli*.

al. (1959) listed isothiocyanates as the main sulphur compounds in cabbage. Although there is some evidence that cauliflower tissue contains inhibitory substances they were not detected in experiments to test the effect of a cauliflower suspension, juice of cauliflower stem and juice of cauliflower sprigs on seven different bacteria, a yeast and two fungi (Mossel & Cornelissen 1960).

D. Ecological survey of bacteria from the prepared raw vegetables

In Table 1 are listed the identified species of bacteria. The following general comments can be made:

(1) In all the products, pseudomonads Sh I and bacteria of the genus *Enterobacter* were found. See Hendrie & Shewan (1979) and Hugh (1981) for identification of pseudomonads, and Brocklehurst & Lund (1981) for occurrence of these organisms on vegetables.

(2) In the standard-processed endive *E. coli* was isolated only once. In GMP-controlled sliced endive the new genus *Kluyvera* and, on one occasion, *Flavobacter meningosepticum*, was identified (Farmer *et al.* 1981; Kleeberger *et al.* 1980). Yeasts and lactobacilli were isolated only from the standard-processed endive.

(3) *Erwinia* organisms were isolated from both the leek products, the GMP-controlled diced carrots and from the sliced endive prepared by the standard industrial method.

TABLE 1

An ecological survey of identified micro-organisms in different sliced and diced raw vegetables

	Vegetable tested					
	Endive		Leek		Carrot	Cauliflower
Organism	(a)*	(b)*	(a)	(b)	(b)	(b)
Pseudomonas Sh. I†	×	×	×	×	×	×
Pseudomonas Sh. II–IV		×				
Pseudomonas maltophilia	×	×				
Xanthomonas	×					
Acinetobacter		×	×			
Citrobacter		×				
Enterobacter	×	×	×	×	×	×
Enterobacter (atypical)	×		×			
Erwinia	×		×	×	×	
Kluyvera		×			×	
Hafnia					×	
Serratia					×	
E. coli	×					
Yersinia	×					
Enteropathogens						
Flavobacter meningosepticum		×				
Lancefield D Streptococci						
Yeasts	×				×	
Lactobacilli	×		×		×	

* (a) Prepared by standard industrial methods; (b) prepared by GMP-controlled methods.
† Sh. I and Sh. II–IV represent groups of Hendrie & Shewan (1979).

(4) *Kluyvera* was also identified among isolates from the GMP-controlled diced carrots.

(5) Lactobacilli and yeasts were isolated from GMP-controlled diced carrots.

(6) Diced cauliflower carried the lowest microbial load.

(7) In this survey pathogenic bacteria seemed to be of no importance.

E. Storage

To prevent growth of bacteria in the period between preparing and consuming the product, it is necessary to store it at low temperatures; an average temperature of 0–1°C is advisable (Anon. 1981). In the plant where our experimental work was done the products were normally stored in the cold and later distributed by vehicles with refrigeration units. In shops the products should, of course, be displayed in refrigerated showcases.

We wish to express our gratitude to Professor D. A. A. Mossel for his support and interest shown during these experiments. In particular we wish to thank the members of his staff, Dr P. van Netten, Dr A. M. Th. Ligtenberg-Merkus and Mrs C. H. M. Klerkx for their valuable co-operation.

5. References

ANEMA, P. J. 1978 'Good Manufacturing Practice' bij de bereiding van voedingsmiddelen. *Voedingsmiddelentechnologie* **11**, 13–15.

ANON, 1981 *Gids voor de Kleinverpakking van Groente en Fruit* (Guide for prepackaging of vegetables and fruit). Bulletin No. 26, 9th revised edn. Wageningen: Sprenger Instituut.

BROCKLEHURST, T. F. & LUND, B. M. 1981 Properties of pseudomonads causing spoilage of vegetables stored at low temperatures. *Journal of Applied Bacteriology* **50**, 259–266.

CLAPP, R. C., LONG, L. & DATEO, G. P. 1959 The volatile isothiocyanates. *Journal of the American Chemical Society* **81**, 6278–6281.

FARMER III, J. J., FANNING, G. R. & HUNTLEY-CARTER, G. P. 1981 *Kluyvera*, a new (redefined) genus in the family Enterobacteriaceae: Identification of *Kluyvera ascorbata* sp. nov. and *Kluyvera cryocrescens* sp. nov. in clinical specimens. *Journal of Clinical Microbiology* **13**, 919–933.

GREEN, S. K., SCHROTH, M. N. & CHO, J. J. 1974 Agricultural plants and soil as a reservoir for *Pseudomonas aeruginosa*. *Applied Microbiology* **28**, 987–991.

HENDRIE, M. S. & SHEWAN, J. M. 1979 The identification of pseudomonads. In *Identification Methods for Microbiologists* 2nd edn, ed. Skinner, F. A. & Lovelock, D. W. pp. 1–14. Society for Applied Bacteriology Technical Series No. 14. London & New York: Academic Press.

HUGH, R. 1981 *Pseudomonas maltophilia* sp. nov. nom. rev. *International Journal of Systematic Bacteriology* **31**, 195.

KLEEBERGER, A., SCHÄFER, K. & BUSSE, M. 1980 Untersuchungen zur Oekologie von Enterobakterien auf Schlachthausfleisch. *Die Fleischwirtschaft* **60**, 1529–1530.

MOSSEL, D. A. A. 1977 *Microbiology of Foods; Occurrence, Prevention and Monitoring of Hazards and Deterioration*, revised edn. Utrecht, The Netherlands: University of Utrecht, Faculty of Veterinary Medicine.

MOSSEL, D. A. A. 1978 Index and indicator organisms—a current assessment of their usefulness and significance. *Food Technology in Australia* **30**, 212–219.

MOSSEL, D. A. A. & CORNELISSEN, M. R. 1960 *Identification Présumptive des Antibiotiques par Voie Microbiologique*. Rapport inédit no. 13046. Institut Central de la Nutrition et de l'Alimentation TNO, Utrecht, Pays-Bas.

MOSSEL, D. A. A. & TAMMINGA, S. K. 1980 *Methoden voor het Microbiologisch Onderzoek van Levensmiddelen* 2nd cdn. Zcist: Noordcrvlict.

MOSSEL, D. A. A. & WESTERDIJK, J. 1949 The physiology of microbial spoilage in foods. *Antonie van Leeuwenhoek* **15**, 190–202.

MOSSEL, D. A. A., EELDERINK, I. & SUTHERLAND, J. P. 1977 Development and use of single 'polytropic' diagnostic tubes for the approximate taxanomic grouping of bacteria isolated from foods, water and medicinal preparations. *Zentralblatt für Bakteriologie und Parasitenkunde, Abt. I. Orig. A* **238**, 66–79.

PEDERSON, C. & FISHER, P. 1944 *The Bactericidal Action of Cabbage and Other Vegetable Juices*. Technical Bulletin no. 273, N.Y. State Agricultural Experimental Station, Geneva, N.Y.

SHEWAN, J. M. 1976 Microbial standards for foods. *Food Technology in Australia* **28**, 493–495.

TAMMINGA, S. K., BEUMER, R. R. & KAMPELMACHER, E. H. 1978 Oriënterende onderzoekingen naar de hygiënische kwaliteit van Nederlandse en geïmporteerde groenten. *Voedingsmiddelentechnologie* **11**, 14–19.

WEBB, T. A. & MUNDT, J. O. 1978 Molds on vegetables at the time of harvest. *Applied and Environmental Microbiology* **35**, 655–658.

YILDIZ, F. & WESTHOFF, D. 1981 Associative growth of lactic acid bacteria in cabbage juice. *Journal of Food Science* **46**, 962–964.

New Prospects and Problems in the Beverage Industry

F. W. BEECH AND R. R. DAVENPORT

*Department of Agriculture and Horticulture, University of Bristol,
Research Station, Long Ashton, Bristol, UK*

Contents

1. Introduction

The prospects for the beverage industry are determined by entrepreneurs, marketing and sales managers, so the techniques for making beverages are always changing to meet the demands of marketing and profitability. The new prospect always brings new problems that must be met by the microbiologist, chemist and engineer. Only those problems of a microbiological nature are considered here and the topic has been further restricted to cider making (Beech 1972; Beech & Carr 1977), with some reference to fruit juices, wines and beers. The quality and stability of the final cider is affected at every stage of the process from the apple to the packaged product (Fig. 1). Even starting with a mixed microflora, clean, fresh-flavoured ciders can be produced consistently using a combination of sound fruit, optimum use of sulphur dioxide (SO_2) (Beech *et al.* 1979), control of pH, exclusion of air, complete fermentation of sugars, good hygiene and large modern fermentation and storage tanks.

FOOD MICROBIOLOGY
ISBN 0 12 589670 0

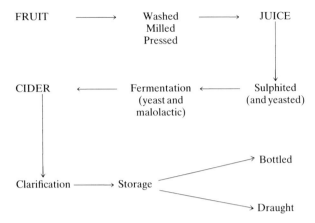

Fig. 1. Flow chart for cider-making.

2. The Orchard as an Ecosystem

In the late 1960s, realizing that their supplies of bitter-sweet cider apples rich in phenolics (Lea & Arnold 1978) were in jeopardy, cider makers planted some 1200 hectares of virus-free trees on dwarfing rootstocks, at a planting density of *ca.* 200/acre instead of the 30/acre used in the traditional orchard (Anon. 1979). The summer of 1976 threw most of the trees into biennial bearing, with massive crops thereafter in alternate years. In such years there is insufficient tank capacity to ferment all the juice and most of it is concentrated to provide raw material against future shortages.

Like all areas of monoculture, each orchard becomes an ecosystem of birds, animals, insects, plants and micro-organisms (Davenport 1976a). The distribution of micro-organisms is related to the time of year, movement of insects, seed-heads, etc. and is affected by humidity, light and chemical sprays (Andrews & Kenerley 1979). The process is complex but is now well documented (Last & Deighton 1965; Davenport 1968, 1976b; Beech & Davenport 1970). Some time after the establishment of an orchard, two groups of micro-organisms can be detected—residents and transients. The residents have certain properties in common: pigment to resist u.v. light, and ability to resist dehydration, to utilize the wide range of nutrients in low concentrations in leaf exudates, and to synthesize and excrete vitamins. Residents include sporing yeasts, such as *Saccharomyces* spp. that colonize mummified fruit, soil bacteria such as *Zymomonas* spp., and the cider spoilage yeast, *Saccharomycodes ludwigii*, that can be isolated from insect frass in nearby oak trees. Transients gradualy die out because they fail to establish themselves in suitable ecological niches.

The microflora, in some instances, has changed as a result of legislation against water pollution. It had been assumed that as the fruit was not eaten directly, slurries of animal wastes could be discharged into these orchards with impunity. This assumption was confounded when bottles of pear juice of pH 4·72, pasteurized at 68°C for 20 min, began to explode after 5 weeks in storage. The juice contained soil-borne organisms with heat-resistant spores. A *Clostridium* sp. producing gas, much butyric and some acetic acid, was isolated from the juice and the sprayed orchard in which the fruit had been gathered (Goverd 1977). Fortunately it is no longer permitted to discharge animal slurries indiscriminately; they must be disposed of as recommended by MAFF/ADAS and all producers of apply and pear juice are advised to decrease the pH value of their juice to less than 3·8, by blending or adding a permitted food acid, before pasteurization or hot filling.

3. Apple Harvesting

Following the reduction in the number of cider apple cultivars being planted from many hundreds to about ten, the harvest season has shortened from four months to less than two. Consequently the crop must be harvested mechanically, resulting in sound fruit being mixed indiscriminately with damaged, mouldy and mummified apples. A proportion of the fruit falls naturally; the rest must be shaken from the trees before harvesting with a pick-up reel.

Harvesting cannot be considered in isolation since the grass sward plays an important role. The trees in the young orchard are planted in herbicide-treated strips with grassed alleyways between the rows. Any fruit falling on to the strip is blown or brushed into the alleyways for picking up, becoming coated with dust or mud containing enterobacteria (Goverd et al. 1979) and spore-forming bacteria able to survive the whole process of cider-making. This problem will remain until the trees reach optimum size when the orchard can be grassed down completely. Apples are difficult to remove from long lush grass, so the type of grass is important. Fescue swards are now being tried because they are slow growing, provide a firm base for heavy machinery and produce a mat that keeps the fruit away from the soil. Grass suppressants reduce the amount of mowing and prevent excessive growth when mowing must be stopped once the fruit starts to fall. Sheep, an alternative to mowing, have to be removed earlier, so that by late November, without suppressants, grass can be very long. Fescue again has an advantage since it survives treatment with suppressants better than coarse grasses.

The fruit on the tree has relatively few organisms on the surface and in the interior (Samish et al. 1963), but their numbers and the complexity of the flora increase at each stage from harvesting onwards. The trailer loads of fruit are dropped on to a concrete base, then elevated and dropped into a lorry, transported, dropped again into a silo and finally transported in water flumes to the mill. Flume water becomes heavily contaminated with spore-forming yeasts and bacteria (Wolford & Berry 1948) and needs straining and chlorinating continuously (Goverd 1977).

Any delay from harvest to milling allows the growth of fungal soft rots in the fruit; this increases the content of oxidizing enzymes and, more importantly, allows a secondary invasion of acetic acid bacteria in the main (Corison *et al.* 1979), and production by them of sulphite-binding compounds such as 5-oxo-fructose and oxo-acids (Table 1). If rots exceed 10%, control of the juice microflora by legally permitted SO_2 levels is impossible. Drosophila flies breed rapidly on rotting fruit and are carriers of acetic acid bacteria. Infection with penicillia and aspergilli sometimes leads to the development of the mycotoxin patulin in the fruit and the greater likelihood of mould growth in bottles of pasteurized apple juice. Fortunately patulin is destroyed during yeast fermentation (Burroughs 1977; Stinson *et al.* 1978).

The production of cider apples has been described in considerable detail to illustrate how a large investment in orcharding has led to new microbiological problems.

TABLE 1

Sulphite-binding compounds produced by bacteria

Sulphite-binding compound	Bacteria producing compounds			
	Gluconobacter spp.	*Chromobacterium* spp.	*Pseudomonas* spp.	*Acetobacter melanogenum*
Dihydroxyacetone	+	−	−	−
5-Oxofructose	+	−	−	−
Acids				
2-Oxogalactonic	+	−	−	−
5-Oxogluconic	+	+	−	−
2,5-Dioxogluconic	+	+	+	−
2-Oxogluconic	+	−	+	−
2-Oxo-3-deoxygluconic	+	−	+	−
2-Oxo-3-deoxygalactonic	+	−	+	−
5-Oxo-4-deoxymucic	+	−	+	−
2-Oxoglutaric	−	−	−	+

+, compound produced.
−, compound not produced.

4. Juice Extraction

Apples are ground to pulp and the juice extracted in one or more of a variety of presses (Bielig 1973), none of which is easy to keep clean, and they can be reservoirs of fermenting yeasts.

Depending on the effectiveness of the cleaning procedures, the juice contains a great variety of organisms. Juice is not sterilized, but the development of spoilage organisms is restricted by using SO_2 which, with its additional control of oxidation (Burroughs 1974), yields a clean, fresh, fruity flavoured cider. The amount of SO_2 added is varied according to juice pH, increasing from 75 mg/1 at pH 3·0 to 150 mg/1 at pH 3·8. Juices above pH 3·8, including all from bitter-sweet cider apples, must be blended with acid juice or be acidified with DL-malic acid before treating with SO_2. Next morning, when the antimicrobial effect of SO_2 on all but organisms capable of fermentation is complete (Carr *et al.* 1976) and the equilibration of sulphite ions and sulphite-binding compounds is largely complete, the amount of free SO_2 remaining is measured. Residual SO_2 values should lie between the two curves in Fig. 2. Values falling below the lower curve indicate the presence of excessive amounts of sulphite-binding compounds, other than the normal glucose, xylose, 1-xylosone, arabinose, and galacturonic acid (Burroughs & Sparks 1973; Table 2), and the need for a further addition of SO_2.

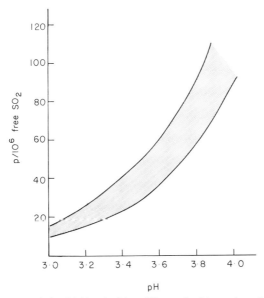

Fig. 2. The minimum antimicrobial level of free SO_2 required in apple and grape juice. The amount of free SO_2 that should be present in a juice after being left standing overnight is shown by the shaded area of the graph.

TABLE 2

Sulphite-binding compounds found naturally in apple juices and ciders

Sulphite-binding compound	Source
Glucose Arabinose	} Juice
Xylosone	Ascorbic acid
Acetaldehyde	Oxidation and yeasts
D-Galacturonic acid D-Glucuronic acid	} Pectin
Pyruvic acid 2-Oxoglutaric acid	} Yeasts

5. Fermentation

The correctly sulphited juice is inoculated with fresh yeast from a pure culture generator; there is no recycling as in brewing. Dried yeast can be used (Reed & Chen 1978) but may not have sufficient reserve of sterols (Aries & Kirsop 1978; Rose 1978) for the complete fermentation of high gravity juices in deep tanks (Whitworth 1978). The yeast chosen should have properties suitable for the process (Rankine 1977), and in cider-making this means it should have good fermentation speed even at relatively low temperature; should settle out rapidly; be alcohol- and SO_2-tolerant; should produce polygalacturonase which converts pectin, demethylated by the fruit enzyme, to galacturonic acid (Pollard & Kieser 1959); have a low oxygen requirement; be a low producer of sulphite-binding compounds (α-oxo-glutarate, pyruvate and acetaldehyde); and should not produce hydrogen sulphide or acetic acid. Sulphite is not formed in cider fermentations as in some wines, because of lower concentrations of growth factors (Eschenbruch & Bonish 1976).

Little attempt has been made to exploit biochemical research on the factors controlling the formation and metabolism of desirable components such as malate, succinate or lactate, or of aroma compounds during fermentation (Beech *et al.* 1968; Williams 1974; Dandt & Ough 1976; Suomalainen & Lehtonen 1979). Such application may have been thought unprofitable because cider and wines are not made by pure culture fermentations (Beech 1969). However, recently a hybrid between a killer and a fermenting yeast, with all the useful fermentation qualities of the latter, suppressed the growth of 'wild' *Saccharomyces* yeasts in a wine fermentation, while yeasts of other genera were suppressed by juice sulphiting (Hara *et al.* 1980).

Unlike in brewing, control of fermentation speed in ciders is not attempted (Ough & Groat 1978) but thiamin and ammonium salts are usually added to prevent premature cessation of fermentation, particularly of diluted concentrates. Concentrated juices whether produced here or abroad are stable at 70° Brix and keep for up to 3 years under cool storage conditions (Daepp 1973). However, unless the concentration plant is kept scrupulously clean, concentrates can become contaminated with osmophilic and osmotolerant yeasts and eventually ferment (Sand *et al.* 1977; Sand 1980). Concentrates, besides requiring less tank capacity, allow fermentations to be started throughout the year and produce ciders with more standard flavours.

6. Storage

After fermentation the ciders are clarified and stored in large, filled air-tight tanks. Now, with year-round fermentations, ciders receive relatively little storage and, while fresh and fruity in flavour, no longer have a mature taste, although that has never been defined in objective terms.

Ciders made from freshly pressed juice and stored from one season to the next usually underwent malolactic fermentation when the temperature rose above 15°C. It is not easy to induce such fermentation when required since there has to be compatability between the strains of fermenting yeasts and the species of lactic acid bacteria (Carr *et al.* 1972; Lafon-Lafourcade 1973). The micro-aerophilic lactic acid bacteria responsible include homo- and hetero-fermenting rods and cocci (*Lactobacillus plantarum*, *L. mali*, *L. collinoides* and *Leuconostoc* spp.) which convert L-malic acid into lactic acid and carbon dioxide, halving the acidity by the loss of one of the two COOH groups. The fermentation is a nuisance if it occurs in most bottled table wines but it is utilized in making special petillant Portuguese wines. Some perry pears also contain citric acid which is metabolized by leuconostocs after all malate has been metabolized, producing acetic acid, acetoin and CO_2. This spoils the flavour of the perry, as does the production in ciders of catechol and acetic acid from quinic and shikimic acids and of ethylcatechol from chlorogenic acid (Carr & Whiting 1971). Mayer & Pause (1973) claim that *Pediococcus cerevisiae* forms biogenic amines in stored wines (see also Woidich *et al.* 1980) but this activity has not been tested in ciders. Ropiness, due to polysaccharide formation by *L. collinoides* (Carr & Davies 1972) and *P. cerevisiae* (Carr 1970) is no longer a problem since the introduction of adequate sulphiting of juice.

Heterofermentative lactic acid bacteria and yeasts of the genus *Bret-*

tanomyces grow in ciders and wines made with ineffective sulphiting and with access of air, producing alkyl-substituted Δ'-piperideine, the source of the so-called 'mousy' taint (Tucknott 1974). *Brettanomyces* spp. are an essential part of the microflora of lambic beer (Spalpen *et al.* 1978) but with adequate thiamin and pediococci present in the beers, produce pleasant fruity, ester-like aromas instead.

Acetic acid bacteria (Passmore & Carr 1975) and film yeasts develop on the surface of inadequately sulphited ciders exposed to air during storage. The former produces acetic acid from ethanol and from malate while the yeasts convert them to carbon dioxide, water and acetaldehyde. The bacteria will also convert fructose and any sorbitol to sorbose and eventually to the sulphite-binding 5-oxo-fructose (Carr & Whiting 1971).

Formerly much naturally sweet cider was produced by terminating the fermentation prematurely. If also of high pH, it was often attacked by the sulphite-resistant bacterium *Zymomonas anaerobia* (Millis 1956; Dadds *et al.* 1973), converting sugars to ethanol and carbon dioxide by the Entner–Douderoff pathway, and acetaldehyde. The ciders acquired an unpleasant sweet pungent aroma and turned milky by reaction between acetaldehyde and polyphenols. The disorder was eliminated by allowing fermentation to go to completion and adding the required sweetening just before bottling. In recent years *Z. anaerobia* has caused spoilage in beers where it produces both acetaldehyde and hydrogen sulphide. Production of H_2S by this organism and some strains of fermenting yeast (Eschenbruch *et al.* 1978) is stimulated by sulphate (Anderson & Howard 1974) and a deficiency of pantothenate.

7. Packaging

As required, ciders are blended, treated with sufficient SO_2 to combine with any free sulphite-binding compounds and to leave up to 50 mg/l in free form to combine with oxygen, either dissolved or in the headspace (Table 3). Some ciders are carbonated artificially beforehand, a limited number by fermentation in a pressure tank. Carbonation on its own (Zschaler 1979), or in combination with benzoic acid and pH adjustment (Back 1981), as in soft drinks (Perigo *et al.* 1964), can inhibit the growth of certain yeasts and bacteria (Witter *et al.* 1959).

After clarification, ciders may be sterilized before packaging using filter sheets (EK or EKS grade) made from cellulose or asbestos, sometimes in combination with membranes, or flash-pasteurized, both processes being followed by aseptic filling (Reintjes 1980). Hot filling is sometimes used or the container may be pasteurized after filling and capping. These conditions

TABLE 3

Average changes in sulphur dioxide and oxygen in 13 bottled white wines

Storage period (months)	Concentration of SO_2 (mg/l)		Concentration of dissolved O_2 (mg/l)
	Free	Total	
0	31	137	3·4
1	22	129	1·3
2	20	126	0·3
10	15	117	0·1

Data from Burroughs (1980).

apply to most beverages, so examples of spoilage will be given from several sources.

It is axiomatic that organisms will colonize any ecosystem that selects those capable of survival (Ayanaba & Alexander 1972; Sand 1976/77; De Freitas & Frederickson 1978). Thus, processes using heat select thermotolerant organisms and/or spore formers (Put *et al.* 1976) while chilled systems select psychrophilic non-spore formers (Morita 1975). Available water, nutrients, oxygen and preservatives provide equally effective selection pressures.

Amongst the services in a factory, water (Thorpe & Everton 1968) can be a source of yeasts and bacteria (Put *et al.* 1972) through contamination of bore holes, or colonization of the cooling or rinsing water supply pipes from an initial infection. Damp rubber hoses harbour bacteria, while plastic dispense pipes, with slightly rough inner surfaces, support a variety of organisms, mainly yeasts (Harper 1980/81). Fermenting yeasts can adapt to exist in lubricating oils and can infect the product from aerosols generated by moving machinery; these yeasts are easily missed in 'trouble-shooting' since it takes about ten days to isolate them from oil on conventional media.

The basic raw material can also provide the initial source of infection, with *Zygosaccharomyces* spp. (Fig. 4) (Warth 1977) occurring in imported citrus concentrates and wines (Rankine & Pilone 1973), or *Saccharomycodes ludwigii* surviving cider fermentations. Bottling more than one beverage in the same plant can cause cross-contamination. The problem is compounded in mixed exotic beverages (Charley 1964), when one component is of high pH or is rich in nutrients or is heat labile, such as egg-yolk, nut-meat, cocoa, vegetable extracts, protein (Luttrell *et al.* 1981), milk (Anon. 1973) or tropical fruit pulps (Bilenker & Dunn 1960), or if the product contains a thickening agent.

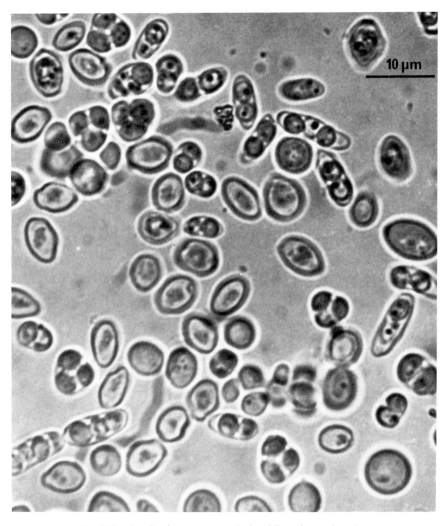

Fig. 3. Sporing *Saccharomyces* spp. isolated from dust on hot pipes.

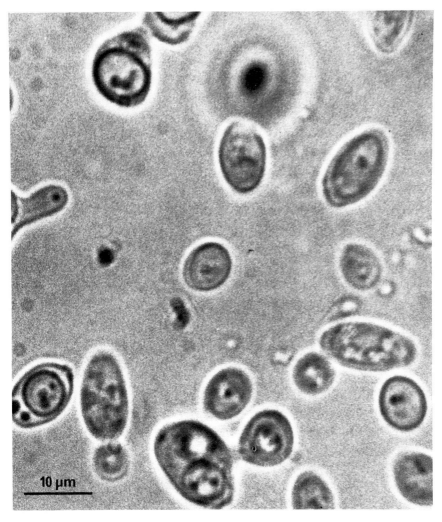

Fig. 4. *Zygosaccharomyces bailii* showing conjugation tube and ascospores.

Equipment must be designed to facilitate sterilization and maintenance in a sterile condition (Jowitt 1980). Non-standard modifications to a filling machine have allowed droplets of the product to accumulate, with subsequent growth of organisms in it and drip transfer to later bottlings. The growth of yeasts in pressure gauges and dead ends of sterilizing filters is a well-known phenomenon, as is the need to avoid sudden pressure surges, which used to disturb the seating of some early models of membrane filters allowing non-sterile product to pass momentarily. Hence it is essential to arrange for continuous sampling of the product immediately before bottling and to ensure that the sampling equipment does not become blocked at any time during a bottling run (Pilone 1977).

Heat-sterilizing systems also need monitoring, to check the accuracy of temperature recorders and the integrity of by-pass valves under sudden pressure changes (Cowland 1975), to ensure freedom from pin-holes in the plates and that the machine is operated at its rated capacity, and even to check for the effect of juice components on the thermal resistance of spoilage yeasts (Juven et al. 1978). Problems with batch pasteurization occur when non-chlorinated, contaminated cooling water is sucked in momentarily before the bottle cap seats properly. Capping machines can be potent sources of infection, those handling corks requiring frequent spraying with a mist of sanitizer during production runs. Sanitizers suitable for sterilizing bottling equipment have been described by Barrett (1979); they must be tested against active spoilage organisms isolated from the plant rather than against 'museum' strains.

Changing an existing process can allow growth of a new group of organisms. Reducing oxygen levels in lager for longer flavour shelf-life allowed the development of hazes due to an anaerobic bacterium described variously as Bacteroides or Pectinatus (Lee et al. 1980). The organism may have been present previously but not detected because samples on isolation media were not incubated anaerobically. Attempts to isolate organisms previously exposed to a hostile environment, e.g. low pH or high alcohol, may fail because the cells may be stressed and not grow readily on conventional media. For example, Z. bailii may take up to 14 days to grow in conventional media unsupplemented with a proportion of the product in which it was originally present.

Even when organisms have grown on media, a decision still has to be made about whether they were the organisms spoiling the product (Zaake 1979), indicators of inadequate sanitation, or purely adventitious and selected by the choice of unsuitable media.

With production running at thousands of bottles, it must be questioned whether or not classical microbiological procedures are appropriate. It is clearly undesirable to hold production quantities of beverages in quarantine for 3 days or more until pronounced sterile. Rapid, automated counting methods may be inadequate (Southern 1979), and predictive (Hachigian 1978) or instrumental methods (Bascomb 1981) are being sought, sometimes in combination with ATP (Hysert *et al.* 1976) or immunofluorescence analyses (Hammond & Jones 1979). Undoubtedly these will be brought into use as soon as they have been fully developed. It is hoped that a combination of such methods with in-line microprocessors will enable the bottling line to be monitored continuously and faults indicated as soon as they occur.

Until then it is both a wonder and a tribute to managers, quality control staff, engineers and equipment manufacturers that so many beverages are produced annually in such vast quantities with only occasional, albeit expensive, spoilage problems.

8. References

ANDERSON, R. J. & HOWARD, G. A. 1974 The production of hydrogen sulphide by yeast and by *Zymomonas anaerobia*. *Journal of the Institute of Brewing* **80**, 245–251.

ANDREWS, J. H. & KENERLEY, C. M. 1979 The effects of a pesticide program on microbial populations from apple leaf litter. *Canadian Journal of Microbiology* **25**, 1331–1344.

ANON. 1973 Natural fruit juices and milk can be mixed without curdling using a suitable stabiliser. *Chemical Processing* **19**, 15, 17.

ANON. 1979 *The Modern Cider Apple Orchard*. Booklet 2072, pp. 13. Pinner, Middlesex: Ministry of Agriculture, Fisheries and Food Publications.

ARIES, V. & KIRSOP, B. 1978 Sterol biosynthesis by strains of *Saccharomyces cerevisiae* in the presence and absence of dissolved oxygen. *Journal of the Institute of Brewing* **84**, 118–122.

AYANABA, A. & ALEXANDER, M. 1972 Changes in nutritional types in bacterial successions. *Canadian Journal of Microbiology* **18**, 1427–1430.

BACK, W. 1981 Microorganisms causing spoilage in fruit juices, fruit nectars and sweet soft drinks. *Brauwelt* **121**, 43–48.

BARRETT, M. 1979 Detergents and sterilants in the brewery. *Brewers' Guardian* **108**, 35, 37–40.

BASCOMB, S. 1981 Application of automation to the general and specific detection of bacteria. *Laboratory Practice* **30**, 461–464.

BEECH, F. W. 1969 The inter-relationships between yeast strain and fermentation conditions in the production of cider from apple juice. *Antonie van Leeuwenhoek* **35s**, F11–12.

BEECH, F. W. 1972 Cider making and cider research: a review. *Journal of the Institute of Brewing* **78**, 477–491.

BEECH, F. W. & CARR, J. G. 1977 Cider and perry. In *Economic Microbiology* Vol. 1, ed. Rose, A. H. pp. 139–313. London & New York: Academic Press.

BEECH, F. W. & DAVENPORT, R. R. 1970 The role of yeasts in cider-making. In *The Yeasts* Vol. 3, ed. Rose, A. H. & Harrison, J. S. London & New York: Academic Press.

BEECH, F. W., POLLARD, A. & WILLIAMS, A. A. 1968 Effect of juice treatments on higher alcohol production in ciders. *Annual Report Long Ashton Research Station 1967*, 72–73.

BEECH, F. W., BURROUGHS, L. F., TIMBERLAKE, C. F. & WHITING, G. C. 1979 Current progress in the chemical aspects and antimicrobial effects of sulphur dioxide (SO_2). *Bulletin de l'O.I.V.* **52**, 1001–1022.

BIELIG, J. H. 1973 *Fruit Juice Processing*. F.A.O. Agricultural Services Bulletin No. 13, pp. 103. London: HMSO.

BILENKER, E. N. & DUNN, C. G. 1960 Growth of food spoilage bacteria in banana purée. *Food Research* **25**, 309–320.

BURROUGHS, L. F. 1974 Browning control in fruit juices and wines. *Chemistry and Industry* No. 18, 718–720.

BURROUGHS, L. F. 1977 Stability of patulin to sulphur dioxide and to yeast fermentation. *Journal of the Association of Official Analytical Chemists* **60**, 100–103.

BURROUGHS, L. F. 1979. Losses of SO$_2$ during processing of white wines. *Annual Report Long Ashton Research Station 1978*, 173–174.

BURROUGHS, L. F. & SPARKS, A. H. 1973 Sulphite-binding power of wines and ciders. *Journal of the Science of Food and Agriculture* **24**, 187–217.

CARR, J. G. 1970 Tetrad-forming cocci in ciders. *Journal of Applied Bacteriology* **33**, 371–379.

CARR, J. G. & DAVIES, P. A. 1972 The ecology and classification of strains of *Lactobacillus collinoides* nov. spec. *Journal of Applied Bacteriology* **35**, 463–471.

CARR, J. G. & WHITING, G. C. 1971 Microbiological aspects of production and spoilage of cider. *Journal of Applied Bacteriology* **34**, 81–93.

CARR, J. G., COGGINS, R. A., DAVIES, P. A. & WHITING, G. C. 1972 The effect of yeasts on malo-lactic fermentation. *Annual Report Long Ashton Research Station 1971*, 184–186.

CARR, J. G., DAVIES, P. A. & SPARKS, A. H. 1976 The toxicity of sulphur dioxide towards certain lactic acid bacteria from fermented apple juice. *Journal of Applied Bacteriology* **40**, 201–212.

CHARLEY, V. L. S. 1964 Fruit juice products—the search for variety. *Report of the Scientific and Technical Commission of the International Fruit Juice Producers* **5**, 43–52. Zurich: Juris-Verlag.

CORISON, C. A., OUGH, C. S., BERG, H. W. & NELSON, K. E. 1979. Must acetic acid and ethyl acetate as mould and rot indicators in grapes. *American Journal of Enology and Viticulture* **30**, 130–134.

COWLAND, T. W. 1975 A source of yeast infection in a bottling line. In *Microbial Spoilage of Fermented Fruit Beverages* ed. Davenport, R. R. & Hammond, S. M. pp. 13–14. Bristol: Long Ashton Research Station.

DADDS, M. J. S., MARTIN, P. A. & CARR, J. G. 1973 The doubtful status of the species *Zymomonas anaerobia* and *Z. mobilis. Journal of Applied Bacteriology* **36**, 531–539.

DAEPP, H. U. 1973 State of the technology of fruit juice concentration. *Report of the Scientific and Technical Commission of the International Federation of Fruit Juice Producers* **13**, 43–56. Zurich: Juris-Verlag.

DANDT, C. E. & OUGH, C. S. 1976 Higher alcohol formation as affected by yeast, temperature, sulphur dioxide and grape variety. *Revista Brasiliera de Tecnologia* **6**, 301–305.

DAVENPORT, R. R. 1968 *The origin of cider yeasts*. Membership Thesis, Institute of Biology, London.

DAVENPORT, R. R. 1976a Ecological concepts in studies on micro-organisms on aerial plant surfaces. In *Microbiology of Aerial Plant Surfaces* ed. Dickinson, C. H. & Preece, T. F. pp. 199–215. London & New York: Academic Press.

DAVENPORT, R. R. 1976b Distribution of yeasts and yeast-like organisms from aerial surfaces of developing apples and grapes. In *Microbiology of Aerial Plant Surfaces* ed. Dickinson, C. H. & Preece, T. F. pp. 325–359. London & New York: Academic Press.

DE FREITAS, M. J. & FREDERICKSON, A. G. 1978 Inhibition as a factor in the maintenance of the diversity of microbial ecosystems. *Journal of General Microbiology* **106**, 307–320.

ESCHENBRUCH, R. & BONISH, P. 1976 Production of sulphite and sulphide by low- and high-sulphite forming wine yeasts. *Archives of Microbiology* **107**, 299–302.

ESCHENBRUCH, R., BONISH, P. & FISHER, B. M. 1978 The production of H$_2$S by pure culture wine yeasts. *Vitis* **17**, 67–74.

GOVERD, K. A. 1977 'Non acid-tolerant' bacteria. *Annual Report Long Ashton Research Station 1976*, 136–138.

GOVERD, K. A., BEECH, F. W., HOBBS, R. P. & SHANNON, R. 1979 The occurrence and survival of coliforms and salmonellas in apple juice and cider. *Journal of Applied Bacteriology* **46**, 521–530.

HACHIGIAN, J. 1978 Computer simulation of partial spoilage data. *Journal of Food Science* **43**, 1741–1748.

HAMMOND, J. R. M. & JONES M. 1979 The immunofluorescent staining technique for the detection of wild yeasts—practical problems. *Journal of the Institute of Brewing* **85**, 26–30.

HARA, S., IIMURA, Y. & OTSUKA, K. 1980 Breeding of useful killer wine yeasts. *American Journal of Enology and Viticulture* **31**, 28–33.

HARPER, D. R. 1980/81 Microbial contamination of draught beer in public houses. *Process Biochemistry* **16**, 2–7.

HYSERT, D. W., KOVECSES, F. & MORRISON, N. M. 1976 A firefly bioluminescence ATP assay method for rapid detection and enumeration of brewery micro-organisms *Journal of the American Society of Brewing Chemists* **34**, 145–150.

JOWITT, R. (ed.) 1980 *Hygienic Design and Operation of Food Plant* pp. 292. Society of Chemical Industry. Chichester: Ellis Horwood Ltd.

JUVEN, B. J., KANNER, J. & WEISSLOWICZ, H. 1978 Influence of orange juice composition on the thermal resistance of spoilage yeasts. *Journal of Food Science* **43**, 1074–1076.

LAFON-LAFOURCADE, S. 1973 De la fermentescibilité malolactique des vins: interaction levures-bactéries. *Connaissance de la Vigne et du Vin* No. 3, 203–207.

LAST, F. T. & DEIGHTON, F. C. 1965 The non-parasitic microflora on the surfaces of living leaves. *Transactions of the British Mycological Society* **48**, 83–99.

LEA, A. G. H. & ARNOLD, G. M. 1978 The phenolics of ciders: bitterness and astringency. *Journal of the Science of Food and Agriculture* **29**, 478–483.

LEE, S. Y., MAYBEE, M. S., JANGAARD, N. O. & HORIUCHI, E. K. 1980 *Pectinatus*, a new genus of bacteria capable of growth in hopped beer. *Journal of the Institute of Brewing* **86**, 28–30.

LUTTRELL, W. R., WEI, L. S., NELSON, A. I. & STEINBERG, M. P. 1981 Cooked flavour in sterile Illinois soybean beverage. *Journal of Food Science* **46**, 373–376.

MAYER, K. & PAUSE, G. 1973 Non-volatile biogenic amines in wine. *Mitteilungen aus dem Gebiet der Lebensmitteluntersuchung und-Hygiene* **64**, 171–179.

MILLIS, N. F. 1956 A study of the cider-sickness bacillus—a new variety of *Zymomonas anaerobia*. *Journal of General Microbiology* **15**, 521–528.

MORITA, R. Y. 1975 Psychrophilic bacteria. *Bacteriological Reviews* **39**, 144–167.

OUGH, C. S. & GROAT, M. L. 1978 Particle nature, yeast strain and temperature interactions on the fermentation rates of grape juice. *Applied and Environmental Microbiology* **35**, 881–885.

PASSMORE, S. M. & CARR, J. G. 1975 The ecology of the acetic acid bacteria with particular reference to cider manufacture. *Journal of Applied Bacteriology* **38**, 151–158.

PERIGO, J. A., GIMBERT, B. L. & BASHFORD, T. E. 1964 The effect of carbonation, benzoic acid and pH on the growth rate of a soft drink spoilage yeast as determined by a turbidostatic continuous culture. *Journal of Applied Bacteriology* **27**, 315–332.

PILONE, G. J. 1977 Continuous monitoring of wine for yeast and bacteria. *American Journal of Enology and Viticulture* **28**, 250–251.

POLLARD, A. & KIESER, M. E. 1959 Pectin changes in cider fermentations. *Journal of the Science of Food and Agriculture* **10**, 253–260.

PUT, H. M. C., VAN DOREN, H., WARNER, W. R. & KRUISWIJK, J. TH. 1972 The mechanism of microbiological leaker spoilage of canned foods: a review. *Journal of Applied Bacteriology* **35**, 7–28.

PUT, H. M. C., DE JONG, J., SAND, F. E. M. J. & VAN GRINSVEN, A. M. 1976 Heat resistance studies on yeast species causing spoilage in soft drinks. *Journal of Applied Bacteriology* **40**, 135–152.

RANKINE, B. C. 1977 Modern developments in selection and use of pure yeast cultures for winemaking. *Australian Wine, Brewing and Spirit Review* **96**, 31–33.

RANKINE, B. C. & PILONE, D. A. 1973 *Saccharomyces bailii*, a resistant yeast causing serious spoilage of bottled table wine. *American Journal of Enology and Viticulture* **24**, 55–58.

REED, G. & CHEN, S. L. 1978 Evaluating commercial active dry wine yeasts by fermentation activity. *American Journal of Enology and Viticulture* **29**, 165–168.

REINTJES, H. J. 1980 Aseptic filling of fruit juices and concentrates in large bags. *Flüssiges Obst* **47**, 254–255.

ROSE, A. H. 1978 The role of oxygen in lipid metabolism and yeast activity during fermentation.

In *European Brewery Convention Monograph V. Fermentation and Storage Symposium*, *Zoeterwoude*. pp. 96–107.

SAMISH, Z., ETINGER-TULCZYNSKA, R. & BLICK, M. 1963 The microflora within the tissue of fruit and vegetables. *Journal of Food Science* 28, 259–266.

SAND, F. E. M. J. 1976/77 The refreshing drink as an ecosystem. *Bios* 7, 27–28; 8, 33–36.

SAND, F. E. M. J. 1980 *Zygosaccharomyces bailii*: an increasing danger for the beverage industries? *Brauwelt* 120, 418–425.

SAND, F. E. M. J., VAN DEN BROCK, W. C. M. & VAN GRINSVEN, A. M. 1977 Yeasts isolated from concentrated orange juice. *Proceedings of the International Special Symposium on Yeasts, Part I, Keszthely, Hungary*. pp. 121–122.

SOUTHERN, JR, P. M. 1979 New developments in automation and rapid methods in microbiology. *Food Technology* 33, 54–56.

SPALPEN, M., VAN OEVELEN, D. & VERACHTERT, H. 1978 Fatty acids and esters produced during the spontaneous fermentation of lambic and gueuze. *Journal of the Institute of Brewing* 84, 278–282.

STINSON, E. E., OSMAN, S. F., HUNTANEN, C. N. & BILLS, D. D. 1978 Disappearance of patulin during alcoholic fermentation. *Applied and Environmental Microbiology* 36, 620–622.

SUOMALAINEN, H. & LEHTONEN, M. 1979 The production of aroma compounds by yeast. *Journal of the Institute of Brewing* 85, 149–156.

THORPE, R. H. & EVERTON, J. R. 1968 Post-process sanitation in canneries. *Technical Manual No. 1*. pp. 187. Chipping Campden: Campden Food Preservation Research Association.

TUCKNOTT, O. G. 1974 Mousy taint in cider. *Annual Report Long Ashton Research Station 1973*, 159–160.

WARTH, A. D. 1977 Mechanism of resistance of *Saccharomyces bailii* to benzoic, sorbic and other weak acids used as food preservatives. *Journal of Applied Bacteriology* 43, 215–230.

WHITWORTH, C. 1978 Technological advances in high gravity fermentation. *European Brewing Convention Monograph V. Fermentation and Storage Symposium, Zoeterwoude*. pp. 155–164.

WILLIAMS, A. A. 1974 Flavour research and the cider industry. *Journal of the Institute of Brewing* 80, 455–470.

WITTER, L. D., BERRY, J. M. & FOLINAZZO, J. F. 1959 Viability of *Escherichia coli* and a spoilage yeast in carbonated beverages. *Food Research* 23, 133–142.

WOIDICH, H., PFANNHAUSER, W., BLAICHER, G. & PECHANEK, U. 1980 Investigations concerning biogenic amines in red and white wines. *Mitteilungen Klosterneuberg* 30, 27–31.

WOLFORD, E. R. & BERRY, J. A. 1948 Bacteriology of slime in a citrus-processing plant, with special reference to coliforms. *Food Research* 13, 340–346.

ZAAKE, S. 1979 Detection and significance of beverage-spoilage yeasts *Monatsschrift für Brauerei* 32, 350–356.

ZSCHALER, R. 1979 Influence of carbon dioxide on the microbiology of beverages. *Antonie van Leeuwenhoek* 45, 158.

Properties of and Prospects for Cultured Dairy Foods

K. M. SHAHANI AND B. A. FRIEND

Department of Food Science and Technology,
University of Nebraska, Lincoln, Nebraska, USA

Contents

1. Introduction

Milk has been used as food since antiquity although it is not known when prehistoric man actually domesticated the cow, sheep or goat for milk production. Archaeologists have found evidence in the Libyan desert that cows were used for milk as far back as 9000 BC (Pederson 1979). Fermentation occurred naturally in surplus milk set aside for later use, and this process served as a primitive method of food preservation. It was no doubt observed that in some cases the milk formed a smooth curd with a pleasing acid odour and flavour, while in other cases the milk became putrid and caused illness when consumed. The preparation of fermented, or cultured, milk products then became an art which was handed down from one generation to another.

Nearly every civilization has consumed cultured milk of one type or another and these products have been, and still are, of extreme importance in the nutrition of people throughout the world. As shown in Table 1, the highest per capita consumption of cultured products is in Finland and other Scandinavian countries. The least per capita consumption occurs in the United States of America, Ireland and the United Kingdom. Although the per capita consumption of milk has not increased in the USA during the past decade (Anon. 1978), the consumption of cultured milk products has increased 300-fold from 0·04 kg per person in 1966 to 1·2 kg per person in 1978.

FOOD MICROBIOLOGY
ISBN 0 12 589670 0

TABLE 1

Per capita consumption of cultured milks

	Per capita consumption (kg) in		
	------	------	------
Country	1966	1974	1978
Finland	12·0	35·2	34·4
Sweden	11·9	17·2	22·2
Denmark	2·8	11·6	16·0
Netherlands	12·9	13·2	15·6
Switzerland	5·5	10·4	13·2
France	4·2	7·3	8·0
Germany	3·4	8·5	7·8
USSR	—	7·0	6·8
Czechoslovakia	—	2·0	4·1
Poland	1·6	3·2	2·7
Japan	7·7	2·9	2·5
UK	0·4	1·5	1·9
Ireland	—	0·8	1·7
USA	0·04	0·7	1·2

Data from Anon. (1980).

Milk is a complex mixture consisting of an oil-in-water emulsion stabilized by phospholipids and protein adsorbed on the surface of the fat globules. It also contains the water-soluble B-vitamins (thiamin, riboflavin, pyridoxine, niacin, pantothenic acid, biotin and folic acid), vitamin C, the fat-soluble vitamins (A, D, E and K), salts, citrates, enzymes (lipase, catalase and phosphatase) and the carbohydrate lactose—all the requirements for the growth of the fastidious lactic acid bacteria (Pederson 1979).

In raw milk, the types of organisms which develop depend on the introduction of organisms into the milk, sanitary conditions, prevailing temperatures and other environmental factors. According to Pederson (1979), the lactic organisms most likely to grow first in milk will be *Streptococcus lactis* and *Strep. cremoris*; in warm climates, however, *Strep. thermophilus* will be the first to predominate. These organisms produce sufficient acid to prevent the growth of acid-intolerant spoilage organisms and to promote the growth of the more acid-tolerant lactic bacteria such as *Lactobacillus bulgaricus* and *L. casei*.

Milk of a variety of animal species, fermented by lactic cultures, is known by different names in different parts of the world. Though all the products listed in Table 2 are produced commercially, the best known of these in the Western world are yoghurt, acidophilus milk, kefir and koumiss. Few people recognize, however, that these products are prepared by bacterial and/or

TABLE 2

Some important cultured milk products

Product	Country	Type of fermentation	Organism
Yoghurt	USA	Moderate acid	*Streptococcus thermophilus* + *Lactobacillus bulgaricus*
Cultured buttermilk	USA	Moderate acid	*Strep. cremoris* *Strep. lactis* *Leuconostoc citrovorum*
Acidophilus milk	USA	High acid	*Lact. acidophilus*
Bioghurt	Europe	High acid	*Strep. lactis* *Lact. acidophilus*
Bulgarican milk	Europe	High acid	*Lact. bulgaricus*
Yakult	Japan	Moderate acid	*Lact. casei*
Dahi	India	Moderate acid	*Strep. lactis* *Strep. cremoris* *Strep. diacetylactis* *Leuconostoc* spp.
Leben	Egypt	High acid	Streptococci Lactobacilli Yeasts
Surati cheese	India	Milk acid	Streptococci
Srikhand	India	Sour and sweet	Dahi microbes
Kefir	Russia	High alcohol	Streptococci *Lact. caucasicus* *Leuconostoc* spp. Yeasts
Koumiss	Russia	High alcohol	*Lact. acidophilus* + *Lact. bulgaricus* + *Saccharomyces lactis*

yeast action and that the characteristic flavours and textures of these products are results of these fermentations.

According to legend, the method of preparing yoghurt was revealed to Abraham by an angel, and he owed his fecundity and longevity to this fermented food (Rosell 1932). Another cultured product of similar and ancient mysterious origin is kefir, which is also known as the 'champagne of milk' and the 'drink of the prophet'. In Russia, Tolstoy conscientiously consumed koumiss for treatment of his tuberculosis (Tolstoy 1971), while Metchnikoff (1908) proposed that the longevity of the Bulgarians was due in part to their ingestion of large quantities of bulgarican milk. Metchnikoff was perhaps the first person, however, to associate the action of specific micro-organisms with the nutritional and therapeutic properties of cultured products.

2. Nutritional Properties

The nutritional value of a food is dependent not only upon its nutritional content but also upon the availability, digestibility and assimilability of the nutrients. Cultured dairy products have the same caloric value as the milk from which they are made, yet are more nutritious because of their ease of digestion, higher concentration of enzymes and increased level of B-vitamins.

Yoghurt proteins have a higher *in vitro* digestibility (Breslaw & Kleyn 1973) and higher biological value (Rasic *et al*. 1971; Simhaee & Keshavarz 1974) than milk proteins. The free amino acid content of yoghurt is also higher than in milk, due to the heat treatment and proteolysis by the yoghurt bacteria (Rasic *et al*. 1971). Several other studies on the proteolytic and lipolytic activity of various lactic cultures (Poznanski *et al*. 1965; Chandan *et al*. 1969*a,b*) demonstrated the relative importance of the lactobacilli in the pre-hydrolysis of milk protein and the importance of streptococci and leuconostocs in the pre-hydrolysis of milk fat.

Hargrove & Alford (1978) also reported that rats had better weight gains and higher food efficiency when fed yoghurt as compared with milk or other fermented or acidified milks. The yoghurt diet maintained superiority even when the other diets were supplemented with vitamins.

Lactose, the carbohydrate in milk, causes nutritionally related problems in some persons who are deficient in the intestinal enzymes lactase or β-galactosidase. These people, therefore, must restrict their dietary intake of milk and milk products. Infants, children and adolescents require the calcium in milk for proper bone growth and development, and a severe restriction in the intake of dairy products thus has serious nutritional consequences. Therefore, it has been suggested that lactose-intolerant persons consume yoghurt and other cultured dairy products because up to half the lactose of the milk is hydrolysed by the lactic organisms during the manufacturing process (Gallagher *et al*. 1974). Such products may also supply additional quantities of the enzyme lactase, which is elaborated by the microbial starter cultures.

Goodenough & Kleyn (1976) noted that dietary yoghurt containing viable organisms greatly enhanced the lactase activity of rat intestinal mucosa. The effect was attributed to inherent lactase activity in the yoghurt, since no comparable effect was noted in rats fed either pasteurized yoghurt or simulated yoghurt containing sucrose or lactose.

Kilara & Shahani (1976) also demonstrated that the level of lactase increases during the incubation of yoghurt cultures. Subsequent *in vitro* digestion of the yoghurt released the enzyme from the culture cells. Since the lactase can be released from the yoghurt by digestion and as it appears to enhance intestinal lactase activity, it is possible that yoghurt could be of benefit to persons with lactose intolerance.

While many lactic organisms require B-vitamins for growth, several cultures are capable of synthesizing vitamins. The extent of biosynthesis or utilization of B-vitamins by the lactic organisms depends on the temperature, length of incubation and other processing parameters (Nilson *et al.* 1965). Reif *et al.* (1976) observed that cottage cheese starter-culture actively synthesized vitamin B_{12} and folic acid during the setting period, resulting in significant increases in these vitamins in the final product (Fig. 1). This is borne out in the B-vitamin composition data for milk and several cultured products shown in Table 3. In addition, cultured products also contain higher levels of B-vitamins, in particular folic acid, than their counterpart products prepared by direct acidification processes.

TABLE 3

B-vitamin content of milk and several cultured products

| Product | Content (μg/100 g) of | | | | | |
	Biotin	B_6	B_{12}	Folic acid	Niacin	Pantothenic acid
Milk	2·9–4·9	17–40	0·27–0·57	0·13–0·73	71–96	330–460
Yoghurt	4·0–5·1	—	0·35–0·52	3·9	130–141	280–381
Cultured cottage cheese	3·2	24–56	0·8–2·1	41	70–257	463
Cultured sour cream	2·6	16	0·3–0·4	10·8	11–67	320–360
Cheddar cheese	0·6–2·5	49–147	—	4·21	13–212	111–711

Adapted from Shahani & Chandan (1979).

3. Therapeutic Properties

In addition to the superior nutritional quality of cultured dairy products, there are other important attributes affecting the health of the consumer. Hypercholesteraemia is considered to be one of the major factors predisposing atherosclerotic heart disease. Supplementation of the diet with cultured dairy foods has demonstrated a favourable trend toward decreasing serum cholesterol levels. Hepner *et al.* (1979) reported a hypocholesteraemic effect

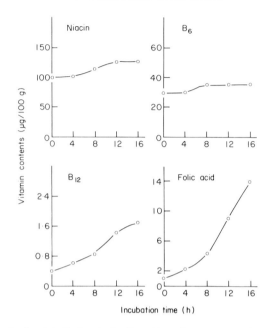

Fig. 1. Biosynthesis of certain B-vitamins by the cottage cheese starter culture.

in human subjects receiving a one-week dietary supplement of yoghurt. Recent investigations in our laboratory (Sinha 1978) demonstrated that unfermented milk containing viable *L. acidophilus* cells significantly reduced the serum cholesterol levels in rats (Table 4).

Several lactic organisms inherently produce natural antibiotics (Fig. 2). For example, *Strep. lactis* produces nisin, *L. bulgaricus* produces bulgarican and *L. acidophilus* produces a number of antibacterial compounds which have been called acidolin, acidophilin and lactocidin. Shahani *et al.* (1976, 1977) have studied the production of acidophilin and bulgarican by *L. acidophilus* and *L. bulgaricus*, respectively. As shown in Table 5, *L. acidophilus* DDS 1 and *L. bulgaricus* DDS 14 are active against a wide variety of Gram positive and Gram negative organisms, including pathogens and non-pathogens. Different strains vary greatly in their production of antibacterial compounds, and factors such as medium, pH, temperature and time of incubation have a pronounced effect on antibiotic production. Milk is a medium that contains essential growth factors, since these organisms failed to produce antibiotic when grown on other synthetic or semi-synthetic media (Shahani *et al.* 1976).

Acidophilin was isolated from milk medium using a combination of methanol–acetone extraction and silica gel or Sephadex chromatography

TABLE 4

*Effect of feeding milk, acidophilus milk and cholesterol on
serum cholesterol of Sprague-Dawley rats*

Treatment	Concentration of serum cholesterol (mg/100 ml) after (days)		
	10	20	30
Milk	146·8	145·7	141·8
Milk + 0·05% cholesterol	167·2	160·7	161·8
Acidophilus milk	132·9	118·3	102·5*
Acidophilus milk + 0·05% cholesterol	151·2	146·5	141·7

From Sinha (1978).
*Significantly different ($P < 0.05$) from other means.

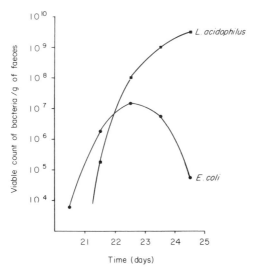

Fig. 2. Antibiotic effect of *Lactobacillus acidophilus* on *Escherichia coli* in the faecal flora of chickens.

TABLE 5

Inhibitory activities of Lactobacillus acidophilus *and*
L. bulgaricus *grown in milk medium, on nine test
organisms*

Test organism	L. acidophilus (DDS 1)	L. bulgaricus (DDS 14)
Bacillus subtilis	+ + +	+ + + +
Escherichia coli	+ +	+ + +
Proteus vulgaris	+ +	+ + +
Pseudomonas aeruginosa	+ +	+ + +
Ps. fluorescens	+ +	+ + +
Sarcina lutea	+ + +	+ + + +
Serratia marcescens	+ +	+ + +
Staphylococcus aureus	+ +	+ + + +
Streptococcus lactis	+ +	+ + +

+ + + + Very strong inhibition of test culture (zone of
    inhibition 9–12 mm wide).
+ + + Strong inhibition (zone 5–8 mm).
+ + Moderate inhibition (zone 3–4 mm).

(Shahani *et al.* 1977). The *in vitro* antibacterial activity of acidophilin is
shown in Table 6. Approximately 30–60 μg (0·2–0·4 units) of acidophilin per
ml of aqueous solution caused a 50% inhibition of the organisms, including
many common pathogens.

Bogdanov *et al.* (1962) were perhaps the first to observe that *L. bulgaricus*
possessed a potent antitumour activity. Later, they isolated three glycopeptides which have promising biological activity against Sarcoma 180 (Bogdanov *et al.* 1975). In a study in collaboration with the Sloan-Kettering
Institute for Cancer Research, several lactobacilli were tested against
Sarcoma 180, and *L. acidophilus* DDS 1 possessed definite antitumour
activity. Additional studies in our laboratory, using Ehrlich ascites tumour
cells, showed that extracts of *L. acidophilus* and yoghurt possessed antitumour activity. When colostrum fermented with *L. acidophilus* DDS 1 was
fed to mice, there was inhibition of tumour proliferation ranging from 13 to
38% (Bailey & Shahani 1976). Feeding *L. acidophilus* cells in a non-fer-
mented milk product (Sinha 1978) also resulted in a significant reduction in
the rate of proliferation of ascites tumour.

Reddy *et al.* (1973) and Farmer *et al.* (1975) found that mice given yoghurt
displayed from 28 to 35% inhibition of the ascites tumour (Fig. 3). Feeding
milk, lactose, lactic acid or cells killed by heat showed no inhibitory effect.
During separation of the antitumour compounds of yoghurt, Ayebo (1980)
demonstrated that the active principles were dialysable and after further
separation of the yoghurt dialysate by ion exchange chromatography, the
anionic fraction contained the active tumour principles (Table 7).

TABLE 6

In vitro *antibacterial activity of acidophilin*

Test organism	Strain	IC_{50} (μg/ml)
Bacillus subtilis	ATCC 6633	30
B. cereus	Difco 902072	29
B. stearothermophilus	ATCC 7954	43
Streptococcus faecalis	ATCC 8043	45
Strep. faecalis var. liquefaciens	ATCC 4532	42
Strep. lactis	NU C_{10}	30
Strep. lactis	NU C_2	38
Lactobacillus lactis	(LY-3 France)	40
L. casei	ATCC 7469	42
L. plantarum	ATCC 8014	60
L. leichmannii	ATCC 7830	59
L. leichmannii	ATCC 4797	58
Sarcina lutea	ATCC 9341	30
Serratia marcescens	NU	29
Proteus vulgaris	NU	30
Escherichia coli	NU	32
Salmonella typhosa	ATCC 167	30
Salm. schottmuelleri	ATCC 417	30
Shigella dysenteriae	ATCC 934	30
Sh. paradysenteriae	ATCC 9580	30
Pseudomonas fluorescens	NU	30
Ps. aeruginosa	(ear infection)	30
Ps. aeruginosa	(green diarrhoea)	60
Staphylococcus aureus	NU (coagulase +ve)	50
Staph. aureus	Phage 80/81	60
Klebsiella pneumoniae	ATCC 9997	60
Vibrio comma	ATCC 9459	30

From Shahani *et al*. (1977).

TABLE 7

Effect of inoculating yoghurt dialysate fractions on proliferation of ascites tumour in mice

Treatment	Tumour cells (10^6/mouse)	Inhibition (%)	Tumour cell DNA (μg/ml)	Inhibition (%)
Saline	46·7	—	521·9	—
Yoghurt dialysate	28·7*	38·5	321·1*	38·5
Cationic fraction	43·2	7·0	482·8	7·5
Anionic fraction	28·6*	38·7	319·8*	38·7
Neutral fraction	47·2	0	527·0	0

Adapted from Ayebo (1980).
*Significantly different ($P < 0.5$) from saline control.

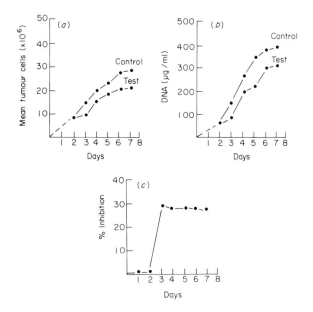

Fig. 3. Progression of Ehrlich ascites tumour *in vivo* as determined by (*a*) daily cell counts; (*b*) DNA determinations; (*c*) per cent inhibition of tumour cell proliferation. Test animals were fed yoghurt *ad libitum*.

4. Prospects

Although the art of preparing cultured dairy foods, such as cheese, yoghurt, kefir and koumiss, has been practised for ages, their science and technology as practised today are of modern origin. Advances made in sciences of microbiology, biochemistry and engineering, particularly during the past two or three decades, coupled with the progress achieved in industrial microbial technology, have led the way for the isolation and identification of more specific cultures and for the development of newer and a wider variety of cultured dairy products.

Historically, fermentation processes involved unpredictable and slow souring milk effected by mixed or 'unknown' cultures or 'back-slop'. However, modern technologies have helped the development and use of 'pure' or known cultures with predictable performance potentials. Use of specific lactic micro-organisms lead to predetermined or specific fermentation under rigorous conditions of pH, temperature and water activity to produce products of superior microbiological, chemical, physical, sanitary and nutritional qualities. Customarily, it was assumed that during the manufacture, the primary function of the culture(s) was to produce lactic acid. However, acid production is only one of the many changes augmented by lactic bacteria. As discussed previously, selected cultures can enhance the nutritional value of the milk products by improving their digestibility, assimilability and B-vitamin content. These cultures may also produce antibiotic or antimicrobial components, anticarcinogens and enzymes so that the ingestion of such fermented products may have considerable therapeutic value as well.

There is a growing interest in the nutritional and health-giving properties of cultured dairy foods. Until recently such claims were considered as having their origin only in folklore or hearsay, and were thought to lack scientific backing. Recent scientifically based claims discussed herein, and those reported by Ayebo & Shahani (1980), Deeth & Tamime (1981), Shahani & Ayebo (1980) and Speck & Katz (1980), appear to have contributed to the recent popularity and greater acceptability of fermented foods such as yoghurt and acidophilus. During the 12-year period 1966–1978 (Table 1), the highest per capita consumption of cultured milk in Finland increased from 12·0 to 34·4 kg, in the UK from 0·4 to 1·9 kg and in the USA from 0·04 to 1·2 kg. These and other figures in Table 1 indicate clearly the growing popularity and improved future prospects for cultured milk. Likewise, the per capita production and consumption of cheese(s) have grown notably in almost all the major countries of Europe and the Americas.

In the case of yoghurt particularly, the increased consumption has been ascribed to its image as a wholesome, high-protein, low-lactose, convenient, healthy, weight-reducing or low-fat food. Concomitantly, there have appeared on the market yoghurt-like or yoghurt-related products like acidophilus milk, bulgaricus milk, kefir, koumiss, etc. Many of these products are being made available in different forms—liquid, semi-solid and solid. The health food industry has contributed heavily to the market availability of cultured products and cultures in the form of whole fermented products as well as extracts and mixtures. The cultures and cultured-products industry has further witnessed the availability of the products as culture powders, capsules and tablets.

A cursory look at the immediate past performance of the fermented food industry suggests that its prospects are indeed bright in that the production and consumption of such products should continue to rise. Traditional cultured products are also expected to maintain their place in the food industry because of their organoleptic, textural, nutritional and therapeutic qualities. Additionally, the appearance of simulated and modified cultured products in the marketplace seem to hold promise, particularly for satisfying the fancies and needs of specific clientele. The past decade heralded the introduction of frozen yoghurt, dietary yoghurt, 'Sweet Acidophilus', acidophilus-yoghurt and yoghurt tablets. During recent times one finds fruit juices fortified with acidophilus, yoghurt salad dressing and yogolaise (yoghurt containing mayonnaise), acidophilus or yoghurt chelated with amino acids, minerals and vitamins, to name a selection.

It is important to keep in mind the considerable strain-to-strain variation among the lactic acid bacteria. Any nutritional or therapeutic benefits linked to the action of one particular strain may not necessarily apply to all other strains of the same organism. For example, screening studies in our laboratory showed that *L. acidophilus* DDS 1 produced the natural antibiotic acidophilin; but no other strain of *L. acidophilus* tested produced significant quantities of this metabolite (Shahani *et al.* 1976). Therefore, care must be taken in ascribing beneficial properties to strains of lactic acid bacteria which have not been properly evaluated in the laboratory. Secondly, commercial preparations of lactic cultures shown to be effective in the laboratory may not contain sufficient numbers of viable organisms to be of any benefit. Finally, while *L. acidophilus* and other lactic bacteria have been recognized to possess many beneficial properties, they also are rather fastidious in their growth and metabolic requirements. Antibiotics, alcohol and other antimicrobials seem to affect their growth rather markedly. Consequently, it is imperative that the ingestion of acidophilus and other beneficial lactic bacteria should not be preceded, coupled or followed by any such incompatible food ingredients

5. References

ANON. 1978 *Milk Facts*. Washington, D.C.: Milk Industry Foundation.

ANON. 1980 Bulletin 119. Brussels: International Dairy Foundation.

AYEBO, A. D. 1980 *Partial purification of anticarcinogenic compounds of yogurt*. Ph.D. Thesis, University of Nebraska, Lincoln.

AYEBO, A. D. & SHAHANI, K. M. 1980 Role of cultured products in the diet. *Cultured Dairy Products Journal* **15** (4), 21–29.

BAILEY, P. J. & SHAHANI, K. M. 1976 Inhibitory effect of acidophilus cultured colostrum and milk upon proliferation of Ascites tumor. *Proceedings of the 71st Annual Meeting of the American Dairy Science Association.* p. 41.

BOGDANOV, I. G., POPKHIRSTOV, P. & MARINOV, L. 1962 Anticancer effect of antibioticum bulgaricum on Sarcoma 180 and on the solid form of Ehrlich carcinoma. *Abstract VIII. International Cancer Congress.* pp. 364–365.

BOGDANOV, I. G., DALEV, P. G., GUREVICH, A. I., KOLOSOV, M. N., MALKOVA, V. P., PLEMYANNIKOVA, L. A. & SOROKINA, I. B. 1975 Antitumour glycopeptides from *Lactobacillus bulgaricus* cell wall. *Federation of European Biochemical Societies Letters* **57**, 259–261.

BRESLAW, E. S. & KLEYN, D. H. 1973 *In vitro* digestibility of protein in yogurt at various stages of processing. *Journal of Food Science* **38**, 1016–1021.

CHANDAN, R. C., ARGYLE, P. J. & JONES, N. 1969*a* Proteolytic activity of lactic cultures. *Journal of Dairy Science* **52**, 894 (Abstract).

CHANDAN, R. C., SEARLES, M. A. & FINCH, J. 1969b Lipase activity of lactic cultures. *Journal of Dairy Science* **52**, 894 (Abstract).

DEETH, H. C. & TAMIME, A. Y. 1981 Yogurt: nutritive and therapeutic aspects. *Journal of Food Protection* **44**, 78–86.

FARMER, R. E., SHAHANI, K. M. & REDDY, G. V. 1975 Inhibitory effect of yogurt components. *Journal of Dairy Science* **58**, 787 (Abstract).

GALLAGHER, C. R., MOLLESON, A. L. & CALDWELL, J. H. 1974 Lactose intolerance and fermented dairy products. *Journal of the American Dietetic Association* **65**, 418–419.

GOODENOUGH, E. R. & KLEYN, D. H. 1976 Influence of viable yogurt microflora on digestion of lactose by the rat. *Journal of Dairy Science* **59**, 601–606.

HARGROVE, R. E. & ALFORD, J. A. 1978 Growth rate and feed efficiency of rats fed yogurt and other fermented milks. *Journal of Dairy Science* **61**, 11–19.

HEPNER, G., FRIED, R., ST. JEOR, S., FUSETI, L. & MORIN, R. 1979 Hypocholesteremic effect of yogurt and milk. *American Journal of Clinical Nutrition* **32**, 19–24.

KILARA, A. & SHAHANI, K. M. 1976 Lactase activity of cultured and acidified dairy products. *Journal of Dairy Science* **59**, 2031–2035.

METCHNIKOFF, E. 1908 *The Prolongation of Life*. New York: G. P. Putnam & Sons.

NILSON, K. M., VAKIL, J. R. & SHAHANI, K. M. 1965 B-complex vitamin content of cheddar cheese. *Journal of Nutrition* **86**, 362–368.

PEDERSON, C. S. 1979 *Microbiology of Food Fermentations*. Westport, Connecticut: AVI Publishing Co.

POZNANSKI, S., LENOIR, J. & MOCQUOT, G. 1965 The proteolysis of casein under the action of certain bacterial endoenzymes. *Le Lait* **45**, 3–26.

RASIC, J., STOJSAVLJEVIC, T. & CURCIC, R. 1971 A study of the amino acids of yogurt. II. Amino acid content and biological value of protein of different kinds of yogurt. *Milchwissenschaft* **26**, 219–224.

REDDY, G. V., SHAHANI, K. M. & BANERJEE, M. R. 1973 Inhibitory effect of yogurt on Ehrlich ascites tumor-cell proliferation. *Journal of the National Cancer Institute* **50**, 815–817.

REIF, G. D., SHAHANI, K. M., VAKIL, J. R. & CROWE, L. K. 1976 Factors affecting the B-complex vitamin content of cottage cheese. *Journal of Dairy Science* **59**, 410–415.

ROSELL, J. M. 1932 Yoghurt and kefir in their relation to health and therapeutics. *Canadian Medical Association Journal* **26**, 341–345.

SHAHANI, K. M. & AYEBO, A. D. 1980 Role of dietary lactobacilli in gastrointestinal microecology. *American Journal of Clinical Nutrition* **33**, 2448–2457.

SHAHANI, K. M. & CHANDAN, R. C. 1979 Nutritional and healthful aspects of cultured and culture-containing dairy foods. *Journal of Dairy Science* **62**, 1685–1694.

SHAHANI, K. M., VAKIL, J. R. & KILARA, A. 1976 Natural antibiotic activity of *Lactobacillus acidophilus* and *bulgaricus*. I. Cultural conditions for the production of antibiosis. *Cultured Dairy Products Journal* **11** (4), 14–17.

SHAHANI, K. M., VAKIL, J. R. & KILARA, A. 1977 Natural antibiotic activity of *Lactobacillus acidophilus* and *bulgaricus*. II. Isolation of acidophilin from *L. acidophilus*. *Cultured Dairy Products Journal* **12** (12), 8–11.

SIMHAEE, I. & KESHAVARZ, E. 1974 Comparison of gross protein value and metabolizable energy of dried skim milk and dried yogurt. *Poultry Science* **53**, 184–191.

SINHA, D. K. 1978 *Development of unfermented acidophilus milk and its properties*. Ph.D. Thesis, University of Nebraska, Lincoln.

SPECK, M. L. & KATZ, R. S. 1980 Nutritive and health values of cultured dairy foods. ACDPI Status Paper. *Cultured Dairy Products Journal* **15** (4), 10–11.

TOLSTOY, T. 1971 *Tolstoy Remembered*. New York: McGraw Hill Book Co. Inc.

Fermented Fish and Meat Products:
The Present Position and Future Possibilities

INGER ERICHSEN

Swedish Meat Research Institute, Kavlinge, Sweden

Contents

1. Introduction

Traditional methods of food preservation were primarily dehydration by sun, wind or smoke, or by treatment with salt which extracted the tissue fluid by osmosis. Other processes have been developed to preserve food for consumption at a later date and they often confer desirable changes in flavour and thus increase the acceptability of the final product. One such process is fermentation, also one of the oldest food preservation methods known to man.

A successful fermentation is one in which a specific microflora has been encouraged to develop in a preferred direction by applying a system of physical, chemical, biochemical and environmental constraints to prevent the growth of all undesirable micro-organisms. Fermentation also implies transformation of organic substances into simpler compounds by the action

FOOD MICROBIOLOGY
ISBN 0 12 589670 0

of enzymes and micro-organisms. In this process most of the nutritive value of the raw material is retained.

Besides helping to preserve the food the fermentation process adds to flavour and texture and makes available a greater variety of food. It can also generate new food components such as vitamins and essential amino acids not present in the original food, thus improving its nutritive value.

2. Fermented Fish Products

A. Tropical types

Fermented foods are important dietary components in many parts of the world especially in South-East Asia, the Near East and in Africa. Millions of people in these areas have remarkably low amounts of total protein in their diets and the animal protein content is even lower.

One of the most important sources of animal protein in these parts of the world is fish, both freshwater and saltwater varieties. Under tropical storage conditions the quality of fresh fish deteriorates rapidly. Many processes have therefore been developed to preserve the fish for consumption at a later date. Much of the fish used for traditional fermentation processes are non-commercial species for which there is little consumer demand. Thus the processes are also important because they make use of fish which would otherwise not be used for human consumption.

All fermentation procedures developed have been subject to local variations and customs and to the fish species used, and this has resulted in a large variety of fermented fish products. Three main types of product have been described (van Veen 1965): fish sauce; fish paste; whole fish, salted and often partially dried.

The manufacturing procedure is as follows: the fish is packed with salt in earthenware pots which are tightly sealed and buried in the ground where they are left for several months during which time an anaerobic fermentation takes place.

In all cases the fermentation process is brought about by autolytic enzymes from the fish itself, the most active being from the viscera, and from micro-organisms, in the presence of high concentrations of salt (Uyenco *et al.* 1953). The microbial activity seems to be essential for the production of characteristic odours and flavours (Nagao & Kimura 1951; Nagasaki & Yamamoto 1954*b*; Zenitani 1955). The precise nature of the final product depends largely on the extent to which dehydration and fermentations are allowed to proceed.

The fermented products described are standard items in the diets in the areas in which they are produced and are used chiefly as condiments for flavouring a monotonous rice diet. The products contain high salt concentrations, usually 20–25% in the finished product, and therefore cannot be consumed in large quantities. If consumption is to be increased it is essential to reduce the concentration of the salt. Because of the high concentration of salt the fermented products from these areas rarely cause food poisoning.

A number of factors have contributed to the success of these fermented products: (a) the methods of production are inexpensive and there is little waste; (b) the techniques are simple and well-understood locally and the products remain stable for long periods; (c) tropical ambient temperatures are particularly suitable for the growth of the micro-organisms responsible for production of the desired product; (d) the products are well established and readily accepted by people of all social classes; (e) no sophisticated packaging and distribution facilities are required.

For further details of existing types of exotic fermented products, of which there are many, and their manufacturing procedures, refer to the excellent review by Mackie et al. (1971) on fermented fish products.

B. European types

The types of fermented fish products produced in tropical areas are of no importance in Europe where other types have been developed.

Salting and pickling of herring is one of the oldest established processes which is still of importance in many European countries. The production process may differ from one country to another as to the species of fish used, the degree of salting, the presence of visceral enzymes, the presence of carbohydrates, maturing temperature and duration of the fermentation process, but the process is essentially the same. This review will therefore be confined to some types of fermented fish products found in Scandinavia.

The most common type of salted, fermented fish products found in the Scandinavian countries are titbits made from herring, and anchovy made from sprats.

The herring used for the production of titbits is usually caught off the coast of Iceland during the summer months. The fish is packed at sea or on land in wooden barrels with a mixture of salt and sugar and allowed to mature in the brine formed in the barrels for several months. At the end of the maturing period the pH value of the brine is 5·6–6·0 and the salt concentration about 13%.

The maturing process is essentially one of protein degradation to peptides and amino acids, with the development of pleasing flavours and odours. The curing is brought about by proteolytic enzymes originating mainly from the fish (Alm 1965).

Bacteria are probably of prime importance in the development of flavours but contribute little to the proteolysis (Omland 1955; Shewan & Hobbs 1967). It has been suggested by some workers that the cured flavour is more likely to be produced by the tissue enzymes of the fish, while other workers claim that although tissue enzymes undoubtedly play a part in the development of cured flavour these are primarily a result of microbial activities (Omland 1955).

The suggestion by Voskresensky (1965) that the tissue enzymes are more active in the earlier stages of the ripening process and that the enzymes produced by micro-organisms are more important in the later stages may be feasible.

The flavours produced in fish fermented with added carbohydrates have not been submitted to any extensive study.

The bacterial flora developing in the barrels consists mainly of halotolerant micrococci, yeasts and Gram positive spore-forming bacteria. After maturation the fish is washed, filletted, cut into pieces and packed in cans in a pickle consisting of acetic acid, lactic acid, sucrose, water and a preservative, usually 0·2% benzoic acid.

In spite of a salt concentration of 8–9%, a pH value ca. 4·6, and the presence of a preservative, the canned product is not bacteriologically stable at ambient temperatures, but should be kept under refrigeration (Erichsen & Molin 1964). The microflora developing in the cans upon storage consists of lactic acid bacteria: pediococci and heterofermentative lactobacilli which cause gas-formation and, ultimately, liquefaction of the product. These bacteria are atypical of fish (Priebe 1962) which implies that their presence is due to secondary infections introduced during processing in the factories. This assumption is supported by the fact that titbits produced in different factories contain different types of gas-forming lactobacilli (Erichsen 1967).

The production of anchovy from sprats is carried out in much the same way. After maturation in barrels the anchovy is packed in cans, either whole or filletted, in a pickle containing salt, sugar, spices and sometimes a preservative. Like herring titbits, anchovies are bacteriologically unstable at ambient temperatures. The borderline between maturation and spoilage is difficult to define in these products and many consumers claim that the best quality titbits or anchovy is obtained after the development of some bacterial spoilage.

Other types of fermented fish products from Scandinavia are the so-called 'surströmming' made in Sweden and 'rakefisk' made in Norway. These products are not manufactured to the same extent as titbits and anchovy but are more of a delicacy, consumed on special occasions.

Swedish 'surströmming' is made from Baltic herring caught in May–June. The whole fish is immersed in a 25% salt brine for 1–2 days. The fish is then eviscerated, with roe and milt retained, packed in barrels in a brine of 17% salt and stored at 15–18°C. After 3–4 weeks the product is mature and the final salt concentration is about 12%. Considerable putrefaction takes place. Some consumers consider the product a delicacy while others find it repulsive because of its strong odour.

Norwegian 'rakefisk' is made from trout which is packed in tubs in layers with salt and stored in a cool place. A salt concentration of 5·6% is considered optimal but salt concentrations as low as 2% have been found. The final product has a buttery consistency and when produced successfully should not smell.

C. Health hazards

Since most fermented fish products are not cooked, but eaten raw, heat does not contribute to the control of food poisoning. The two main factors controlling food-poisoning bacteria in fermented fish products are high salt concentrations and low pH. Sometimes added preservatives may contribute as well. Where high salt concentrations and high acidity are relied upon to control food poisoning it is important that production is properly controlled. Salt must penetrate rapidly into the fish flesh and the pH must be reduced rapidly.

The two types of food-poisoning micro-organisms requiring special attention in connection with salted, fermented fish products are *Clostridium botulinum* and *Staphylococcus aureus*, since they are both more resistant to salt and acids than other food poisoning bacteria.

Clostridium botulinum, particularly type E, is consistently associated with fresh fish (Bott *et al.* 1967; Cann *et al.* 1967; Nickerson *et al.* 1967; Johannsen 1953) while *Staph. aureus* more commonly contaminates the fish during handling and processing. All types of *Cl. botulinum* are inhibited by 10–12% salt and generally by a pH below 4·5. *Clostridium botulinum* types B, E and F are able to grow at 8–10°C but are inhibited below 4°C. Toxin production is very slow below 10°C. *Staphylococcus aureus* does not grow below 6·5°C, and is inhibited by 15–20% salt and a pH below 4·5–5·0.

Food poisoning caused by these two types of micro-organisms has never been reported in connection with Scandinavian types of salted, fermented herring or anchovy products. No food poisoning has been reported from Swedish surströmming but several outbreaks of botulism type B and E have been reported from consumption of the Norwegian rakefisk (Nordbö & Valland 1967; Skulberg 1958; Yndestad 1970; Forfang & Skulberg 1964). This is due primarily to the fact that this type of product is often prepared in the home without proper care and knowledge, frequently resulting in salt concentrations which are too low to prevent the growth of food-poisoning bacteria.

D. Fish ensilage

Freshly caught fish spoils more rapidly than tissue from warm-blooded animals because it is less acid *post mortem*: having lower glycogen levels there is less accumulation of acid in the fish muscle. Also, fish is usually heavily loaded with psychrotrophic micro-organisms growing and causing spoilage even when the fish is stored in ice.

Fresh fish may be caught periodically in quantities exceeding the local processing or freezing capacities. Similar situations occur both in industrialized and in developing countries. A convenient way of salvaging this food would be by preserving the surplus fish by ensilage. Silage is a preservation process traditionally used in connection with green forage, which has been adopted for whole fish or parts of fish. There are two ways by which fish can be preserved as silage (James *et al.* 1977):

(1) preservation by addition of organic/inorganic acids which lowers the pH sufficiently to prevent microbial spoilage;
(2) preservation through fermentation by lactic acid bacteria.

In the latter case fish is minced or chopped and mixed with a fermentable sugar which favours the growth of lactic acid bacteria. These bacteria may be present either naturally on the fish or they can be added in the form of a starter culture. Molasses and cereal meals are often used as fermentable carbohydrates to initiate the lactic acid fermentation which prevents putrefactive and other undesirable changes taking place. The most successful process results in a rapid reduction of pH to 4·5 within 48–50 h, reaching a final pH of 4·0.

The method of preserving fish by means of an ensiling process is not well known and has not been put into practice to any significant extent. In Denmark ensilage of fish has been carried out for the past 30 years and there is a silage industry in Poland. In Norway silage is produced commercially from guts and other waste products from the fish industry. In South-East Asia fish silage production is currently being introduced as a means of utilizing waste and surplus fish. The fish silage produced so far has been used mainly for animal feed. Feeding experiments have shown that the nutritive quality is very good (Krishnaswamy 1965).

There exists, however, a growing interest in the fermented silage process for fish because it utilizes excess and waste fish for human consumption thereby contributing to protein needs, particularly in the developing countries.

3. Fermented Meat Products

Fermented meat products have constituted an important part of the human diet since ancient times. Possibly as long ago as 1500 BC, people learned that meat would not spoil if finely chopped, mixed with salt and spices and allowed to dry in rolls. This gave a product with good keeping qualities and good flavour. The product was a forerunner of our present fermented sausage: in fact salami is thought to be named after the city of Salamis on Cyprus, which was destroyed more than 2000 years ago.

Traditionally this type of fermented meat became an accepted and important food because it offered variety in the diet during the warm summer months, when lack of refrigeration and other means of preservation severely limited the meat items which could be consumed with any degree of safety. By using the proper techniques sausage could be prepared during the winter months, when game and domestic animals could be slaughtered, and consumed during the following summer (hence the name 'summer sausage'). The meat products thus produced were dehydrated to varying degrees depending upon the duration and conditions of storage.

With the passage of time various geographical regions developed their own characteristic types of product. Local climatic conditions, temperature, spicing, storage conditions and type of meat used were some of the factors contributing to wide variations in flavour of the products and in time certain localities became known for the particular flavour of local products.

People of the warm Mediterranean countries desired a heavily seasoned, dried type of sausage while people further north desired a product which was slightly spiced, heavily smoked, moist and well salted. Traditionally fermentations were dependent upon chance or random contamination with 'wild' organisms, often resulting in products with variable qualities. However, in many production places a sort of house flora of desirable micro-organisms had built up through the years, impregnating equipment, shelves, tables and even the air itself. This house flora infected the meat and usually gained control of the fermentation process thereby imparting the characteristic flavour to the product and preventing growth of undesirable types. The presence of micro-organisms and the observations that certain bacteria caused desirable changes in food products did not become known until hundreds of years later.

A. Types of product

Today fermented sausage products are the main types of fermented meats produced industrially.

Fermented meat products can be divided into two main groups:

(1) fermented, salted meat products which have been exposed to varying degree of drying;
(2) marinated products which have been kept in an acid marinade, often with sugar added.

Most of the questions discussed in connection with fermented sausage are applicable also to the whole field of fermented meat products and the following presentation will therefore deal mainly with fermented sausage products.

Fermented sausage, as is produced today, can be divided into two main types:

(1) Dry fermented sausage which has an ageing time of several months and has been subject to considerable drying resulting in a low a_w, the water content being below 40%. This type of product has a very good keeping quality owing to the low a_w. It is usually a heavily spiced product, but seldom smoked. Salami (Hungarian and Italian) comes into this category of fermented sausage.
(2) Semi-dry, fermented sausage which usually has a short processing and ageing time lasting from 3 days to 3 weeks. The a_w is high in these products (0·93–0·98) and water content is 40–60%. They are usually mildly spiced and in most cases subjected to a smoking process. They do not have good keeping qualities and should be kept under refrigeration. Having a high a_w these products rely mainly on biologically produced acid for keeping quality, even though salt concentration and smoking are also important factors. Many German types of sausage and most Swedish types of fermented sausages belong to this category.

B. Manufacturing process

The manufacturing process of fermented sausages consists of grinding or chopping the meat, mixing it with salt, nitrate or nitrite, spices and other seasonings and sugar, and stuffing the mixture into casings. The sausages are then transferred to a 'green' room where fermentation takes place. Temperature and relative humidity in the green room are carefully controlled and the time varies depending on the type of sausage desired. If it is a smoked variety the smoking usually takes place at the same time.

The first indication of acid production occurs after 8–16 h in the green room. Factors affecting the initiation of acid production are the types and the numbers of bacteria present and the heating schedule employed. The initial pH values of the meat mixes are usually in the range 5·8–6·4, whereas the pH values of finished products of good quality vary from 4·5 to 5·3. Sausages having pH values higher than 5·3 often possess an inferior soft texture.

C. Fermentation process

Successful manufacture of fermented sausages is dependent on the types of micro-organisms developing in the raw comminuted meat. The microbial load of raw meat is usually about 10^3–10^4/g. Chopping the meat increases the meat surface and may increase the microbial load by up to 10^5/g. Many different types of micro-organisms are represented on the meat, mainly Gram negative types (Barnes & Ingram 1956; Gardner *et al.* 1967; Hall & Angelotti 1965; Seideman *et al.* 1976; Snijders & Gerats 1976; Vanderzant & Nickelson 1969). The major microbiological changes occurring during natural fermentation of meat products are the shift of the fresh, refrigerated or frozen meat flora from Gram negative, catalase-positive, highly aerobic flora to a Gram positive, catalase-negative and mostly micro-aerophilic flora. These changes are attributed partly to the decrease in the redox potential within the sausage to the point where aerobic organisms cannot grow, paving the way for the growth of micro-aerophilic types of bacteria, and partly to the selective inhibitory effect of the curing salts added. Other parameters important for the development of a special microflora from a mixed population are temperature and gas composition of the environment, e.g. CO_2. The bacteria types usually developing under such conditions are lactic acid bacteria which utilize carbohydrates, producing acid and lowering the pH in the product causing suppression of undesirable bacteria, including pathogens.

Lactic acid bacteria, mainly *Lactobacillus* spp., have been shown to proliferate rapidly during fermentation of meat products, reaching numbers of 10^7/g. Other types that have been shown to contribute to fermentation are pediococci and leuconostocs. Not all types of lactic acid bacteria are desirable in the fermentation process. Many heterofermentative types cause gas- and slime formation and discolouration of the products while others take part in the fermentation process of certain types of sausage like Genoa salamis. Homofermentative types producing primarily lactic acid tend to give products with a more uniform quality.

Other types of micro-organism considered by some to be of importance during the initial stages of fermentation are micrococci. Because of their ability to reduce nitrates to nitrites they contribute to the development of flavour and to the stabilization of colour. Current trends to shorten fermentation time, and to add nitrites directly, tend to limit the role of micrococci.

D. Use of starter cultures

Most sausage producers have long relied upon naturally occurring lactic acid bacteria in the ingredients for successful fermentation. This must be considered a poor approach to fermented meat production since it offers minimal safety, lack of uniformity in the finished product and extensive losses due to spoilage. It was not until about 1940 that there arose the idea to use starter cultures in the manufacture of meat products, i.e. the addition of pure cultures of lactic acid bacteria in such amounts as to speed up the fermentation process and take control of the development of the microflora, thus increasing uniformity of the finished product, minimizing losses and improving flavour.

Through their ability to form organic acids and macromolecular antibiotics many types of lactic acid bacteria have also been shown to offer protection against growth of undesirable micro-organisms in foods (De Klerk & Smith 1967; Upreti & Hinsdill 1975). For example, some lactic acid bacteria lacking a catalase system can produce toxic levels of hydrogen peroxide which has been shown to inhibit *Pseudomonas* spp., the most active spoilage micro-organisms in connection with meats (Price & Lee 1970).

An attempt was made first to use various lactic acid bacteria of the types used in the dairy industry. The majority of these strains did not grow in the meat mixtures, presumably due to their lack of tolerance to salt and/or nitrite. Later, attempts were made to utilize various lactobacilli that constituted the predominate flora of good-quality marketed meat products. In these products pediococci were found quite often, but lactobacilli usually predominated. In the 1950s the first starter cultures for meats were made from such isolated cultures.

The development of starter cultures for use in fermented sausages followed separate ways in Europe and in the USA. In the beginning European producers favoured cultures that would reduce nitrate, such as strains of *Micrococcus* (Niinivaara 1955). Later this view was changed when it was observed that the addition of lactobacilli along with micrococci gave better colour development and a somewhat more acid product. The result was a starter culture composed of *L. plantarum* and *Micrococcus* sp. (Niinvaara 1958; Nurmi 1966).

In the USA sausage producers utilized nitrite cures and thus needed to add only lactic acid bacteria. *Pediococcus cerevisiae* was the bacterial strain considered most suitable for fermented sausage products common in the USA (Niven *et al.* 1958). Pediococci grow readily in meat products and produce less acid than some of the *Lactobacillus* species, which can produce excessive amounts of acid. In recent years the *Pediococcus* type of starter culture has also gained in popularity in Europe.

In Sweden a new starter culture containing *P. pentosaceus* has been developed. This strain has a lower optimum growth temperature than *P. cerevisiae*, making it more suitable for the production of most Swedish types of semi-dry sausages, which are fermented at temperatures between 25–30°C.

Many different types of starter culture or starter culture combinations are now available commercially and they are all combinations of the following three main types of bacteria: *Lactobacillus* spp.; *Pediococcus* spp. and *Micrococcus* spp. Originally the starters were available in lyophilized form but frozen concentrates of starters have become increasingly popular since it was observed that the frozen type of starter is more active and has a shorter lag phase.

E. Health hazards

Dry and semi-dry fermented sausages have rarely been implicated in food poisoning. However, some outbreaks have been reported in connection with salami (Anon. 1971*a*,*b*, 1975). Prerequisites for pathogenic bacteria to cause problems in connection with fermented food products are: (1) that they are present in large numbers in the raw material from the beginning; (2) that they are given the opportunity to grow during the initial stages of the fermentation process.

The organisms most commonly discussed as potential risks in connection with fermented food products are *Staphylococcus aureus*, *Salmonella* spp. and *Clostridium* spp.

(i) Staphylococcus aureus

In recent years *Staph. aureus* has attracted considerable attention with regard to the safety of fermented meats. These organisms are widely distributed in meats and grow better than other pathogens under what would be considered adverse conditions like those prevailing in fermented sausage. *Staphylococcus aureus* added during the manufacture of fermented sausage may grow rapidly during the first 2–3 days of fermentation (Daly *et al.* 1973).

Different types of lactic acid bacteria have, however, been shown to exert a strong limiting effect on both growth and toxin production by *Staph. aureus* (Graves & Frazier 1963; Haines & Harmon 1973; Genigeorgis 1976). *Pediococcus* spp. were found to inhibit *Staph. aureus* more effectively than lactobacilli due to the rapid fall of pH caused by these bacteria during the early stages of fermentation (Daly *et al.* 1973; Genigeorgis 1976). Practical experiments have shown that if lactic acid bacteria are added together with *Staph. aureus*, the lactic acid bacteria will suppress and completely inhibit the activity of *Staph. aureus* (Daly *et al.* 1973; Niskanen & Nurmi 1975). Excessive growth of *Staph. aureus* in fermented sausage usually reflects faulty acidulation. According to Daly *et al.* (1973) the use of commercial starter cultures shows a 99·6–99·9% inhibition of *Staph. aureus* compared with uninoculated controls.

(ii) Salmonella *spp.*

Salmonella infections caused by the intake of fermented sausages are rare (Smith *et al.* 1975*a,b*). Growth of *Salmonella* in fermented sausage is inhibited by the activities of lactic acid bacteria as well as by the salt concentration, the low water activity and, in some instances, also by the smoking process (Genigeorgis 1976). Possibilities for the growth of *Salmonella* under such conditions are very small. Reports exist, however, on survival of *Salm. typhimurium* in artificially contaminated fermented sausage (Östlund & Regner 1968) but their numbers declined upon subsequent refrigeration storage of the finished product (Smith *et al.* 1975*a,b*).

The addition of starter cultures like *P. cerevisiae* and *Lactobacillus plantarum* drastically decrease the survival of *Salmonella* in fermented meat products (Smith 1975*a,b*). The reason for this decrease, besides the rapid decline in pH, is that many strains of *Pediococcus* spp. and *Lactobacillus* spp. exert a pronounced antagonistic effect on *Salmonella* spp.

(iii) Clostridium botulinum

Risk of botulism from fermented sausages which contain nitrite is very small (Kueper & Trelease 1974; Christiansen *et al.* 1975). Moreover, in the absence of nitrite, toxin was not produced by *Cl. botulinum* (Type A and B) (Kueper & Trelease 1974).

4. Future Possibilities

In most developing countries large quantities of surplus food, especially fish, go to waste because tropical temperatures are high and inadequate technological development does not allow preservation procedures like canning, freezing and refrigeration. Many of these countries are rich in a variety of natural resources for food and should have no problems in feeding their population provided foodstuffs can be preserved. The greatest demand today is perhaps not only for the production of more food but for methods of preserving food already available.

One important means of aiding the developing countries would be for the industrialized countries to devote more of their resources to applied and basic research on new fermentation processes which could be transferred and applied in the developing countries. The only preservation method for food used in the developing countries today is the fermentation of fish by the addition of high concentrations of salt to stop putrefaction. However, the high salt content of the resulting products, in most cases, make them unfit for human consumption in quantities necessary to cover the needs of the population for high grade proteins. Therefore, alternative fermentation processes should be considered by means of which fish, meat and vegetables could be converted to edible and acceptable preserved foods.

Of the alternative processes available, the technique of ensilage using lactic acid bacteria as the major preserving agent seems to be the most promising. To initiate the fermentation cheap local sources of carbohydrates like rice or molasses could be used. Any such type of fermentation process is well suited for the developing countries because it is inexpensive and requires little energy.

So far little is known about how traditional fermented products of fish and meat acquire their acceptable qualities and typical flavours, and how they remain safe for consumption. More extensive studies by the industrialized countries into the microbiological side of the fermentation processes would be desirable to gain more knowledge about the micro-organisms involved in the process, their actual role in the changes taking place during fermentation, and in the production of special flavours and other desirable characteristics of the products. Such studies could lead to the development of different starter cultures of lactic acid bacteria with the ability to speed up the fermentation process and perhaps to produce new products with improved flavours and textures, properties making them more acceptable for human consumption. Lyophilization of such starter cultures and their introduction and distribution to the developing countries constitutes no major problem and could, together with further research into the technology of the process, lead to the solution of at least some of the malnutrition problems in the world.

In the industrialized countries with their access to effective but energy-requiring preservative processing methods like heat sterilization and freezing, the development of fermentation products has been directed more towards creating new 'fancy' foods rather than those that are merely nutritional. Considering the present energy crisis a situation may arise in which even the industrialized countries could be forced to look for alternative, more energy-saving processes for food preservation. In such a situation biological preservation in these countries also appear to be the best alternative. New products will have to be developed which again call for further research into the technology and microbiology of new fermentation processes.

5. References

ALM, F. 1965 Scandinavian anchovies and titbits. In *Fish as Food*, Vol. 3, ed. Borgström, G. p. 195. London & New York: Academic Press.

ANON. 1971a Gastroenteritis associated with salami—New York. *Morbidity and Mortality Weekly Report* 20, 253, 258.

ANON. 1971b Gastroenteritis attributed to Hormel San Remo stick Genoa salami—New York. *Morbidity and Mortality Weekly Report* 20, 370.

ANON. 1975 Staphylococcal food poisoning associated with Italian dry salami—New York. *Morbidity and Mortality Weekly Report* 24, 374, 379.

BARNES, E. & INGRAM, M. 1956 The effect of redox potential on the growth of *Clostridium welchii* strains isolated from horse muscle. *Journal of Applied Bacteriology* 19, 117–128.

BOTT, T. L., DEFFNER, J. S. & FOSTER, E. M. 1967 Occurrence of *Clostridium botulinum* type E in fish from the Great Lakes with special reference to certain large bays. In *Botulism 1966* ed. Ingram, M. & Roberts, T. A. pp. 21–24. Proceedings of the 5th International Symposium on Food Microbiology, Moscow, July 1966. London: Chapman & Hall.

CANN, D. C., WILSON, B. B., HOBBS, G. & SHEWAN, J. M. 1967 *Clostridium botulinum* in the marine environment of Great Britain. In *Botulism 1966*, ed. Ingram, M. & Roberts, T. A. pp. 62–65. Proceedings of the 5th International Symposium on Food Microbiology, Moscow, July 1966. London: Chapman & Hall.

CHRISTIANSEN, L. N., TOMPKIN, R. B. & SHAPARIS, A. B. 1975 Effect of sodium nitrite and nitrate on *Clostridium botulinum* growth and toxin production in a summerstyle sausage. *Journal of Food Science* 40, 488–490.

DALY, C., LA CHANCE, M., SANDINE, W. E. & ELLIKER, P. R. 1973 Control of *Staphylococcus aureus* in sausage by starter cultures and chemical acidulation. *Journal of Food Science* 38, 426–430.

DE KLERK, H. C. & SMITH, J. A. 1967 Properties of *Lactobacillus fermentii* bacteriocin. *Journal of General Microbiology* 48, 309–316.

ERICHSEN, I. 1967 The microflora of semi-preserved fish products. III. Principal groups of bacteria occurring in titbits. *Antonie van Leeuwenhoek* 33, 107–112.

ERICHSEN, I. & MOLIN, N. 1964 The microflora of semi-preserved fish products. II. The effect of the quality of the raw materials, added materials and storage conditions. *Antonie van Leeuwenhoek* 20, 197–208.

FORFANG, K. & SKULBERG, A. 1964 Botulism type E fra rakefisk. *Tidsskrift for den Norske Legeforening* 84, 973–977.

GARDNER, G. A., CARSON, A. W. & PATTON, J. 1967 Bacteriology of prepackaged pork with reference to the gas composition within the pack. *Journal of Applied Bacteriology* 30, 321–333.

GENIGEORGIS, C. A. 1976 Quality control of fermented meats. *Journal of the American Veterinary Medical Association* 169, 1220–1228.

GRAVES, R. R. & FRAZIER, W. C. 1963 Food microorganisms influencing the growth of *Staphylococcus aureus*. *Applied Microbiology* 11, 513–516.

HAINES, W. & HARMON, L. G. 1973 Effect of variations in conditions of incubation upon inhibition of *Staphylococcus aureus* by *Pediococcus cerevisiae* and *Streptococcus lactis*. *Applied Microbiology* 25, 169–171.

HALL, H. E. & ANGELOTTI. R. 1965 *Clostridium perfringens* in meat and meat products. *Applied Microbiology* **13**, 352–357.

JAMES, M. A., IYER, K. M. & NAIR, M. R. 1977 Comparative study of fish ensilage prepared by microbial fermentation and formic acid ensilage. *Proceedings of the Conference on the Handling, Processing and Marketing of Tropical Fish, Tropical Products Institute, London*, 273–278.

JOHANNSEN, A. 1953 *Clostridium botulinum* in Sweden and adjacent waters. *Journal of Applied Bacteriology* **26**, 43–47.

KRISHNASWAMY, M. A., KADKOL, S. B. & REVANKAR, G. D. 1965 Nutritional evaluation of an ensiled product from fish. *Canadian Journal of Biochemistry* **43**, 1879–1883.

KUEPER, T. V. & TRELEASE, R. D. 1974 Variables affecting botulinum toxin development and nitrosamine formation in fermented sausage. *Proceedings of the Meat Industry Research Conference*, pp. 69–74.

MACKIE, I. M., HARDY, R. & HOBBS, G. 1971 *Fermented Fish Products*. FAO Fisheries Report No. 100, Rome: FAO.

NAGAO, S. & KIMURA, T. 1951 Bacteriological studies on "shiokara". *Bulletin-Faculty of Fisheries, Hokkaido University* **1**, 21–32.

NAGASAKI, S. & YAMAMOTO, T. 1954 Studies on the influence of salt in microbial metabolism. IV. Some chemical observations on the course of ripening of "Ika-Shiokara". *Japanese Society of Scientific Fisheries Bulletin* **20**, 617–620.

NICKERSON, J. T. R., GOLDBLITH, S. A., DIGIOIA, G. E. & BISHOP, W. W. 1967 The presence of *Clostridium botulinum* type E in fish and mud taken from the Gulf of Maine. In *Botulism 1966*, ed. Ingram, M. & Roberts, T. A. pp. 25–33. Proceedings of the 5th International Symposium on Food Microbiology, Moscow, July 1966. London: Chapman & Hall.

NIINIVAARA, F. P. 1955 The influence of pure cultures of bacteria on the maturing and redenning of raw sausage. *Acta Agral Fennica* **85**, 95–101.

NIINIVAARA, F. P. 1958 Observations on the significance of a *Micrococcus* strain in meat processing. *Proceedings of the 2nd International Symposium on Food Microbiology, London*, 187–189.

NISKANEN, A. & NURMI, E. 1975 The effect of starter culture on staphylococcus enterotoxin and thermonuclease production in dry sausage. *Applied and Environmental Microbiology* **31**, 11–20.

NIVEN, C. F., DEIBEL, R. H. & WILSON, G. D. 1958 The AMIF sausage starter culture. *American Meat Institute Foundation Bulletin* **41**, 5–25.

NORDBÖ, E. & VALLAND, M. 1967 Bacteriological and toxiological investigation on a case of human botulism type E by consumption of home-prepared fermented fish (rakefisk). *Nordisk Veterinaer Medicin* **19**, 536–539.

NURMI, E. 1966 Effect of bacterial inoculations on characteristics and microbial flora of dry sausage. *Acta Agral Fennica* **108**, 1–77.

OMLAND, D. 1955 Litt om salt sild. *Fiskets Gang* pp. 495–503.

ÖSTLUND, K. & REGNER, B. 1968 Undersökningar rörande mikrofloran i isterband. *Nordisk Veterinaer Medicin* **20**, 527–542.

PRICE, R. J. & LEE, J. S. 1970 Inhibition of *Pseudomonas* species by hydrogen peroxide producing lactobacilli. *Journal of Milk and Food Technology* **33**, 13–18.

PRIEBE, K. 1962 Zur Frage der Herkunft von Betabakterien bei Heringsmarinaden. *Archiv für Lebensmittelhygiene* **13**, 278–281.

SEIDEMAN, S. C., VANDERZANT, C., SMITH, G. C., HANNA, M. O. & CARPENTER, Z. L. 1976 Effect of degree of vacuum and length of storage on the microflora of vacuum packaged beef wholesale cuts. *Journal of Food Science* **41**, 738–742.

SHEWAN, J. M. & HOBBS, G. 1967 The bacteriology of fish spoilage and preservation. In *Progress in Industrial Microbiology* Vol. 6, ed. Hockenhull, D. J. D. pp. 169–208. London: Heywood Books Ltd.

SKULBERG, A. 1958 A case of fish-borne type B botulism. *Nordisk Hygienisk tidsskrift* **39**, 133–136.

SMITH, J. L., HUHTANEN, C. N., KISSINGER, J. C. & PALUMBO, S. A. 1975*a* Survival of *Salmonella* during pepperoni manufacture. *Applied Microbiology* **30**, 759–763.

SMITH, J. L., PALUMBO, S. A., KISSINGER, J. C. & HUHTANEN, C. N. 1975b Survival of *Salmonella dublin* and *Salmonella typhimurium* in Lebanon Bologna. *Journal of Milk and Food Technology* **38**, 150–154.

SNIJDERS, J. M. A. & GERATS, G. E. 1976 Hygiene bei der Schlachtung von Schweinen *Die Fleischwirtschaft* **56**, 717–721.

UPRETI, G. C. & HINSDILL, R. D. 1975 Production and mode of action of lactocin 27: Bacteriocin from a homofermentative *Lactobacillus*. *Antimicrobial Agents and Chemotherapy* **7**, 139–145.

UYENCO, V., LAWAS, I., BRIONES, P. R. & TARUC, R. S. 1953 Mechanics of "bagoong" (fish paste) and "patis" (fish sauce) processing. *Proceedings Indo Pacific Fisheries Council*, Section II, **4**, 210–218.

VANDERZANT, C. & NICKELSON, R. 1969 A microbial examination of muscle tissue of beef, pork and lamb carcasses. *Journal of Milk and Food Technology* **32**, 357–361.

VAN VEEN, A. G. 1965 Fermented and dried seafood products in South-East Asia. In *Fish as Food* Vol. 3, ed. Borgström, G. pp. 227–250. New York & London: Academic Press.

VOSKRESENSKY, N. A. 1965 Salting of herring. In *Fish as Food* Vol. 3, ed. Borgström, G. pp. 107–133. New York & London: Academic Press.

YNDESTAD, M. 1970 A case of botulism, type E after consumption of rakefisk. *Norsk Veterinaer Tidsskrift* **82**, 321–323.

ZENITANI, B. 1955 Studies on fish-fermentation products. I. On the aerobic bacteria in "Shiokara". *Japanese Society of Scientific Fisheries Bulletin* **21**, 280–283.

Genetic Engineering for Food and Additives

J. R. PELLON AND A. J. SINSKEY

*Department of Nutrition and Food Science,
Massachusetts Institute of Technology,
Cambridge, Massachusetts, USA*

Contents

1. Introduction

In the past, a primary concern of food microbiologists has been food-borne micro-organisms which are able to cause infections and intoxications in humans or which affect food quality. In the first case, efforts were directed towards elimination of the public health risks associated with food-borne diseases. In the second case, the microbial spoilage of foods has been, and still is, important economically. Besides supporting the above two areas, genetic engineering offers unique opportunities to the food industry, especially on the more positive aspects of food science. The design of improved microbial strains and the production of flavouring agents, preservatives, sweeteners, colourants and new proteins through the rational manipulation of genes is today a feasible goal and in some cases a reality. The food processing industries, together with the pharmaceutical and chemical industries, should be the first to perceive the commercial effects of the application of genetic engineering. Whether or not this happens represents a unique challenge to food researchers.

Genetic engineering is a technology at the laboratory level used to increase or modify the genetic information contained in micro-organisms. Most important, it allows for control of biological information processing. Its applications have reached the pilot plant scale. Genetic engineering can

FOOD MICROBIOLOGY
ISBN 0 12 589670 0

be applied to the food processing industry in different ways. For example,

(a) to design micro-organisms which are more efficient in their desired interactions with food. Genetic engineered organisms that have increased productivities and yields in various fermentation processes would be significant accomplishments;
(b) to design micro-organisms to produce additives useful as food ingredients;
(c) to design micro-organisms which are more sensitive to detect mutagenic agents in food;
(d) to develop novel and very specific microbial detection procedures based on DNA hybridization technology.

In general, the utilization of genetic engineering techniques to improve food processing requires a sequence of studies (Table 1). The main problem limiting the application of genetic engineering techniques to food microorganisms is that the basic biochemical and genetic knowledge of the microbial characteristics that could improve food is lacking. Hence the application of genetic engineering in the food processing industry will be seen not as having an industry-wide impact but as a crystallization of realities around certain more easily definable problems.

What are typical cases where genetic engineering has already produced tangible results or where developments are expected in the near future? Examples can be drawn from the following areas: single cell protein production, food processing, and the production of additives.

TABLE 1

Application of genetic engineering in food processing requirements

---

1. Determination of the microbial species fundamental to the process.
2. Define the biochemistry of the microbial functions of interest.
3. Analysis of the genetic determinants of the biochemical functions of interest.
4. Study and development of methods of genetic manipulation in the microbial species of interest.
5. Study of environmental effects on the expression of the desired microbial function(s).

---

2. Single Cell Protein Production

The term single cell protein (SCP) refers to protein in microbial cells, which can be used for animal feed or for human consumption. This term was coined to replace the old term 'microbial protein' which at the end of the sixties was considered less aesthetic. The typical protein content is 60–70% in bacteria, 45–65% in yeasts and 35–40% in moulds. The world production of SCP is estimated at 2 million tons per year (Anon. 1981a).

The need to produce SCP is based on the scarcity and unbalanced geographical distribution of the protein produced today on the planet, together with a deficiency in essential amino acids of some other protein sources. The perceived advantages and disadvantages of SCP are summarized in Table 2.

The substrates used as raw materials for conversion vary widely from petroleum-based hydrocarbons to carbohydrates or to agricultural, industrial and municipal residues. Today, most SCP produced comes from cane and beet molasses, wood wastes, corn trash and papermill wastes (Anon. 1981a).

The application of genetic engineering techniques to obtain more efficient strains for the production of SCP from non-edible substrates is exemplified by the process developed by Imperial Chemical Industries (ICI). The story began in the late 1960s when methane was utilized for production of SCP, but there was a problem in that methane as a growth substrate resulted in poor yields of cells and its limited solubility and the safety factor were troublesome. Using methanol as a substrate, the yields were significantly higher. Furthermore, the oxidation state of methanol is closer to that of cells

TABLE 2

Advantages of SCP production	Disadvantages of SCP production
1. The process requires less time due to the efficiency of micro-organisms in producing protein, compared to the production of vegetable or animal protein.	1. High content in nucleic acids which, in humans, results in uric acid as an end product. High levels of uric acid in humans may predispose to gout.
2. Higher nutritive value of SCP with respect to vegetable protein (Wang 1968).	2. Some compounds present in the cell wall are non-digestible and toxic.
3. Minimal soil surface needed for SCP production.	3. Fluctuations in price and availability of substrates. The substrate constitutes about 40–50% of the SCP manufacturing costs (Cooney et al. 1980).
4. Climate does not affect production.	4. Lack of functionality as a food ingredient.
5. Flexibility in process design, due to multiplicity of substrates and micro-organisms which may be employed.	

than is methane and less oxygen is required, which means lower fermentation costs.

In the early 1970s, process development research and environmental manipulations resulted in the development of a large-scale process. Around 1977, the scientists in charge of the research development programme realized that further improvements in fermentation yields were not likely to occur unless engineering techniques were employed.

The micro-organism utilized in the ICI process is the bacterium *Methylophilus methylotrophus* strain AS1, which in this case utilizes methanol as a source of carbon and energy, and ammonia as a source of nitrogen. The pathway of ammonia assimilation in *M. methylotrophus* starts with two reactions catalysed by the enzymes glutamine synthetase (GS) and glutamate synthase (GOGAT) (Hopwood 1981). This pathway requires the expenditure of ATP in the GS reaction (1 mol of ATP per mol of ammonia assimilated):

$$\text{glutamate} + NH_3 + ATP \xrightarrow{\text{GS}} \text{glutamine} + ADP + P_i$$

$$\alpha\text{-ketoglutarate} + \text{glutamine} + NAD(P)H \xrightarrow{\text{GOGAT}} 2\,\text{glutamate} + NAD(P).$$

In other micro-organisms, including *Escherichia coli*, the ammonia assimilation pathway depends on only one enzyme, glutamate dehydrogenase (GDH), which does not require ATP:

$$\alpha\text{-ketoglutarate} + NH_3 + NAD(P)H \underset{\text{GDH}}{\xrightleftharpoons{\hspace{2cm}}} \text{glutamate} + NAD(P).$$

The ICI investigators' approach was to replace the original assimilation pathway by the *gdh* gene which codes for GDH. The genetic engineering strategy employed consisted of (Senior & Windass 1980; Hopwood 1981) (Fig. 1):

(1) Identification and isolation of a donor of the *gdh* gene (*E. coli*).
(2) Construction of vector plasmids to transfer the gene *gdh* to *M. methylotrophus* AS1. This was done by splicing chromosomal fragments of an *E. coli gdh*⁺ into a plasmid vector, into an *E. coli gdh*⁻. The desired clone is able to utilize ammonia while the rest of the *E. coli gdh*⁻ cells are not. Next, the plasmid DNA is isolated.
(3) Isolation of GOGAT⁻ mutants of *M. methylotrophus* AS1 cannot utilize ammonia as a nitrogen source (auxotrophs).
(4) Transfer of the plasmid vector with the *gdh* gene to the *M. methylotrophus* auxotroph.

The resulting strain was able to grow on a mixture of methanol, ammonia and salts, and fermentation studies showed enhanced yields, suggesting that

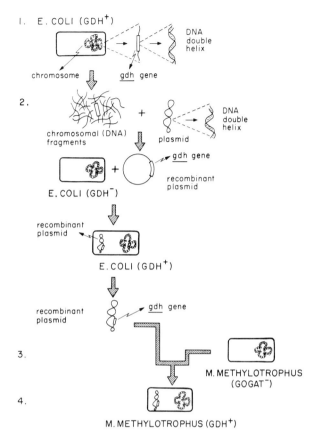

Fig. 1. Genetic engineering improvement of *Methylophilus methylotrophus*.

energy production was limiting growth. Whether this was the case in large-scale processes has not been reported. If the yield of cells with methanol is not improved, then the study has indicated that other factors are resulting in the less than theoretical yields.

The brand name of ICI's SCP is 'Pruteen' and its main problem commercially is the lower price of the main competing source of protein, soy-beans. 'Pruteen' is used as animal feed, but it may be used also in human food after further purification and adjustment of its organoleptic and rheological properties. In 1979, the price of 1 kg of SCP for animal feed was 40–50c; while for human consumption, it was $1–$1.30 (Anon. 1981*a*).

The above example is probably one of the first instances of the application of genetic engineering techniques to an industrial scale process.

Other applications can be envisaged. Future developments may improve the essential amino acid content of SCP sources. In addition, we have described how temperature-sensitive mutants of yeast may be utilized to improve the physical and functional properties of SCP (Miyasaka *et al.* 1980). Applications of genetic engineering to improve functionality of SCP is now a challenging strategy worth pursuing.

3. Food Processing

In the near future, the application of genetic engineering techniques in food processing has particular possibilities in the brewing and dairy industries.

In the brewing industry, genetic engineering techniques have been employed in the yeast *Saccharomyces cerevisiae* to produce beer with a lower carbohydrate content (Rose 1981). The dextrins, which constitute 22% of the carbohydrates in malt wort, are not fermented by *S. cerevisiae* and so they appear in the final product (beer). A related species, *S. diastaticus*, is able to ferment dextrins due to the extracellular enzyme glucoamylase, which hydrolyses glucose units from the non-reducing ends of these large molecular-weight polysaccharides. Two genes coding for the production of glucoamylase have been identified in *S. diastaticus* (Stewart 1981). The obvious aim was to transfer the genetic information for dextrin utilization from *S. diastaticus* to *S. cerevisiae*.

The genetic transfer to *S. cerevisiae* was successful but it was soon realized that in transferring the dextrin utilization genes transfer also occurred of undesirable genes which produce an unpleasant phenolic flavour in beer due to the compound 4-vinyl guaiacol (Rose 1981). The genes responsible for the appearance of 4-vinyl guaiacol were thought to be linked closely to those responsible for dextrin breakdown, but it has been shown now that the two groups of genes do not appear together when more refined genetic engineering techniques are used to tailor these micro-organisms (Rose 1981).

The genetic manipulation of *S. cerevisiae* will enable improved strains to be designed (i.e. higher ethanol tolerance). Two techniques show promise at the present time: transformation and spheroplast fusion (Stewart 1981). These techniques will be employed in the near future to produce yeast strains that will be used in the commercial production of 'light' beer.

In the dairy processing industry, genetic engineering may solve different types of problems, including the shortage of mammalian rennin, instability in group N streptococci of metabolic functions fundamental to their commercial utilization, or the infection of the bacterial cultures by bacteriophages.

Rennin (chymosin in recent nomenclature) is a proteolytic enzyme utilized in cheesemaking to coagulate milk. Its world market is about $40 million per year (Anon. 1981*a*). The enzyme is obtained from the fourth stomach of the unweaned calf, but the shortage of rennin is forcing the cheesemaking industry to look for cheaper and more reliable sources. One solution is the microbial rennins extracted from fungi, but their biochemical properties are not the same as calf rennin.

In the future the best solution would seem to be the transfer of the genes coding for chymosin (rennin) to bacteria. The first step, the identification of the mRNA coding for prochymosin, the precursor of chymosin, has been accomplished (Anon. 1981*b*). Next the mRNA will serve as a template to synthesize the complementary DNA (cDNA), which in turn will be spliced in some genetic vector (i.e. plasmid) and incorporated into a suitable bacterium or yeast to produce prochymosin. The conversion of prochymosin to the active chymosin (rennin) may be effected by acid pH. Figure 2 outlines the strategy being employed by several research investigators.

In the dairy industries, the lactic acid bacteria are involved in the production of a variety of food items. Among then, the micro-organisms of the genus *Streptococcus* pertaining to the antigenic group N are especially

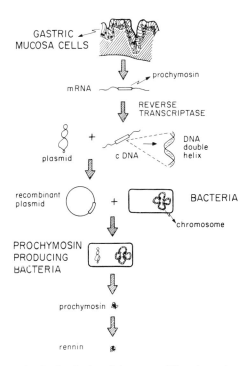

Fig. 2. Rennin production by cloning of the responsible eukaryotic genes in bacteria.

important (Rose 1981). Group N streptococci is composed of two species: *Streptococcus lactis* and *Strep. cremoris*, which are the lactic acid bacteria most studied at the biochemical and genetic levels.

The main problems to be solved with respect to the industrial application of group N streptococci are the instability of some fundamental metabolic functions and the infection of the cultures with phages. In addition, the development of strains that result in higher yields and productivities would be most desirable. This could be accomplished by altering fermentation substrate utilization patterns, as well as by increasing gene dosage of enzymes in existing metabolic pathways.

The growth of group N streptococci in milk to cell densities found in coagulated cultures requires the transformation, by proteases, of casein in small oligopeptides and amino acids (Mills & Thomas 1981) and the conversion of lactose in lactic acid. The result is the presence in the medium of amino acids utilized directly by the micro-organisms (nitrogen source) which at the same time obtain energy from the lactose fermentation. Both functions, and some others commercially interesting (i.e. citrate utilization), are unstable. Approximately 1–3% of colonies isolated from an unstable culture lose their proteinase activity or their ability to ferment lactose (McKay 1978). The reason for the instability of these functions has remained unclear until recent years. Today, mostly through the efforts of the group headed by Dr L. L. McKay of the University of Minnesota, we know that the instability is due to the loss of plasmids which carry genes coding for enzymes needed for the proteinase- and lactose-fermenting activities. Lactose fermentation is a multi-step process and some genes involved are plasmid encoded. For example, in *S. cremoris* B_1 the enzymes EII-*lac* and FIII-*lac*, which are part of the phosphotransferase uptake system, and phospho-β-D-galactosidase, also involved in lactose metabolism, are encoded in a plasmid (Anderson & McKay 1977).

Plasmids can replicate autonomously, physically separate from the chromosome. The stability of plasmids is affected by different factors including:

(a) host characteristics (i.e. growth rate with and without plasmids, physiological consequences of plasmid gene expression);
(b) plasmid characteristics (i.e. stringent or relaxed replication control, copy number, size, genetic complement encoded);
(c) environmental characteristics (i.e. necessary expression of plasmid genes).

The instability of plasmid-coded characters may be due to different mechanisms which bring about the loss of the responsible genes: segregation, recombination with the chromosome, deletion, various rearrangements, mutation in the plasmid replication origin, etc. (Jackson 1981).

The sizes of the plasmids determining commercially important functions in the dairy industry vary from 5.5×10^6 daltons to 6×10^7 daltons. The copy number is usually high (10–30 copies) for plasmids smaller than 7×10^6 daltons and low (1 or 2 copies) in the case of large plasmids (Clewell 1981; Walsh & McKay 1981). Because plasmids with molecular weights lower than 25×10^6 daltons usually do not contain genes coding for self-transfer, it is thought that the small plasmids are likely to be transferred together with large self-transferable plasmids (Kempler & McKay 1981).

The methods available today to transfer genetic material in group N streptococci include protoplast fusion, transduction and conjugation (Davies & Gasson 1981). Some of the transconjugants exhibit a cell-aggregation phenotype and donate *lac* at a high frequency (Walsh & McKay 1981). This could be related to the production of sex pheromones which could induce the aggregation response and the preparation for plasmid transfer, the nature of which is not known (Clewell 1981).

The use of recombinant DNA techniques is limited by the lack of a transformation system in group N streptococci (Kempler & McKay 1981). In this respect, a plasmid (ρAMβ) conferring resistance to erythromycin has been shown to be transmissible by conjugation between different strains of lactic streptococci, and to promote the mobilization of chromosomal genes and the transfer of non-transmissible plasmids (Davies & Gasson 1981). This plasmid, together with the use of protoplasts to develop a transformation system, will allow the introduction and selection for hybrid plasmids in group N streptococci. This in turn will allow the use of DNA cloning techniques to improve strains in the dairy industry.

It is already known (McKay & Baldwin 1978) that some strains rendered stable in the *lac* trait appear to have inserted the genes responsible in the chromosome. To obtain stable strains, the genes involved were transferred from one strain to the chromosome of the recipient strain by a transducing phage (McKay & Baldwin 1978). This strategy may allow for the stabilization of other commercially important traits (e.g. citrate utilization).

The infection of starter cultures by bacteriophages causes slow acid production which has tremendous economic importance in the cheesemaking industry. The plasmids mentioned previously and the recombinant DNA techniques may soon help to obtain strains resistant to phages by incorporating restriction endonucleases which recognize and degrade foreign DNA molecules.

4. Production of Additives

The term additive is employed here to indicate any substance added to the food during its processing, packaging or storage. In general, additives are added to food to facilitate processing, handling, or distribution; to control physical, chemical or microbiological changes; or to improve nutritional, rheological and organoleptic properties.

The commercially important additives produced by micro-organisms fall into four categories: (a) enzymes employed in food processing, (b) primary metabolites or compounds essential for the micro-organisms' growth, (c) secondary metabolites or compounds not required for the growth of the micro-organisms, and (d) additives produced by introducing chemically synthesized genes into the micro-organism chosen as the 'factory'. The economic importance of additives is exemplified by the world market of amino acids (primary metabolites) which is about $1700 million/year.

In the production of enzymes, genetic engineering techniques can be applied at different levels: (1) introducing multiple copies of the gene codifying for the enzyme (genetic amplification) to increase its yield, (2) manipulating the regulatory mechanisms to maximize gene expression or to control gene expression, and (3) manipulating the genes responsible for the formation of the cellular membrane in order to create leaky mutants. Recent applications of microbial enzymes are the use of proteinases in brewing and meat tenderizing, amylases in baking and brewing, α-amylase, glucamylase and glucose isomerase in the production of high-fructose corn syrups from starch and rennin in cheesemaking. The production of cheaper and purer enzymes through genetic manipulation will undoubtedly help to develop new applications in the food processing industry. An example of genetic amplification is the 500-fold increase in the yield of DNA ligase by inserting multiple copies of the ligase gene into *E. coli* (Eveleigh 1981). DNA ligase is one of the fundamental enzymes in recombinant DNA technology.

Micro-organisms, due mainly to their fast metabolism and the specificity of reactions which they are able to carry out, are tiny chemical 'factories' specializing in the production of small molecules such as amino acids and other primary and secondary metabolites. An advantage of fermentation over chemical synthesis is the use of renewable raw materials.

The use of genetic engineering techniques in the production of primary and secondary metabolites employed as food additives holds particular promise in amino acid production. For example, methionine is one of the eight essential amino acids (i.e. it cannot be synthesized by man) and hence must be supplied in the diet. Because most cereal grains are deficient in it, methionine is synthesized chemically (105,000 tons/year). The microbial production in high enough yields has been a failure because the strain development programmes were based on mutation and selection, and these methods cannot change the biochemical pathways in the strains of interest. Using genetic engineering techniques, we should see in the near future the introduction of the necessary new pathways in the appropriate strains.

Some amino acids have already been produced in laboratory scale quantities employing recombinant DNA technology, e.g. the essential amino acid threonine. Ajinomoto, the company producing it by fermentation, has patented the process in Japan (Morris 1981). Proline has also been produced by recombinant DNA technology by Bethesda Research Laboratories (USA) (Morris 1981).

The most representative example of the use of genetic engineering technology in the production of a food additive is probably the case of aspartame, a low-calorie nutritive dipeptide sweetener (L-aspartyl-L-phenylalanine methyl ester), which is 200 times more potent than sucrose. It was approved by the Food and Drug Administration (FDA; USA) in July 1981, and its use is already permitted in Belgium, Brazil, France, Mexico, the Philippines and Switzerland. Currently precursors such as aspartic acid are produced by fermentation. However, at the laboratory level, the

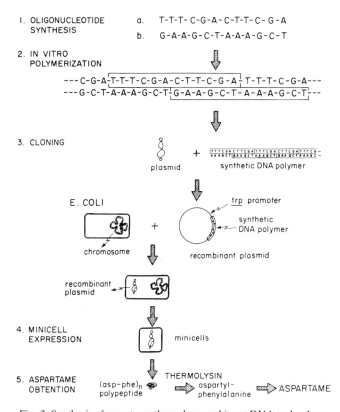

Fig. 3. Synthesis of aspartame through recombinant DNA technology.

production of the dipeptide aspartyl-phenylalanine using the most sophisticated techniques in the recombinant DNA arsenal was reported in 1980
(Doel *et al.* 1980). The strategy (Fig. 3) was as follows:

(1) Chemical synthesis (phosphotriester methodology) of the oligonuc-
    leotide T-T-C-G-A-C, which codes for phenylalanine-aspartic acid.
    Two oligonucleotides (dodecanucleotides) partially complementary
    and capable of overlapping self-hybridization were synthesized.
(2) *In vitro* polymerization of the synthetic oligonucleotides.
(3) Cloning of the synthetic DNA in *E. coli* using a plasmid as vector. The
    plasmid (pWT121) has upstream of the cloned DNA a promoter (*trp*)
    that controls its expression.
(4) Development of minicells able to express the synthetic DNA. Lysis of
    the minicells.
(5) Enzymatic treatment of the synthetic polypeptide with the enzyme
    thermolysin. This results in the obtention of the dipeptide aspartyl-
    phenylalanine, which in turn can be methylated to yield aspartame.

Aspartame is the only new sweetener approved by the FDA in almost 25
years. It may be in the stores at the end of this year with the trade name
'Nutrasweet', commercialized by the company that discovered and produced
it (Searle) together with other licensed firms (i.e. Ajinomoto). The market
for aspartame is estimated to be $450 million/year in 1985, which is almost
one-third of this year's total amino acid market. Future developments that
may be expected are, for example, the introduction of the plasmid containing
the synthetic DNA in a food-grade micro-organism.

5. Conclusions

The programme of genetic improvement of strains involved in the production of food and food additives have, until very recently, been based on
random mutation and selection of the mutants showing an improvement in
the desired product or commercial characteristic. These programmes were
costly, labour intensive and time consuming. Genetic engineering allows us
to design improved strains through the rational manipulation and recombination of genes. This new technology speeds up the programmes for strain
improvements. We have described here the realities of applying genetic
engineering techniques to food production processes together with an
advancement of future developments.

The future impact of genetic engineering in the food processing industry
will not be felt over the whole industry, but in a piecemeal fashion largely as
a result of the lack of biochemical and genetic knowledge of the commer-

cially interesting functions, to the production of low profit items and to the conservativeness of the food industry to invest in research and development. The latter reason suggests that in the next decade we will see a number of companies with a high technology base entering the food processing field. Also, some food processing industries will invest in genetic engineering firms to strengthen their lack of in-house research capabilities.

For the application of genetic engineering we will have to find alternatives for *E. coli*, which today is used for most genetic manipulations. *Escherichia coli* is the best-understood bacterium at the biochemical and genetic level, but it has the disadvantage of producing endotoxic byproducts and being potentially pathogenic. Micro-organisms generally regarded as safe (GRAS) are needed as hosts for genetic engineering. *Bacillus subtilis* and the yeast *Saccharomyces cerevisiae* are likely candidates. *Bacillus subtilis* is non-pathogenic, does not produce toxins, is relatively well-known genetically, excretes proteins easily and its large-scale fermentation requirements are well known (Sherwood & Atkinson 1981).

In the near future, we will see new genetic engineering techniques, like site-directed mutagenesis, which adds a new dimension to genetic engineering, being applied in the development of strains to be used in food processing. This means that genetic engineering is going to continue helping in a very humanitarian field—feeding all of us.

6. References

ANDERSON, D. G. & McKAY, L. L. 1977 Plasmids, loss of lactose metabolism, and appearance of partial and full lactose-fermenting revertants in *Streptococcus cremoris* B. *Journal of Bacteriology* 129, 367–377.

ANON. 1981a *Impacts of Applied Genetics*. OTA-HR-132, Office of Technology Assessment, 20402. Washington D.C.: U.S. Government Printing Office.

ANON. 1981b *Chemical Week* May 20, p. 33.

CLEWELL, D. B. 1981 Plasmids, drug resistance, and gene transfer in the genus *Streptococcus*. *Microbiological Reviews* 45, 409–436.

COONEY, C. L., RHA, C. & TANNENBAUM, S. R. 1980 Single cell protein: engineering, economics and utilization in foods. *Advances in Food Research* 26, 1–52.

DAVIES, F. L. & GASSON, M. J. 1981 Reviews of the progress of dairy sciences: genetics of lactic acid. *Journal of Dairy Research* 48, 363–376.

DOEL, M. T., EATON, M., COOK, E. A., LEWIS, H., PATEL, T. & CAREY, N. H. 1980 The expression in *E. coli* of synthetic repeating polymeric genes coding for poly(L-aspartyl-L-phenylalanine). *Nucleic Acids Research* 8, 4575–4592.

EVELEIGH, D. E. 1981 The microbiological production of industrial chemicals. *Scientific American* 245, 120–130.

HOPWOOD, D. A. 1981 The genetic programming of industrial micro-organisms. *Scientific American* 245, 66–78.

JACKSON, D. A. 1981 *Trends in the Biology of Fermentations for Fuels and Chemicals* ed. Hollaender, A. pp. 187–200. New York: Plenum Press.

KEMPLER, G. M. & McKAY, L. L. 1981 Biochemistry and genetics of citrate utilization in *Streptococcus lactis* subsp. *diacetyllactis*. *Journal of Dairy Science* **64**, 1527–1539.

McKAY, L. L. 1978 Micro-organisms and their instability in milk and milk products. *Food Technology* **32**, 181–185.

McKAY, L. L. & BALDWIN, K. A. 1978 Stabilization of lactose metabolism in *Streptococcus lactis* C2. *Applied and Environmental Microbiology* **36**, 360–367.

MILLS, O. E. & THOMAS, T. D. 1981 Nitrogen sources for growth of lactic streptococci in milk. *New Zealand Journal of Dairy Science and Technology* **16**, 43–55.

MIYASAKA, Y., RHA, C. & SINSKEY, A. J. 1980 Application of temperature-sensitive mutants for single-cell protein production. *Biotechnology and Bioengineering* **22**, 2065–2079.

MORRIS, C. E. 1981 Swift restructures, prepares for future. *Food Engineering* May, 57–69.

ROSE, A. H. 1981 The microbiological production of food and drink. *Scientific American* **245**, 95–104.

SENIOR, P. J. & WINDASS, J. 1980 Biotechnology: a hidden past, a shining future. In *Proceedings of the 13th International TNO Conference, Rotterdam*, ed. Verbraeck, A. pp. 97–101. The Hague, Netherlands: Netherlands Central Organization for Applied Scientific Research.

SHERWOOD, R. & ATKINSON, A. 1981 New microbes for genetic engineers. *New Scientist* **91**, 665–668.

STEWART, G. G. 1981 The genetic manipulation of industrial yeast strains. *Canadian Journal of Microbiology* **27**, 973–990.

WANG, D. I. C. 1968 Proteins from petroleum. *Chemical Engineering* **75**, 99–108.

WALSH, P. M. & McKAY, L. L. 1981 Recombinant plasmid associated with cell aggregation and high frequency conjugation of *Streptococcus lactis* ML3. *Journal of Bacteriology* **146**, 937–944.

The Potential for Fermentation Processes in the Food Supply

D. C. BULL AND G. L. SOLOMONS

*RHM Research Limited, The Lord Rank Research Centre,
High Wycombe, Buckinghamshire, UK*

Contents

1. Introduction

Fermentation has been used for the production of food and beverages for thousands of years. Traditional processes are used for the production of many foodstuffs such as beer, bread and cheese. Only relatively recently has the fermentation process been considered for producing other foodstuffs on a commercial scale. The legislative requirements, high cost of substrate and capital intensity of the process inevitably lead to the conclusion that only high value products will be economically viable.

Microbial protein will only be an economic success if it can be sold in direct competition with meat or other products of high value. Vitamins, colourings and flavourings are other areas of existing or potential use for the process. However, each new food product made by fermentation will be subjected to

FOOD MICROBIOLOGY
ISBN 0 12 589670 0

close scrutiny by the legislative authority before clearance is given for sale to the public, and this may be an expensive and time-consuming exercise.

Current engineering knowledge has enabled very large fermenters to be designed, built and operated, and there is wide scope for organism, equipment and product development, although none of these are likely to provide cheap products, either for the Western or Third Worlds.

2. Traditional Fermentation Processes

Man has been using the fermentation process for thousands of years, both by accident and design, for the production of various foodstuffs and beverages. Today we usually ferment by design and for the food and drink industry it is big business.

In the UK alone, we brew about 20 million litres of beer per day, ferment almost 250,000 tonnes of cheese per annum, 100,000 tonnes of bakers' yeast p.a. and ferment and bake more than 1 million tonnes of bread p.a. by the traditional (non-Chorleywood) process. The total annual sales value of these items alone is in excess of £6000 million (or well over £100 p.a. per man, woman and child in the UK).

Considerable research and development work has been and is being done to improve the yields and efficiencies of these processes and hence reduce or stabilize the costs associated with production. An example is the work on the continuous production of beer using tower fermenters. So far these fermentations appear to have produced an adequate alcohol concentration but, unfortunately, the products have yet to match the flavour and organoleptic properties of traditional batch-brewed beers. This attempt to make the fermentation process continuous is a common approach, but while there may be improvements in economics, the technical problems may be considerable.

Other improvement areas which may be considered include alternative substrate sources, better fermenter designs in terms of mass and heat transfer, and more efficient product collection. Recycling of trace metals and vitamins can also improve the overall economics. Looking further into the future, genetic engineering may give improved strains of organism, again, marginally improving overall economics, but it is unlikely that any of these modifications will give startling reductions in costs.

3. Costs of New Processes

Only during relatively recent times has man deliberately considered fermentation for creating something new from readily available foodstuffs. If we are to expand fermentation in the food supply business, what areas should we consider—what products should we make; or perhaps, what products could we afford to make?

Assuming that the legislators will only permit the use of food grade materials to make foodstuffs, one of the most plentiful substrates available will be carbohydrates, the bulk prices of which vary quite widely (Table 1).

TABLE 1

Bulk prices for carbohydrate sources

Source	Price (£/tonne)
Sucrose	350
Glucose	300
Starches	200–300
Molasses	90

Sucrose and glucose can be used directly, but the starches generally have to be solubilized and hydrolysed and the extra costs involved bring the substrate cost almost up to glucose prices. Molasses needs significant pre-treatment before use and may give colour problems with the finished product. Substrate, therefore, will cost approximately £300/tonne. As a rule of thumb, for a biomass product, a tonne of product will be produced from two tonnes of substrate, so the total substrate cost will be around £600/tonne of biomass produced. Production costs, capital charges, etc. will add, probably, another £300–£400/tonne to the cost, which brings the selling price of a simple product to approximately £1000/tonne.

4. Potential Fermented Products

Table 2 lists approximate food and allied product prices in 1980. Meat prices are 'on the bone', and butchered; 'off the bone' prices would be considerably higher (e.g. stewing steak, £1600/tonne; fillet steak, £4000/tonne).

Fermentation processes can produce products similar to many of those in Table 2 (apart from vitamin B_{12}, cheese, yoghurt and bread, which are of course produced by fermentation).

A. Lipids

Some yeasts and fungi can produce 50% or more of their cell mass as lipid. However, the oils listed in Table 2 are all relatively cheap, with predicted

TABLE 2

Approximate wholesale food and allied product prices (1980)

Product	Price (£/tonne)
Vitamin B_{12}	2,300,000
Natural colourings	50,000
Artificial colourings	10,000
Flavourings	5000
Haddock fillets	2500
Beef	1700
Lamb	1600
Cheese	1300
Pork	1000
Chicken	900
Extruded soya	700
Yoghurt	600
Castor oil	500
Bread	300
Palm oil	300
Soya oil	300
Tallow	300
Milk	200

Data from Anon. (1981) and Dunnill (1981).

future over-production both in developed and underdeveloped countries (viz. European farmers are growing more and more rapeseed, and the underdeveloped countries are producing more and more palm oil). With the high cost of substrates, low yields and high processing costs, their production by fermentation may never be economic.

B. Proteins

Products in the price range of £1000/tonne and above include meats and fish. The markets here are well established and the scope for market penetration is considerable. Meat consumption in the UK is in excess of 4 million tonnes p.a. and 23 million tonnes p.a. in the EEC.

RHM has now developed a process for producing meat-like analogues, and are able to simulate 'chicken' and 'ham' pieces, which are virtually indistinguishable from the real thing.

The process has been developed over a period of about 12 years and is now at a stage where clearance has been obtained from the Ministry of Agriculture, Fisheries and Food for carrying out test market surveys on the finished product.

It has always been our intention to produce high-value products aimed at the replacement market rather than the extender market, although in the very early days, we were hopeful that the process could be used in the underdeveloped countries to convert surplus carbohydrate materials into valuable protein for the undernourished. Clearly, economics dictate against this when there is competition with proteins such as soya and skim-milk powder in the £200–700/tonne price range.

Our basic process is the continuous production of *Fusarium graminearum* on a glucose substrate in submerged culture. The additional processes required to produce a meat- or fish-like product are expensive and include steps to reduce the RNA content to acceptable levels for human food intake, as well as texturization and the addition of flavouring and/or colouring.

By far the major part of the work has been in satisfying the legislative authority of the toxicological safety of our product. The submission to MAFF extended to 24 volumes of data concerning every aspect of the process from raw material specification to mycotoxin assay methods used, from the pH control system used to the number of our volunteers who actually enjoyed eating the product.

This was a very expensive exercise and carried the additional penalty of our being the first to submit such a product to MAFF for clearance. Any company manufacturing a new product of fermentation for human consumption will have to satisfy the legislative authority in this manner, although now that the first submission has been covered, the machinery is available for handling future submissions.

C. Flavours and colours

At the top of the price range are food flavourings and colourings. Considerable work is being done in the fermentation preparation of flavours and colours, where the tendency is for synthetic products to lose the approval of legislative authorities and of the consumer. It has to be borne in mind, however, that every new product will have to satisfy the authorities from a toxicity viewpoint, and the costs of testing may prove prohibitive, particularly for relatively small markets (measured in tens, rather than thousands, of tons).

Nevertheless, the Japanese are currently investigating a red pigment from the microfungus species *Monascus*. Traditionally, *Monascus purpureus* has been grown on polished rice. The fermented rice is deep red in colour and it is the intensity of this colour, and the added fact that it has been consumed for hundreds of years, that has attracted the attention of the investigator. The problems with these natural products are mainly concerned with colour stability under heat, light and differing pH conditions.

D. Edible gums

Gums are also important food additives, and, here again, the products are valuable. Traditionally, gums are obtained from plant extracts, tree exudates and seaweeds. Some 20,000 tonnes of gum tragacanth are used annually within the EEC, with a sales value of some £40 million. Tate & Lyle have recently installed a plant for the commercial production of xanthan gum from *Xanthomonas campestris*. The process is reported to give finished products of high and consistent quality.

In summary, as the food supply stands today, the fermentation process will only be economically viable for: (1) traditional foodstuffs' manufacture employing fermentation; (2) high-value, 'up-market', novel foods; (3) high-value, small tonnage food additives.

5. Strengths and Weaknesses of the Fermentation Process

The fermentation process as currently operated has a number of strengths and weaknesses in commercial use.

A. Renewable resources for feedstocks

Using plentiful carbohydrates such as sugars and starches means that acceptable foodstuffs can be made *now* using fermentation techniques. However, with substrate costs currently constituting over half of the finished product cost, cheaper substrates will have to be found for the expansion of fermentation into general markets. Cellulose is a possibility; the world's vegetation produces over 10^{11} tonnes p.a. of cellulose (Vlitos 1981) much of which is wasted. Nearly half of the 9–13 million tonnes p.a. of straw produced in the UK is burned. New techniques in chemical engineering and biotechnology will probably lead to the conversion of this waste, economically, into fermentable sugars.

B. Produce materials needing very little further purification

High quality foodstuffs obtained by fermentation processes will be produced in axenic culture in which only the desired micro-organisms will be grown. In this way, products of the fermentation can be predicted and will require little or no purification after separation.

In the case of metabolite production, organisms can be mutated, and even genetically engineered, to maximize yields of the desired product.

C. Scope for future process improvements

Compared with most food processes, aseptic fermentation is in its infancy. There is considerable scope for development of organisms, equipment and products, although each major change in products will probably require further clearance from the legislative bodies.

D. Capital intensive operations

Commercial fermenters are generally large, and the surfaces in contact with foodstuffs are generally made of stainless steel. The size of the plant is dictated by the fact that fermentation usually takes place only in dilute solutions (Table 3).

TABLE 3

Concentration of fermentation products in the broth

Product	Concentration (% dry wt/vol)
Lactic acid	13
Ethanol	8
Bakers' yeast	5
Microbial protein	2
Vitamin B_{12}	0·002

Data from Dunnill (1981).

As biomass concentrations increase so do problems of heat and mass transfer. Specific heat exchange surface areas therefore increase and gas/liquid mixing design has to be improved.

Separation of biomass from fermentation broths can be difficult and centrifuges or vacuum filters are generally used. However, because the specific gravity of micro-organisms is close to that of water, and a suspension of micro-organisms in water tends to be slimy and viscous, low centrifugation and filtration rates are to be expected. This leads to large equipment being specified; equipment that must be cleaned quickly and easily to food-handling standards.

Large volumes of highly biodegradable effluent are produced which in the short term can be further processed to give animal feed or biogas, or in the longer term recycled to the fermenter.

E. Contamination

The majority of new, continuous fermentation processes being developed are carried out aseptically, since the fermentation conditions used may favour the growth of undesirable organisms. Traditional batch fermentation, such as bakers' yeast and beer, are usually produced non-aseptically. The problems associated with building very large plants and operating them aseptically were daunting, but with the developments in rotating shaft seal design, the use of air lift mixing, and continuous development in absolute air and liquid filters, the problems appear to have been largely overcome; RHM regularly operate $1\cdot3$ m^3 vessels under continuous culture, growing *F. graminearum* in thousand-hour runs. We now expect less than one contamination per year of fermenter operation. This one failure per year is generally a result of operator error rather than equipment failure.

F. Yields

Generally speaking, in biomass production, only one-half of the substrate is converted to a useful product. The remainder is released as carbon dioxide.

In metabolite production, however, yields range from almost zero in producing products like vitamins, to 90% in lactic acid manufacture. Low yields may be improved (e.g. amino acids and vitamins) by gene amplification and blocking metabolic pathways. These facts again illustrate the need to aim at high-value products in the food business as the theoretical yield on high-cost substrate is likely to be low.

6. Conclusion

We do not, therefore, foresee the fermentation process as being able to provide cheap protein for human consumption either in the Western world or in the developing countries. Its use will be limited to providing alternative high-value foods or food additives as long as the use of high-cost food grade substrates is required by legislative authorities.

7. References

ANON. 1981 *Farmers Weekly* 1st May.
DUNNILL, P. 1981 Biotechnology and industry. *Chemistry and Industry* No. 7, 204–217.
VLITOS, A. 1981 Natural products as feedstocks. *Chemistry and Industry* No. 9, 303–310.

Sampling Programmes for the Microbiological Analysis of Food

D. C. KILSBY AND A. C. BAIRD-PARKER

Unilever Research, Colworth Laboratory, Colworth House,
Sharnbrook, Bedfordshire, UK

Contents

1. General Considerations

The food microbiologist has struggled to control the variability in his counting procedures since he first began to practise his science, indeed the need to achieve reproducibility has taken up a large amount of available experimentation time and facilities. The development of food microbiology techniques has always been influenced and retarded by the need to compare new systems with inexact, existing methods (Sharpe 1980). In order to handle new developments, the microbiologist has been forced increasingly to adopt more sophisticated methods for comparing counting procedures, until now, when no comparison of methods is regarded as valid unless endorsed by a full statistical analysis. Unfortunately such analyses only confirm the inexactitude of the methodology, a fact of which the micro-biologist was already too well aware. His only recourse, therefore, is to produce more sophisticated and expensive experiments in an attempt to obtain a meaningful conclusion. This has led to a move away from straight comparisons within the laboratory, to comparisons based upon variation between analysts and between laboratories. This is the unfortunate, but inevitable, result of our inability to apply the available methods universally, with success, i.e. the methods rely too much upon the skill of the technician.

This then is the position in which the food microbiologist finds himself, with the consequence that he has no faith in the analyses he carries out. This position is very clear when dealing with microbiological criteria for foods, to

FOOD MICROBIOLOGY
ISBN 0 12 589670 0

which very few people give credence. Under such circumstances it is probably wise to ask why such a situation has developed. Most of the microbiologist's effort is taken up in trying to control his analytical technique, yet very little effort is made in investigating the natural variability in numbers of micro-organisms in the materials analysed.

The basic unit of interest to a food microbiologist is a batch, or lot. This term, as used by food-microbiologists, is not defined here and is only important if it has an effect upon the accuracy of analyses carried out as measurements representative of the batch. It is important in this context to consider the variability of analytical results obtained for samples from a batch. There is often a significant, and in many cases large, difference between replicate analyses (e.g. microbial counts) from a batch of material tested.

Any attempt to set a microbiological specification must therefore take account of this variability. Any system based upon a single sample unit analysis must, consequently, set a limit sufficiently high to allow for this observed variability. Many food microbiologists find it unacceptable to operate with the inevitable high limit required, which seems sensible because the operation of such a limit does not provide useful differentiation between batches of good and bad quality. That is, a batch yielding a result above the limit set may well be unacceptable, but a batch providing a sample with a result below the limit may, or may not, be acceptable. One alternative is to set the limit such that the reverse is true, i.e. a result below the limit is probably from a satisfactory batch, but a count above may or may not be from an unsatisfactory batch. Both approaches are unacceptable because the first accepts a high proportion of unacceptable batches and the second rejects a high proportion of acceptable batches, hence an alternative system of handling the variability is required.

A suitable approach adopts multiple analyses for one batch. This is generally an unpopular method because of the increase in analytical work required. Its strength, however, is in its ability to accommodate variability between replicate sample units. The International Commission for Microbiological Specifications for Foods (ICMSF) has proposed two such schemes (Anon. 1974). Both of their schemes recognize each individual sample unit analysis as belonging to a particular concentration range. Because belonging to this range is an attribute of the sample unit, these schemes are, therefore, attribute sampling plans. These plans are very much more powerful than those using a single limit value, and in many of the specific plans given by ICMSF, tolerances are given (i.e. a proportion of the sample units comprising the sample are allowed to exceed a specified limit). In this way the variability of the batch is allowed for. Basically these schemes accept or reject batches depending upon the proportion of units in the batch falling within each attribute. Table 1 gives an example of such a scheme where the group of five sample units is examined ($n = 5$) and up to two of those five sample units can have the attribute of lying between the two limits (m) and (M). None of the sample units may possess the attribute of having a concentration in excess of (M). This scheme defines three attributes (less than (m), between (m) and (M), and greater than (M)) and is therefore a 3-class attribute plan. In Table 1, the performance of such a scheme can be seen with results from three batches. By using this scheme batches 1 and 3 are rejected and batch 2 accepted. Observation of the individual sets of concentrations for each batch will show very little justification for making such a distinction between the three sets of results. This is because the attributes schemes have no way of allowing for the concentration measured (i.e. in the example given a concentration of 1.3×10^5/g in batch 1 is assigned to the same class as the concentration 8.5×10^5/g in batch 2).

This disadvantage can be overcome by using a variables sampling plan, in which each sample unit contributes to the total plan according to its actual concentration. In order to do this, however, it is necessary to recognize and describe the distribution of micro-organisms within units of a batch.

TABLE 1

Operation of the ICMSF 3-class attribute sampling plans

Scheme: 5 sample units per batch
$m = 10^5$/g $M = 10^6$/g
Allow 2 or less sample units between m and M; no sample unit to exceed M.

Batch 1	Batch 2	Batch 3
8.0×10^4	3.0×10^4	7.1×10^4
9.6×10^4	6.2×10^4	8.6×10^4
1.3×10^5	9.3×10^4	9.7×10^4
2.0×10^5	3.0×10^5	2.0×10^5
2.6×10^5	8.5×10^5	1.1×10^6

Action: Reject batches 1 and 3; accept batch 2.

2. The Log-normal Distribution

One distribution which might be used to describe the relationship between concentrations within units is 'log-normal'. There is no theoretical justification for any distribution, but the log-normal distribution appears to fit observations. Table 2 shows the results of a test for normality (Fillibens 1975) carried out on the logarithms of concentrations for total viable counts obtained for batches of foods. By chance in the particular analysis used, 5% of batches should appear to be non-log-normal if the underlying distribution is in fact log-normal. The actual average value is 7·8%, with a maximum of 12·7% for powdered products and a minimum of 2·4% for frozen vegetable products. The distribution is therefore very close to log-normal.

Using statistics related to the log-normal distribution, it is possible to produce a variables sampling plan (Burr 1976) which can be adapted for schemes with low sample unit numbers (Kilsby et al. 1979). An example of such a scheme is given in Table 3 (the k values used are obtained from statistical tables). The use of a variables sample scheme allows for a much clearer definition of a specification (see Table 3) and is only sensitive to the proportion of the batch which exceeds the limit. A secondary advantage of the variables sampling plan is its ability to be modified to be used for operating good manufacturing practice (GMP) standards (Kilsby et al. 1979). Table 4 compares the ICMSF sampling plan and the variables sampling plan showing that the attributes scheme fails to give a consistent probability of rejection as the proportion of the batch exceeding M increases.

There are other important consequences to the food microbiologist of a log-normal distribution of micro-organisms with batches tested. These result from the relationships of means for such a distribution. Table 5 gives a comparison of some means for a sample. Three means are given, the log mid interval value (MIV), the mean log and the log mean. The log MIV and the mean log are clearly very similar, the log mean is, however, larger than both. It matters which mean is selected. The food microbiologist is interested

TABLE 2

Log normality of total viable count in batches of foods

Commodity group	No. of suppliers	No. of batches not log-normal	No. of batches examined	% of batches not log-normal
Frozen fish products	2	35	518	6·7
Frozen crustacea	1	31	393	7·9
Frozen meat products	2	13	159	8·2
Frozen vegetable products	1	1	41	2·4
Frozen dairy products	1	5	52	9·6
Powdered products	2	15	118	12·7
OVERALL	6	100	1281	7·8

in the number of micro-organisms within the batch tested and the log mean is the value of interest. The value obtained, however, is an estimate of the mean log (Kilsby & Pugh 1981) and these two values are related as indicated in Table 6. The count obtained usually underestimates the true count required, the amount depending upon the variance (s^2) of the log concentrations. The variance for batches of foods may be very large, Table 7 gives a range of typical standard deviations. The counting error can, therefore, be very large.

TABLE 3

The application of a variables scheme

Standard: 90% sure of rejection of a batch if more than 10% exceeds 10^7/g

$$k_1 = 2 \cdot 8 \, (\bar{x} + k_1 s < 7)$$

Sample: sample unit	1	300,000/g
	2	125,000/g
	3	1,200,000/g
	4	500,000/g
	5	1,000,000/g

$$\bar{x} = 5 \cdot 67 \qquad\qquad s = 0 \cdot 40$$

Calculation: $\bar{X} + k_1 s = 6 \cdot 79$

Accept the batch

TABLE 4

Comparison of an attributes and variables scheme

Probability of rejecting a batch using 5 sample units:

variables scheme $P = 0 \cdot 99, p = 50\%$
attributes scheme $n = 5, c = 3$

Proportion	V or M*	0·10	0·20	0·30	0·40	0·50
Pr (variables)		0·81	0·90	0·95	0·98	0·99
Pr (attributes)	0·1†	0·41	0·67	0·83	0·92	0·97
	0·3	0·48	0·70	0·85	0·94	1·00
	0·5	0·68	0·80	0·93	0·98	1·00
	0·7	0·94	0·96	1·00	1·00	1·00

*V the upper limit for the operation of the variables sampling plan; M the upper limit defined by ICMSF.
†Proportion of the batch between m and M.

TABLE 5

Comparison of means for a sample of five sample unit counts from a batch

Count	Log count	Log MIV*	Mean log	Log mean
$2 \cdot 0 \times 10^3$	3·301			
$9 \cdot 9 \times 10^3$	3·996			
$5 \cdot 2 \times 10^3$	3·716	3·556	3·530	3·642
$1 \cdot 2 \times 10^3$	3·079	$(3 \cdot 6 \times 10^3)$	$(3 \cdot 4 \times 10^3)$	$(4 \cdot 4 \times 10^3)$
$3 \cdot 6 \times 10^3$	3·556			

*Log MIV, log of the mid interval value (median).

TABLE 6

Relationship between arithmetic and geometric means

$$\log A = \overline{X} + a.s^2$$

where
A is the arithmetic mean ($\log A = \log$ mean)
$\overline{X}$ is the mean logarithm (to the base 10)
$a = \ln.10/2$
$s^2 = $ variance

TABLE 7

Some examples of mean logs and their standard deviation for a range of food product total counts

	Mean log	Standard deviation*
Frozen beef	5·483	1·363
Fish	4·912	0·385
Comminuted beef	6·024	0·267
Skim milk powder	3·604	0·093

*Variance = [standard deviation]².

There are many implications to the food-microbiologist of the log-normal distribution of micro-organisms within a batch of food (Kilsby & Pugh 1981), all affecting the precision of the estimate of concentration obtained. Many of the apparent inaccuracies observed at the moment may be explained by this distribution, and recognizing it enables the food microbiologist to introduce more precise methods of measurement.

3. Conclusion

The application of microbiological criteria (guidelines, specifications and standards: see Ch. 19, this volume) will always be an inefficient and imprecise method of exercising microbiological control over batches. Such control is best exercised by the monitoring of critical control points, identified by risk analysis of a particular food operation, and controlled by non-microbiological methods and procedures wherever possible, e.g. by measurement of temperature and/or relative humidity, by design of food plant and the implementation of properly validated cleaning procedures. Where the application of microbiological sampling for monitoring and/or control is the only option, without doubt the use of standard variables level sampling plans will give the most precise data. However, such plans should only be used with an understanding of the degree of precision with which decisions are taken concerning a particular batch of product or the status of a critical control point. If the degree of confidence with which a decision is made is fully appreciated by the person making a decision on the results of a microbiological analysis, such analysis will continue to be a valuable adjunct to other control procedures used to assess quality and safety of foods.

4. References

ANON. 1974 *Micro-organisms in Foods Vol. 2. Sampling for Microbiological Analysis: Principles and Specific Applications.* An ICMSF publication. Toronto: University of Toronto Press.

BURR, I. W. 1976 *Statistical Quality Control Methods.* New York & Basel: Marcel Dekker.

FILLIBENS, J. J. 1975 The probability plot correlation coefficient test for normality. *Technometrics* 17, 111–117.

KILSBY, D. C. & PUGH, M. E. 1981 The relevance of the distribution of micro-organisms within batches of food to the control of microbiological hazards from foods. *Journal of Applied Bacteriology* 51, 345–354.

KILSBY, D. C., ASPINALL, L. J. & BAIRD-PARKER, A. C. 1979 A system for setting numerical microbiological specifications for foods. *Journal of Applied Bacteriology* 46, 591–599.

SHARPF, A. N. 1980. *Food Microbiology, A Framework for the Future.* Springfield, Illinois, USA: Charles C. Thomas.

Guidelines, Specifications and Standards for Foods

B. SIMONSEN

Danish Meat Products Laboratory, Ministry of Agriculture,
Copenhagen, Denmark

Contents

1. Introduction

It is probably correct to presume that everybody interested in applied food microbiology keeps a file with reprints, photocopies etc. on microbiological criteria for foods, a file which constantly increases in volume. So much has been said about microbiological criteria that one might soon expect to see the end of it, but this is unlikely to happen. There has been—and still is—much confusion about definitions of criteria, and of their establishment and application, and much remains to be settled about numerical limits and sampling plans.

FOOD MICROBIOLOGY
ISBN 0 12 589670 0

This paper does not attempt to give the final answers to the questions raised over the years, but will try to summarize recent activities on definitions and establishment of microbiological criteria for foods.

Guidelines, specifications and standards are now three words that are generally accepted to designate three different applications of microbiological criteria. Other words employed include: recommended limits, limit values, norms, and the recently introduced term, reference values (Mossel 1977). Although all these words may have a specific meaning in the English language, much confusion arises when they are translated into other European languages in the European Community.

In this paper the generally accepted international definitions of the different criteria will be presented, and the elements necessary for the establishment and application of microbiological criteria for foods will be discussed.

2. Definitions of Microbiological Criteria

The first generally accepted definitions of microbiological criteria for foods were given by an NAC/NRC Committee (Anon. 1964). The word 'guideline' was not employed, but the term 'recommended limit' was used instead and defined as: "the suggested maximum acceptable number of micro-organisms or of specific types of micro-organisms, as determined by prescribed methods in food". This definition makes no reference to the application of this particular criterion.

A 'microbiological specification' was defined by the committee as "the maximum acceptable number of micro-organisms or of specific types of micro-organisms, as determined by prescribed methods, in a food being purchased by a firm or agency for its own use". This clearly refers to a commercial situation rather than to its use by an official agency having jurisdiction.

Finally a microbiological standard was defined as "that part of a law or administrative regulation designating the maximum acceptable number of micro-organisms or of specific types of micro-organisms, as determined by prescribed methods, in a food produced, packed or stored, or imported into the area of jurisdiction of an enforcement agency". This definition has undergone only slight changes over the years. It is, however, deplorable that the word 'standard' is often given other meanings, such as the more neutral, summarizing term 'criterion', and sometimes meaning a guideline or a specification—or just a numerical limit.

The word 'guideline' was introduced by Elliott (1970) as "that level of bacteria in a final product, or in a shipped product, that requires identity and correction of causative factors in current and future production, or in handling after production". The definition omits to specify the enforcement agency (e.g. an official agency or a manufacturer's quality control department) but it explains clearly the object of a guideline, which is not to reject food from the particular batch being examined, but to take corrective steps to improve the microbiological condition of future productions of that specific food. Olson (1971) simplified the definition and indicated more clearly that a 'guideline' could be used by an official agency: "(It) . . . specifies microbial limits which, when coupled with plant inspection showing substantial insanitary conditions, provide a basis for subsequent actions".

Although being very active in the evolution of microbiological criteria and sampling plans for a number of foods the International Commission on Microbiological Specifications for Foods (Anon. 1974) has added nothing new to the definitions from the NAC/NRC Committee, and it does not define a guideline. It is noteworthy that this commission, which dates from 1962, employs the term 'specifications' in its title. The term 'criteria' seems more appropriate today in relation to its activities.

The term 'reference value' was introduced in food microbiology by Mossel (1977) because of the often observed emotional rection to the word 'standard'. He does not define the term directly, but the following definition would probably cover his intentions: "A reference value is a value somewhat above the 95th percentile in the distribution of data derived from adequate surveys assessing what is technologically attainable under good manufacturing practice".

This definition does not include any reference to the status of the criterion—whether it is mandatory or advisory, whether it should be used for commercial or for official purposes. As will be discussed later the definition contains the rather imprecise and undefined term 'good manufacturing practice', which may lead to difficulties in establishing reference values.

The definitions of microbiological criteria that now are most widely accepted internationally have been developed by two working groups set up by the Food and Agriculture Organization of the United Nations and the World Health Organization (Anon. 1977a, 1979). In the Codex Alimentarius Food Hygiene Committee a need was felt to have definitions for Codex purposes of mandatory and advisory criteria, as well as setting down the principles for the establishment and application of microbiological criteria. The final version of these definitions and principles appear in the report of the Codex Alimentarius Commission (Anon. 1981).

Although these definitions apply specifically to Codex Alimentarius—i.e. international—purposes, these may, with slight modifications, be useful for national official purposes. They may not be directly applicable to commercial purposes, for which the above-mentioned definition of 'specification' still seems valid.

For Codex Alimentarius purposes criteria are divided into two groups: advisory and mandatory. There are two advisory criteria, viz. a microbiological guideline and a microbiological end-product specification. A microbiological guideline is applied at the establishment at a specified point during, or after, processing to monitor hygiene. It is intended to guide the manufacturer and is not intended for official control purposes.

Thus, this definition is comparable with those proposed by Elliott (1970) and Olson (1971) quoted above. It may be used by an official agency for monitoring purposes, but not for control purposes.

The second advisory criterion is a microbiological end-product specification (not to be confused with 'commercial specification').

A microbiological end-product specification is intended to increase assurance that the provisions of hygienic significance in a code have been met. It may include micro-organisms which are not of direct public health significance.

As the name indicates this criterion applies to the final product, either immediately after manufacture, or—which would often be the case in a Codex Alimentarius situation—at the port of entry in international trade. For a number of food products in international trade the Food Hygiene Committee, the Committee on Processed Meat and Poultry Products, the Committee on Fish and Fishery Products and other Codex Alimentarius Committees have elaborated Recommended International Codes of Practice or Codes of Hygienic Practice. The need for microbiological criteria attached to such commodity codes has been felt, and an end-product specification, which is advisory (as is a code of hygienic practice) is the proper criterion for this purpose.

Under the Codex Alimentarius, Recommended International Standards for a number of food commodities have been elaborated. If accepted by the member countries of the Codex Alimentarius these standards become mandatory. Such standards often contain a hygiene section, and theoretically a microbiological standard could be appended to such a section. So far this has, however, not been the case. To make it possible to have such a criterion attached to a food standard, the Codex Alimentarius has adopted the following definition:

A microbiological standard is a criterion contained in a Codex Alimentarius Standard. Wherever possible it should contain limits only for pathogenic micro-organisms of public health significance in the food concerned. Limits for non-pathogenic micro-organisms may be necessary . . . A microbiological standard shall not be introduced de novo, but shall be derived from microbiological end-product specifications which have accompanied Codes of Practice through the Codex Procedure and which have been extensively applied to the food.

In case a member country accepts a Codex Standard, and if this contains a microbiological standard, the criterion will most likely have a status not very different from that contained in a microbiological standard as defined originally by the NAC/NRC Committee (see above). As the concern over a public health hazard is generally higher than the concern over a quality deterioration as a consequence of microbial growth, a microbiological standard would normally only specify limits for pathogenic organisms. It might also specify limits for non-pathogenic organisms. The development of a microbiological standard is important. It should have been derived from extensively applied end-product specifications, and not be developed directly for a Codex Standard. This would ensure especially that the cost/benefit of having a microbiological standard has been extensively evaluated.

A microbiological criterion has in most cases been synonymous with a numerical limit for a given organism or group of organisms. There is, however, more than just a numerical limit in a Codex Alimentarius criterion, whether it be a guideline, an end-product specification or a standard.

For Codex purposes, a microbiological criterion consists of five elements:

(1) A statement of the micro-organisms and parasites of concern and/or their toxins. For this purpose, micro-organisms include bacteria, viruses, yeasts and moulds.
(2) The analytical methods for their detection and quantification.
(3) A plan defining the number of field samples to be taken, the size of the sample unit, and where and when, if appropriate, the samples are to be taken.
(4) The microbiological limits considered appropriate to the food.
(5) The number of sample units that should conform to these limits.

Whether it is for Codex, national or trade purposes it is important to keep in mind that a microbiological criterion must consist of these five elements; otherwise the criterion is incomplete and unsatisfactory. Later these five elements will be discussed together with the other elements which should be considered for the establishment of microbiological criteria.

3. Previous Approaches to the Establishment and Application of Microbiological Criteria for Foods

Elliott (1970) quoted E. M. Foster of the Food Research Institute of Wisconsin, USA for the three ways microbiological criteria can be established: the bad, the better and the best way. The bad way is for a group of people with little or no knowledge of food microbiology and technology to 'pull a figure out of the air'.

A better way is to collect data on final products of more or less unknown history with respect to the manufacturing practice, and to settle on a figure based upon the distribution of those data. According to Foster (and Elliott) the best way is to relate the microbial levels to specific conditions of good sanitary practice. Before sampling products from a number of companies an inspector–microbiologist team should inspect a large number of plants preparing the same type of food. As there is for many foods a good correlation between microbial levels and hygienic observations, micro-biological criteria are based on products prepared according to good manu-facturing practice.

Provided we have an unambiguous understanding and definition of good manufacturing practice, the third way of establishing criteria is the best. The problem is that we do not yet have a clear agreement between micro-biologists and food inspectors of what constitutes good manufacturing practice. An inspector would probably concentrate on the hygiene and cleanliness of the plant, before or during operations. He would also satisfy himself that cleaning and disinfection is carried out properly, that clean and unclean operations are separated and that personnel observe the rules of hygiene. A microbiologist would think that the microbiology of the raw material, or the time–temperature conditions to which the food is exposed during processing, are important. Additionally, the really critical control point may be something quite different, e.g. a defective can-closing machine.

For simple food operations with few, clearly defined and regularly moni-tored critical control points the manufacturing practice could be charac-terized as good or bad. Our present knowledge on more complicated food processing operations does not allow us to make this characterization. For that reason we shall probably have to follow what Foster called the better way: compilation of data on final products, irrespective of their origin.

There have been, and still are, differences of opinion on where and when to apply a microbiological criterion. Generally a criterion is applied at the point where the final consumer will buy the food, i.e. in the retail shop. Experiences, however, from the state of Oregon, where microbiological standards were established and applied at the point of sale of ground beef, showed that the innocent had to suffer for the guilty, because the retailer was punished for exceeding an *Escherichia coli*-standard, although the raw meat was contaminated with that organism on arrival in the shop.

In Sweden there are several microbiological guidelines for a wide range of foods. These foods are examined on the 'sell-by' date, i.e. the criterion applies at the end of the estimated shelf life. For some foods, where the micro-organisms have entered the death phase before the food is organolep-tically unacceptable, the application of such a criterion does not seem to be logical.

The ICMSF (Anon. 1974) has so far no detailed policy concerning the establishment and application of numerical limits in microbiological criteria. The limits were based on the information available to the commission in the early 1970s, and it seems evident that not all that information had been based on a systematic collection of data. For this the ICMSF cannot be blamed, as such information was not available at that time, and will hardly be available in this decade. But another element in a microbiological criterion, the sampling plan, was studied intensely by the ICMSF, resulting in a clearer understanding of the most important factors for the stringency of a sampling plan, and resulting in a number of 'cases', that would apply to virtually all foods. The two main factors determining the case are the type of hazard, i.e. no direct health hazard, low-indirect, moderate-direct and severe-direct hazard, and the conditions in which the food are expected to be handled and consumed after sampling, i.e. whether the handling will reduce the degree of hazard, cause no change or increase the hazard. The more severe the hazard (dependent on the pathogenicity of an organism), and the higher the risk that handling before consumption will increase this hazard, the higher case number is given to the food and, in a 3-class sampling plan, the more samples should be taken, or fewer of the samples taken can be accepted as marginally defective. The ICMSF does not specify clearly when and where its recommended criteria, including sampling plans and numerical limits, should apply, but as the criteria seem more applicable to large consignments than to small batches in a retail shop, the ICMSF criteria should preferably be used shortly after production or at a port of entry in international trade.

4. FAO/WHO Approach to the Establishment and Application of Microbiological Criteria for Foods

The two working groups set up by the FAO/WHO to define microbiological criteria also described the general principles to be observed for the establishment and application of microbiological criteria. These general principles are to be followed in the Codex Alimentarius in case a microbiological criterion is to be attached to a Code of Hygienic Practice or to a Codex Alimentarius Standard. The principles are primarily intended for Codex purposes, but they are as useful for other international, national or regional agencies concerned with microbiological control of foods.

   Before establishment and application of a microbiological criterion there are a number of elements to be considered. These elements will be considered below in the form of questions, for which the consensus of answers should decide whether or not a criterion should be established.

A. Is there a need for a microbiological criterion?

A microbiological criterion should be established and applied only where there is a definite need for it. Such need is demonstrated by epidemiological evidence that a particular food is a public health hazard, or where an assurance is required that the provisions of hygienic significance have been adhered to.

This would necessitate a hazard analysis of the particular food, i.e. an assessment of hazards associated with growing, harvesting, processing/manufacturing, marketing, preparation and/or use of a given raw material or food product. In this context 'hazards' include contamination of food with unacceptable levels of food-borne, disease-causing micro-organisms and/or contamination with spoilage organisms to the extent that hazards occur within the expected shelf-life or use of the product.

It is evident that a hazard to human health is the most important demonstration of a need for a microbiological criterion, but depending upon the interpretation of 'hazard' there may be a need for assurance that hygiene provisions have been adhered to. In the latter case the criterion would probably be a guideline to be used in connection with microbiological monitoring. This would serve two purposes, one is to assure the inspector in an objective, numerical way that existing provisions have been met, the other to check whether changes in technology of manufacture may have any unexpected microbiologicl effect.

B. Will the microbiology of the raw material make a criterion meaningful?

Where a criterion is proposed for a non- or low-processed food such as raw meat, poultry, fish and vegetables, the microbiology of the raw material, and especially quantitative and qualitative fluctuations of the microflora, should be taken into account. Slaughtering or harvesting under the best hygienic conditions, according to detailed hygiene directives can do almost nothing to improve the microbiology of soiled live animals or vegetables, and it would be futile to institute a criterion for raw food, where that food originates from animals or vegetables with a heterogenous composition of the microflora. Meat should not contain salmonellas, but what if animals before slaughter are regular carriers of salmonellas?

C. Will the effect of further processing and handling increase hazards?

Most processing procedures for foods will reduce microbiological hazards, but there are situations where processing has an effect on the intrinsic factors of the food such that microbial growth potentials may be enhanced. Some forms of preservation may eliminate a harmless but competitive microflora, whereby the processed food becomes more sensitive to post-process contamination with a pathogen. The need for a criterion also increases if the normal or occasional handling conditions after production will increase hazards. No stringent criterion can safeguard a food that is handled eventually under a most unlikely set of conditions, like a TV-dinner being thawed in an oven and left overnight for consumption the following day. Where occasional abuse conditions are known to happen, a criterion may be meaningful to safeguard against a hazard, e.g. where it can be expected that a dried milk preparation will be reconstituted in a way that would allow proliferation of pathogenic micro-organisms.

D. Can inspection of processing make a criterion unnecessary?

An adequate inspection programme, and especially one based on the hazard analysis critical control point (HACCP) concept, can make a microbiological criterion on a final product superfluous. The HACCP-concept in itself, however, contains a monitoring programme, and microbiological monitoring could very well be useful to assist the inspector. For this purpose micro-biological guidelines may be necessary. Some leniency with respect to sampling, microbiological examination and evaluation can be tolerated under these circumstances, where a criterion is a direct supplement to the inspector's observations—as compared to a criterion used for a final product, where information on the inspection and the observations is inadequate or lacking.

E. What is the cost/benefit ratio of a microbiological criterion?

The term cost/benefit or cost/effectiveness of microbiological control has, in its simplest form, been defined as the benefit produced by microbiological control of food as the difference between the cost of examination and the cost of the expected consequences if the testing had not been carried out.

There are few, if any, thorough cost/benefit analyses of microbiological control programmes, and it may be assumed that had there been such analyses in the past many approaches to the protection of public health would have been different. The cost/benefit ratio for *Salmonella* examina-tions in slaughtered poultry is now so high, that there is really no point in having a stringent *Salmonella* criterion for poultry (FAO/WHO 1979). If, at the start of industrialized poultry production after the Second World War, efforts had been concentrated on feedingstuff control, on prevention of cross-contamination in hatcheries, etc. we would today have a much smaller *Salmonella* problem with a much lower cost/benefit ratio.

F. Is a microbiological criterion effective?

Microbiological control, including the use of microbiological criteria, should be effective: otherwise there is no point in having it. Efficiency should be demonstrated by a reduction in bacterial counts and public health hazards, and an improvement of quality. The situation is difficult, because it is hard to prove anything until it has been tested, and microbiological control should not be applied until its efficiency has been proved.

The previously mentioned standards for minced meat in the state of Oregon were a failure. They had been in force for some years when a committee was set up by the Oregon Department of Agriculture to examine their impact (Anon. 1977b). The examination resulted in their being revoked without delay. The committee asked, and answered, a number of questions, the most important being the following.

(a) Had the standards a significant role in improving sanitation and handling? Yes, the application of the standards have an overall effect of improving sanitation and handling.
(b) Had the standards a meaningful role in reducing the number of bacteria? No, there was no evidence to show that there had been a significant change in the number of bacteria present.
(c) Had the standards resulted in a reduction of the risk of food-borne disease? No, there was no evidence to show the standards had resulted in less risk of food-borne disease.
(d) Had the standards a significant purpose in reducing the number of incidents of food-borne disease? No, there was no documented evidence to suggest that the standards have altered, or could significantly alter, the safety record. In this context the committee pointed to the US Center for Disease Control where it had been demonstrated that most food-borne disease resulted from mishandling of food in the food service industry and in the home.
(e) Had the standards any meaningful role in improving the quality of meat? It was concluded that there had probably been no significant change in quality.

Such questions can only be reasonably answered after a criterion has been applied for a period of time, as was the case in Oregon. Before instituting a criterion questions like: "Is it reasonable to assume that the criterion will reduce the number of bacteria?" should be asked, and answers computed from any relevant information available.

G. Is a criterion attainable by good manufacturing practice?

The FAO/WHO working groups emphasized that "the criterion should be attainable by good manufacturing practice . . .". Ideally, as was mentioned for the establishment of reference values, a criterion in force should always reflect what is obtainable by good manufacturing practice, by what the more advanced processors can do. The problem is that there is often no definition of good manufacturing practice. Virtually no food hygiene legislation contains the HACCP-concept, and therefore does not distinguish between the important and the less important points in manufacturing practice, and does not identify the critical points. There seems to be clear need to identify the critical point in each processing and handling step in food production and to define the measures that can be considered as 'good manufacturing practice'. For many foods we are probably only able to define clearly 'bad manufacturing practice', and perhaps a criterion should be set that will reject food produced under bad manufacturing practice.

H. Will a criterion encourage the use of objectionable treatments?

Good hygiene can be, and often is, assisted by other measures that cause a reduction in microbial counts or inhibit or retard growth. Some of these measures are fully acceptable, others are under heavy attack—from toxicologists, consumer groups, etc. Nitrite is an example of current interest. Defining what is acceptable and what is objectionable is a little outside this topic, but food hygienists and microbiologists should not impose such requirements that would force the food industry to take measures that could be considered of doubtful acceptability.

I. What action to take, if a criterion is exceeded?

A major concern, especially among producers, is what will happen if a food product does not meet criteria, especially the official ones. For that reason it must be clearly stated in the criterion what would happen should this situation arise. The Codex Alimentarius document recommends that exceeding a limit in a guideline should not necessarily result in rejection of a product, but should in general lead to the identification and correction of causative factors. Similarly for the advisory end-product specification an appropriate action should be taken to rectify the causative factor. It is optional whether any further action is taken. If the criterion is a microbiological standard the product concerned must be rejected as unfit for its intended use.

The term 'rejected' is interpreted in the following way: when a product is rejected there are, in principle, several options to be taken. These include sorting, reprocessing (e.g. by heating) and destruction. The option should be specified in the criterion. When deciding the option the major consideration is to keep to a minimum the risk that unacceptable food reaches the consumer. On the other hand food must not needlessly be destroyed nor declared unfit for human consumption.

J. What micro-organisms are relevant in a criterion?

Microbiological criteria often specify limits for total counts, an indicator organism and a relevant pathogen, e.g. in most of the criteria proposed by the ICMSF (Anon. 1974). This list is, however, often extended considerably, and the same groups of organisms are sought in foods of different nature. Such an extended analytical programme is often explained by the fact, that "once the dilutions are made, it is no problem adding one or two extra media". This explanation, however, ignores the fact that media, and especially media for specific organisms, are expensive, and that the preparation of these extra media adds a tremendous burden to the media kitchen.

Too little is yet known about correlation of different microbial counts. If it is assumed that an examination of a specific food for, say, six different types of organisms gives 100% information of the microbiological condition of that food, what would be the reduction in the percentage of information obtained, if one, two or three media are omitted? If the omission of one medium reduces the obtained information to 98%, the cost of using this medium would often be much higher than the benefit lost.

The statement by Mossel (1977) should be re-emphasized:

> The number of criteria used in monitoring the microbiological quality of foods should be limited to a strict minimum. This helps saving the available laboratory capacity for the examination of more samples rather than applying endless series of tests to fewer samples.

K. Is there a sufficiently precise methodology?

Especially for use in standards or end-product specifications, internationally elaborated and tested methods should be preferred and they should be as precise as possible. By 'precision' is meant the closeness of agreement between the results obtained by applying the experimental procedure several times under prescribed conditions. In other words, several laboratories, or several technicians in the same laboratory, should get practically the same result by examining field samples from the same batch of food—taking the heterogeneity of distribution into consideration.

Accuracy is also an important parameter. This is defined as "the closeness of agreement between the true value and the mean result which would be obtained by applying the experimental procedure a very large number of times". In other words, the method will detect all, or nearly all, organisms, even those present in very small numbers. A word of warning may be necessary here. For chemical contaminants in food the level accepted by toxicologists is often the detection level. By refining the analytical procedure the detection level is lowered, and so is the accepted concentration. If a similar situation exists for a microbiological test, especially one requiring absence of a pathogen, we may gradually—after improving the accuracy and sensitivity of our methods—come into a situation, where food is rejected because of the demonstration of a pathogen in a quantity far below the minimum effective dose.

For guidelines there is less need for precise and accurate methods, and more for rapid screening methods to be used for in-plant control. A number of such methods have been developed and tested in laboratories, but have been tested only occasionally under practical circumstances. There is a definite need for application of such methods in food processing establishments.

L. Which numerical limit(s) are to be included in a criterion?

In most criteria a numerical limit (or limits) form an important part. For pathogenic organisms there is normally only one limit equal to the detection level of the method. For non-pathogenic organisms, and especially such from foods with a quantitatively heterogenous microflora, there is an argument for having two numerical limits (in a three-class plan). The ICMSF (Anon. 1974) has advocated this approach, and designates the value 'm' for that number of organisms separating the acceptable and the marginally acceptable microbiological quality, while 'M' separates the marginally acceptable from the unacceptable quality. A three-class plan allows for some samples to fall into the grey zone between 'm' and 'M', but none to exceed 'M'.

Numerical limits for standards and end-product specifications should be based on data gathered at various stages of production, storage and distribution—either (as discussed above) on products produced under good manufacturing practice, if this could be defined, or on a representative number of samples from processors irrespective of the state of manufacturing practice.

For guidelines there need not necessarily be one (or two) numerical limits. It is sufficient to state in the criterion that a critical process, or the whole process, or a change in processing, should not cause any increase in microbial counts, nor a significant change in the composition of the microbial flora, unless this change is desirable.

M. Which sampling plans should be used?

Sampling plans should be administratively and economically feasible, but should also give as much information as possible on the lot or the production run. They could be attribute sampling plans, like those proposed by the

ICMSF (Anon. 1974), or variable sampling plans. The choice between the two types would depend on the food, the micro-organisms sought and the acceptance criteria. For standards or end-product specifications the taking of multiple samples is to be preferred, while for guidelines a fewer samples, or often single samples, can be used. As referred to previously, the number of tests to be performed will be decisive for the number of samples taken, and it is preferable to test more samples with few microbiological tests rather than to perform many tests on few samples.

N. Should there be a mechanism for revision of the criteria?

The constant changes in food technology, a changing health hazard situation, or improvements in hygiene, necessitate a regular reviewing, and if necessary, revision of the criteria. A clause should be included in a microbiological criterion requiring the criterion to be reviewed, and possibly revised, e.g. every three years.

5. Conclusions

It is proposed that for purposes other than those connected with the activities of the Codex Alimentarius, similar definitions of microbiological criteria, advisory and mandatory, be adopted. With slight amendments such criteria could be used for national or regional purposes. Work on the adaptation of these criteria is currently being carried out at the level of the Commission of the European Communities. The establishment and application of microbiological criteria should not, however, be enforced until satisfactory answers to the questions raised above have been given.

6. References

ANON. 1964 Food Protection Committee. An Evaluation of Public Health Hazards from Microbiological Contamination of Foods. Publication No. 1195. Washington D.C.: National Academy of Sciences, National Research Council.
ANON. 1974 *Micro-organisms in Foods 2. Sampling for Microbiological Analysis: Principles and Specific Applications.* International Commission on Microbiological Specifications for Foods. University of Toronto Press.
ANON. 1977a Microbiological Specifications for Foods. Report of the Second Joint FAO/WHO Expert Consultation Held in Geneva, 21 February–2 March 1977. EC/Microbiol/77/Report 2. Food and Agriculture Organization of the United Nations.
ANON. 1977b Report of the Meat Bacterial Standards Review Committee. Department of Agriculture, State of Oregon.
ANON. 1979 Report of an FAO/WHO Working Group on Microbiological Criteria for Foods. Geneva, 20–26 February 1979. WG/Microbiol/79/1.

ANON. 1981 Codex Alimentarius Commission. Report of the Seventeenth Session of the Codex Committee on Food Hygiene. Washington D.C., 17–21 November. Alinorm 81/13.

ELLIOTT, R. P. 1970 Microbiological criteria in USDA regulatory programs for meat and poultry. *Journal of Milk and Food Technology* **33**, 173–177.

MOSSEL, D. A. A. 1977 *Microbiology of Foods. Occurrence, Prevention and Monitoring of Hazards and Deterioration* revised edn. Utrecht: The University of Utrecht, Faculty of Veterinary Medicine.

OLSON, JR., J. C. 1971 Microbiology in FDA—new horizons. *Food Drug Cosmetic Law Journal* **26**, 28–36.

Food Microbiology into the Twenty-first Century— a Delphi Forecast

B. JARVIS

Leatherhead Food RA, Leatherhead, Surrey, UK

Contents

FOOD MICROBIOLOGY
ISBN 0 12 589670 0

1. Introduction

Since earliest times man has been concerned not only with today but with what will happen tomorrow, next week, or next year. Seeking information on the future via the soothsayer, the necromancer, the fortune-teller, has long been traditional and even in these modern times the readers of the tabloids and the various weekly and monthly magazines have an opportunity of finding 'what the stars foretell'. It is perhaps not surprising, therefore, that over the years a considerable amount of effort has been devoted to the development of technological forecasting techniques, particularly for use in business, economics, etc. Technological forecasting, as we know it today, probably had its origins in the late 1950s and early 1960s since when there has been a bushfire spread of forecasting techniques, particularly amongst American firms. Jantsch (1967a) estimated that more than 600 firms in the USA were using technological forecasting techniques in planning their future activities and Earl (1969) noted that in 1966 about 90% of US firms used some kind of formal planning compared with only about 20% in 1947. This accelerated trend towards the planning of future activities has clearly been continued in many other areas, particularly those where technological development is at a critical state.

For the latter reason it was felt timely to attempt to forecast the way in which food microbiology and related topics might develop over the next twenty years. Timely, not only because food microbiology has now become a respected branch of microbiology which has itself existed for about a hundred years, but also because in a recession many people look to the future to see how their company, their science and their technology, will develop. When a forecast was first proposed, I had severe reservations about both its potential usefulness and my being able to undertake the necessary work. In retrospect my qualms were at least in part justified but the outcome of the exercise has been illuminating and has certainly helped to sharpen my personal perception of what may occur in the foreseeable future. In relation to the use of the Delphi method for forecasting future market profiles, Parker (1970) commented that there are those who murmur "Tout ça change—tout c'est la même chose" and continued that in spite of the sophistication of modern forecasting techniques, it is nevertheless essential to treat the outcome with a measure of hard-headed reserve.

The objective of the survey was to take the consensus views of a panel and thence to consider to what extent these views are pertinent to one's own situation. It is not suggested that because a group of people has made a conjoint forecast that this event or that event will or will not occur as indicated by the group. Indeed, using a forecasting technique may change the way in which developments occur because people may stop to consider what is happening in a particular area.

A. The choice of forecasting techniques

The standard work on technological forecasting in Jantsch's (1967b) survey of technological forecasting for the O.E.C.D. The reviews by Earl (1969) and by Griffin (1968) both provide biased assessments of technological forecasting, the latter from an industrial viewpoint. There are about twenty different basic methods for technological forecasting with several hundred variations which have been worked out in detail (Jantsch, 1967b). These techniques can be divided roughly into four groups: intuitive forecasts; trend extrapolation; morphological research; and normative relevance trees.

B. Intuitive forecasts

The intuitive forecast is probably most familiar and is based upon the individual forecast of one 'genius' or, in the absence of a genius, a committee consisting of a group of specialists working together to produce a consensus. However, committee forecasting can rarely be a proper consensus since one or two dominant individuals will sway the opinions of their colleagues. The Rand Corporation's Delphi Technique (Dalkey & Helmer, 1963; Dalkey, 1967) is designed to overcome such problems. Essentially the Delphi technique is an iterative procedure for obtaining the consensus views of a group of specialists within a particular area of technology (q.v.).

An improvement on the Delphi technique, scenario writing, starts with the present state of affairs and seeks to develop systematically a logical train of events which may indicate a possible future state of affairs. By identifying the points at which events may take one turn or another, it is feasible to depict a series of alternative futures depending on which route is taken. This technique has been used successfully in social and economic forecasting and to some extent also within the space industry. The scenario approach has the advantage of taking various options and depicting possible future developments from them.

C. Trend extrapolation

Carrying out a time series analysis on past data is familiar to most of us. There is no doubt that in certain situations trend prediction can be of considerable relevance and value: e.g. in predicting microbiological shelf life and stability of food stored under various conditions. Nevertheless, prediction of future trends in a subject based upon historical events is beset by hazards of technological innovation. A frequently quoted example (Quinn, 1967) is that of the successful manufacturer of piston engines who ignored the development of turbo jets because his executives considered incorrectly that engine efficiency and fuel economy mitigated against jets ever being substituted for piston engines in commercial aircraft. Trend extrapolation over a relatively short period offers the forecaster considerable potential but the choice of parameters, the interactions between them, the interpretation of trends and the aspect of technology innovation makes application of trend extrapolation unreliable.

D. Normative relevance trees and morphological research

This method may be described as the orderly examination of the developments required to reach specific goals and objectives, commencing at the goal and tracing back via all the possible routes identified to indicate deficiencies in specific technologies. It involves the application of critical path analyses and networks to optimize programmes taking account of time, cost, etc. The technique has been applied to systems analyses via think-tank groups and has been used extensively in what is known as 'impact' forecasting. In many instances the forecaster is able to isolate a series of hypothetical problems and from this can pinpoint accurately the best lines of development to achieve technical solutions. The forecast studies the likely impact of any new technology in a specific technological field. This technique was applied successfully to developments in the field of automation of six favoured postulates relating to mass communication, computer hardware, etc.; all were shown subsequently to have been correct (Ozbekhan 1967). The morphological approach to technological forecasting (Zwicky 1962) involves exploration without prejudice of all possible solutions to any given problem as well as the interrelations with problems in surrounding areas of interest, so that nothing, as far as is humanly possible, is overlooked. In essence, the morphological approach is a more detailed investigation of the scenario or normative relevance tree approaches.

E. The Delphi oracle

The choice of the Delphi oracle method was based upon several factors. Firstly, it was not necessary for the panel of experts to be convened in any one place. Secondly, it provided a technique which could be handled on a sequential basis and the mathematical analysis was relatively straightforward. Thirdly, it enabled a wide cross-section of expertise to be incorporated into the panel. The major disadvantage of the Delphi system is that it is a consensus. Because of this, the views of the man of vision may well be overlooked because the majority of the panel is over-cautious or does not take account of certain factors in a particular area.

None the less, the Delphi approach has been used with some reasonable success for profiling the British Chemical Industry of the 1980s (Parker, 1969, 1970) and in various food-related areas. Following a general Delphi exercise concerned with trends in the food industry over the next 20 years, surveys were done on trends in the meat and fish products industry, in the fresh meat and poultry industry, in the chocolate and confectionery products industry and in the edible oils and fat industry. All of these various Delphi forecasts have been surveyed by Steiner (1975). In addition, Hudson (1972) studied the potential application of new protein foods in the UK and Katzenstein (1975) forecast future trends in US food technology developments. In microbiology the Delphi system has been used to predict preventive treatments for tuberculosis infections (Koplan & Farer, 1980).

2. The Approach to the Present Forecast

Initially a list was compiled of some 120 persons known to be working in food microbiology and belonging to appropriate learned societies. The list was compiled subjectively and took no account of whether individuals were employed within a university, research institute, industry or consultancy, nor whether they had medical, veterinary or other scientific qualifications. Random number tables were then used to select a panel consisting of 30 people, each of whom was invited by letter to participate in the study. Only one of that 30 declined.

The first-round questionnaire invited each participant to identify up to five topics of food microbiology within each of the areas of: methodology; control of quality and safety of food; the effects of technological change; public health food microbiology; biotechnology in the food industry; and miscellaneous topics.

On receipt of the various responses, amounting to nearly 1000 individual suggestions, a second-round questionnaire was developed combining or otherwise modifying individual responses to try to avoid overlap, ambiguity and non-definitive questions. In so doing it is probable that some specific nuances were lost. It is also certain that in some cases ambiguity and lack of definition of the questions was not eliminated and in retrospect I believe that more definite answers would have been obtained in some instances by a better phraseology of the round-two questions. The second questionnaire covered the first five of the original sections identified for round one and also incorporated a small section relating to trends in food consumption. The breakdown of the questions into the different sections and the organization of the questions clearly reflected my own personal bias as to the most relevant section.

Participants were asked to read the questionnaire thoroughly before attempting to complete it. They were asked to state for each date the probability that an event would occur using the following scale: 1·0, certain; 0·8, highly probable; 0·6, probable; 0·4, possible; 0·2, unlikely; 0, certain *not* to occur. Intermediate values were requested rather than just these specified values. In addition, each participant was asked to provide a self estimate of personal expertise in each area using a scale ranging from 1 (no specialized knowledge of the area) through 3 (average knowledge) to 5 (expert knowledge in the area concerned). These values were subsequently to be used to weight the probability of responses. Section F (trends in food consumption) was handled similarly except that the participants were asked to estimate the earliest, the most probable and the latest date at which the indicated event might be expected to occur. The survey was done during autumn 1980 and spring 1981.

The questionnaires were analysed to provide (a) distribution data for the probability responses, and (b) the median, upper and lower quartile responses.

In only a few instances were the personal weightings used in the analyses; in these cases the overall responses differed only marginally before and after weighting. The results presented are the unweighted responses to each individual question.

Having completed the analysis of the round-two questionnaire the median responses were circulated to all participants who were asked to return their

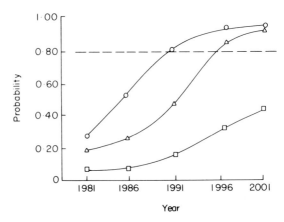

Fig. 1. Changes in median (△), lower quartile (□) and upper quartile (○) responses by year. Data transformed for tables by taking year 'high probability' response ($P = 0·80$) attained.

round-three forms indicating only those areas wherein they wished to modify their earlier responses. In only a few instances did this result in a significant change compared with the round-two results. In the majority of cases panellists were content to leave alone their original predictions, indicating that consensus was less important in this Delphi forecast than in forecasts reported elsewhere (e.g. Steiner 1975).

A. Data presentation

For simplicity the data are presented as tables, each of which shows the earliest, most likely and latest dates for specific events to occur with a probability of $P = 0.80$. These were derived from the quartile responses by date, as shown in Fig. 1. In addition the probability profiles of the panel responses are shown for the year 2001.

3. Methodology for Food Microbiology (Table 1, A1)

The results indicate a high probability that by 1991 there will be international agreement on statistically based sampling programmes, on microbiological criteria for acceptability of foods and on reference methods for analysis of foods. More than 75% of the panel considered that these events would occur before the end of the century. The high degree of consensus between panel members for these questions contrasts with other responses. For instance, there was no consensus that international agreement would be reached on methods of food analysis for routine purposes. This is perhaps not surprising. The upper quartile indicates that the earliest date for such agreement is 1991, with a most probable date of the year 2000.

A. Enumeration of microbes (Table 1, A2–A3)

The panel was unanimous that traditional methods for enumeration of organisms in food, i.e. colony count methods, will be superseded by automated and mechanized procedures for colony counts before the end of the century. By contrast, the panel were far from unanimous on the acceptability of alternative methodologies. For instance, only 25% of the panel considered that methods based on estimation of microbial products using biochemical techniques were likely to be used by 1996 and the most probable date was 2001. Microbial activity estimates, for instance impediometry, were more favoured: 25% considered this to be highly probable by 1991 and 50% thought that it would occur by the end of the century; however, the remainder were unconvinced. The panel was even

TABLE 1

Methodology for food microbiology

	Year event most likely to occur (P = 0·80)			Panel response for year 2001 (probability × 100)		
	Upper quartile	Median	Lower quartile	Lower quartile	Median	Upper quartile
Enumeration						
A1. For specific foods there will be international agreement on:						
(A) Statistical sampling programmes;	1991	1991	2001	80	90	100
(B) Microbiological criteria for acceptability;	1991	1996	2000	84	88	100
(C) Methods of analysis for:						
(i) routine purposes;	1991	2000	*	60	80	100
(ii) reference purposes.	1988	1990	1996	88	100	100
A2. Traditional methods for enumeration will be suppressed by:						
(A) Automated colony count procedures;	1989	1991	2001	80	100	100
(B) Estimation of microbial metabolites;	1996	2000	*	60	80	100
(C) Estimation of microbial activity;	1991	*	*	46	78	100
(D) Separation of microbes followed by (B) or (C).	1996	*	*	40	70	100
A3. (A) Reliance on colony counts will be superseded by estimated of microbial activity;	1996	2000	*	60	80	100
(B) Relationships between indirect indices of microbial populations and the quality or safety of foods will become generally accepted.	1996	2000	*	60	80	100

Detection

A4. The detection of specific organisms in mixed populations will be achieved by:

(A) Detection of specific metabolic products;	1991	1999	*	66	84	100
(B) Use of new selective media in automated systems;	1991	1996	*	51	89	100
(C) Indirect methods, e.g. metabolism of radiochemically labelled substrates;	1966	*	*	40	69	90
(D) Development of new concepts for quantitative estimation of sublethally damaged organisms.	1996	*	*	40	70	100
A5. Methods for food-borne viruses will be commonplace.	1996	*	*	39	60	90
A6. Rapid methods for microbial toxins will be commonplace.	1986	1991	1996	95	100	100
A7. *Identification* will be based on:						
(A) Metabolism of specific substrates and use of appropriate detection systems;	1991	1996	2001	80	90	100
(B) Use of batteries of miniaturized biochemical tests, available as commercial standard units;	1985	1986	1989	97	100	100
(C) Use of on-line computer data banks on:						
(i) a national basis;	1991	1996	2001	80	99	100
(ii) an international basis.	1996	2001	*	50	80	100
(D) Direct analysis of microbial composition e.g. g.c.-pyrolysis.	1996	2001	*	66	80	100

*Probability level of $P = 0.8$ not reached.

less convinced of the application of physical separation of organisms from food, followed by an appropriate chemical or physical analysis. Less than 50% believed that this was likely to occur by the year 2001. The lower quartile response was only 0·45; that is to say that 25% of the panel considered such techniques are only a possibility.

Relatively little support was received for suggestions that reliance on colony counts as indices of quality would be superseded by application of chemical, biochemical or physical activity measurements, and that the relationship between such indirect measurement of microbial populations and the quality or safety of foods will become generally accepted. The upper quartile responses indicated that the earliest probable date was around 1996 with a median response of 2001. In both instances the spread of probability responses was identical and, whilst there was clearly an increased probability trend over the years 1981 to 2001, the overall response for the year 2001 indicated a lower quartile probability of only 0·6. At least 25% of the panel were unconvinced that this indirect technique would become acceptable.

B. Detection of specific microbes (Table 1, A4–A6)

It was considered that detection of specific pathogens and spoilage organisms in mixed populations may be done by the detection of specific metabolic products. The median response time was 1999. An alternative, but better accepted methodology concept, was the development of new, selective, culture media suitable for use in automated systems. By contrast, indirect methods, e.g. metabolism of specific radiochemically labelled substrates and the development of totally new concepts for quantitative estimation of sublethally damaged organisms received scant support, as did the suggestion that methods for detection and isolation of food-borne viruses would become commonplace. The greatest consensus was that rapid serological methods for detection, quantification and identification of microbial toxins will become commonplace. The upper quartile response of 1986, median response of 1991 and lower quartile response of 1996 indicates a high degree of probability that such methodology will be widely used in the foreseeable future. The probability range from lower to upper quartile for the year 2001 was from 0·95 to 1·0, whereas the ranges of probability responses for other questions in this group were very wide (from 0·4 through to 0·9 or 1 in the year 2001).

C. Identification (Table 1, A7)

The panel was almost unanimous in believing that metabolism of specific substrates and the use of appropriate chemical, biochemical or physical detection systems (such as radiochemistry, h.p.l.c., g.l.c., etc.) will be used extensively for identification of microbes. Indeed, over three-quarters of the panel *believed* that by 1996 the application of such techniques was highly probable. Of even greater probability was the use of batteries of miniaturized biochemical tests which would be available as standard commercial units. The panel median response time for this was 1986. It is recognized, of course, that this technique is already in existence but perhaps not used as widely as it will be in the future. Having obtained appropriate data for identification, how is the information to be used? The panel was unanimous that national data banks operated as on-line computer systems would be in operation by the end of the century but they were less optimistic about the use of international data banks. By contrast, direct analysis of microbial composition, for instance through g.c.-pyrolysis methods linked to computer data banks, elicited a somewhat less than enthusiastic response from the panel.

4. Microbiological Control of Quality and Safety
(Table 2, B1–B9)

There was a high degree of consensus amongst panel members that control of quality will become of greater importance because of pressure from consumers for tougher legislation, including the introduction of legislative 'standards'. The panel was unconvinced that economic pressures would restrict consumer concern for food safety and quality in the face of a need for low-cost foods. The implication is that, in the developed world, consumer pressure will continue regardless of the cost of food and that this will increase the pressures upon manufacturers and suppliers to meet more vigorous criteria for quality and safety.

A very wide response was obtained from the panel members as to whether "there will be a general reluctance to replace traditional bacteriological methods, i.e. a misplaced faith in the value of traditional colony count method will continue for some time to come. However, attitudes will change gradually and new concepts in methodology will become widely accepted for quality control purposes". The intended question B3 was "will new concepts in methodology become widely accepted . . ." but it was poorly phrased and, I believe, may have been misinterpreted. If one assumes that the question was answered in the way intended, about 25% of the panel believe that changes in methodology will be accepted within the next five years; the median response suggested that it will take 15 years and there was no consensus opinion that it would be likely to occur by the end of the century.

TABLE 2

Microbiological control of quality and safety

	Year event most likely to occur ($P = 0.80$)			Panel response for year 2001 (probability $\times$ 100)		
	Upper quartile	Median	Lower quartile	Lower quartile	Median	Upper quartile
Quality and safety						
B1. Control of quality will become of greater importance because of consumer pressure for legislative 'standards'.	1987	1993	1996	81	89	100
B2. Economic pressures will restrict consumer concern for food safety and quality in the face of the need for 'low-cost' foods.	1981	*	*	20	40	80
B3. New concepts in methodology will become widely accepted for QC purposes.	1986	1996	*	59	90	100
B4. Microbiological criteria will be used to define the requirements for product safety.	1988	1991	2001	80	94	100
B6. Statistical methods will be used in establishing microbiological criteria.	1986	1991	1996	80	100	100
B7. Variables sampling plans will be developed for microbiological testing of foods.	1986	1991	1996	80	100	100
B8. Food control criteria will be concerned more with the presence/absence of specific organisms than with the total microbial numbers.	1991	2001	*	40	80	100
B9. Risk/hazard analysis will be commonplace and based on use of computer data banks of:						
(A) Critical data relating to conditions for growth of spoilage and pathogenic organisms;	1991	1996	*	61	80	100
(B) Compositional and physical data for various foods;	1991	1996	*	70	88	100
(C) Process data necessary for L.I.P.S.	1996	1996	*	61	70	100

Quality assurance

B10. Traditional QC will be replaced by QA based on HACCP, including:	1991	1996	*	60	80	100
(A) Control of raw material quality rather than end product analysis;	1990	1996	*	71	80	100
(B) Measurement of microbial activity in products held in 'ab use' conditions;	1996	*	*	39	60	90
(C) Compositional analysis of foods in relation to specific criteria for microbial inhibition;	1991	2001	*	64	80	100
(D) Use of computer control of processes to ensure compliance with agreed processes specifications;	1991	1996	2001	80	90	100
(E) Use of 'in-line' rapid microbiological methods for estimation of microbial activity;	1990	2001	*	55	80	100
(F) Methods for rapid assessment of hygienic efficacy of cleaning regimes.	1986	1996	2001	80	89	100

Shelf life of foods

B11. Microbial enzymes will be shown to be a significant cause of chemical spoilage.	1989	1991	*	59	100	100
B12. Shelf life will be assessed by:						
(A) Automated monitoring of microbiological profiles during distribution;	1996	*	*	44	59	90
(B) Automated monitoring of changes in product composition during distribution;	1996	*	*	40	65	100
(C) Prediction, based on measurement of microbiological activity and use of data banks on growth of spoilage organisms;	1993	*	*	40	65	96
(D) Automated monitoring of physical parameters during distribution of food products.	1997	2001	*	34	80	100

*Probability level of $P = 0.8$ not reached.

I do not believe that this is so and can only assume that different people answered the question in different ways.

As might be supposed, there was a high degree of consensus that micro-biological criteria will be used increasingly to define the requirements for product safety, but not that present-day use of criteria will be considered to be non-cost-effective and misleading to the consumer. There was also a high degree of consensus that the statistical basis of microbiological methods will be better understood, that statistical methods will be used to establish microbiological criteria and sampling plans for foods and that 'variables sampling plans' will be developed for microbiological testing of foods. In contrast, there was a low degree of consensus that food control criteria will be concerned more with the presence or absence of specific organisms than with total microbial numbers, and that the 'freedom from pathogens' concept will become of greater importance in establishing food control criteria. In analysing the responses to this question retrospectively, it occurred to me that there might be a different response from panellists with medical or veterinary qualifications, or with public health involvement, than with the industrial or scientifically qualified panellists. In fact, no such differentiation could be detected, indicating that there is considerable diversity of opinion within the 'profession of food microbiology' concerning tests for the presence or absence of specific organisms.

There was reasonable consensus on the desirability of basing risk hazard analysis on computer data banks of critical data for inhibition or growth of selected spoilage and pathogenic organisms, and the composition and physical data for various groups of foods. Greater consensus was attained on the use of process data for 'longitudinal integration of processing for safety'. However, for none of these questions was a high degree of probability given by the panellists although more than 75% suggested that it was probable, if not highly probable, that such risk hazard analysis would become common-place by the end of the century.

A. Quality assurance (Table 2, B10)

"Traditional quality control will be replaced by quality assurance based on principles of hazard associated critical control points". At least 50% of the panellists believed that this would occur before the end of the century although the lower quartile response rate for 2001 reached only a probability level of 0·6. A similar response was obtained in relation to control of raw material quality rather than end product analysis. In general, panellists were less convinced as to: the value of microbial activity measurements on products held in abuse conditions, rather than estimation of micro-organisms *per se*; compositional analysis of foods in relation to specific criteria for microbial inhibition; the use of in-line rapid microbiological methods for estimation of microbial activity. There was, however, a high degree of consensus that by the end of the century the food process operations would be controlled by 'computers', to ensure compliance with agreed process specifications. Similarly, it was agreed that methods for rapid assessment of the efficiency of process plant cleaning regimes would become commonplace.

In some ways the response on compositional analysis of foods is contradictory to that concerning risk/hazard analysis (q.v.). Also, the relatively low probability forecast for measurement of microbial activity on products held in abuse conditions is at variance with the earlier suggestion of the acceptability of microbial activity measurements on food products as alternatives to colony count methods.

These disappointing responses indicate either a lack of awareness of current developments in methodology or a defeatist attitude that possibilities for early detection of unsound goods during manufacture will not be widely accepted within the food industry.

B. Shelf life of foods (Table 2, B11–B12)

The panel was asked to indicate the probability for the assessment of shelf life by automated monitoring of microbiological profiles or automated monitoring of changes occurring in composition during storage and distribution of the product. In both instances 25% of the panellists thought that these approaches might be introduced by 1996; however, there was a wide divergence of opinion within the panel, probability responses ranging from 0·44 to 0·9 and 0·4 to 1, respectively for 2001. In each case the median probability response was of the order of 0·6, indicating that introduction of such techniques was possible but not highly probable.

Suggestions that shelf life might be assessed by computer prediction based upon measurement of microbiological activity and the use of data banks of information on growth of spoilage organisms was again poorly received by the panel and confirmed the earlier conclusion, although the upper quartile response indicated a high probability that this would occur in about 12 years' time. For 2001 the probability ranged again from 0·4 to 0·96 for the lower and upper quartiles, respectively. This is not as wide a range as that obtained in relation to automated monitoring of physical parameters during storage and distribution where for 2001 the probability ranged from 0·34 to 1 with a median response of 0·8.

The one question in this area which elicited a reasonably high degree of consensus related to the suggestion that microbial enzymes will be shown to be a significant cause of chemical spoilage in foods. That is to say that many forms of spoilage which at the moment are described as being chemical will, in time, be shown to be caused by microbial enzymes. The panel generally felt that there was little probability that this would occur during the next five years but by 2001 more than half believed it was highly probable and all thought it was at least probable.

5. Effects of Technological Change (Table 3, C1–C2)

The panel was asked its opinion concerning probable changes in food technology practice. The panel agreed that whilst there was a high degree of probability of greater use to be made of food fermentations, there was a slightly lower probability that rapid freezing will replace traditional processes or that irradiation will be used for pasteurization and sterilization of foods. There was only a 50:50 chance that UHT processes and aseptic packaging will replace traditional processes totally or that new biocidal processes will be developed for foods.

There was again a high degree of consensus that by the end of the century controlled atmosphere packaging will be used for a much wider range of foods, particularly meat and poultry, to control microbial and/or enzymic changes; and that vacuum packaging will become commonplace for low temperature storage of retail packs of meat. Some members of the panel were of the opinion that the use of controlled atmosphere packaging will introduce new microbiological problems by selection for organisms which previously have not been of significance in this type of product.

It was generally agreed that liquid foods will be marketed increasingly in concentrated form, to reduce the bulk for transportation, but that the present-day upsurge in the use of domestic microwave ovens will not necessarily result in an increase in unit packs of meat and other foods, before the end of the century. There was a little support for the idea that improved cold chain distribution will reliably maintain products at less than 2°C though, by and large, the panel did not believe this to be likely. Thermostable plastics are unlikely, in the view of the panel, to replace traditional food containers.

It was considered that the consequences of a continuing need for energy conservation will lead to an increased use of short-time, high temperature processes, of combined preservation methods as alternatives to freezing and sterilization, and to the development of new ambient temperature, shelf-stable foods (e.g. intermediate moisture foods) which will not require refrigeration, freezing or thermal processing. It was considered unlikely that the need for energy conservation will itself lead to increased use of irradiation or of processes based on microbial antagonism. It was also felt that there is little likelihood of a reduction in the extent of thermal processing of foods before the end of the century.

A. Legislative changes affecting food manufacture and distribution (Table 3, C3)

The panel was unanimous in considering that product liability legislation will intensify pressures for manufacturers to produce sound wholesome foods and to ensure satisfactory storage throughout the distribution chain. It was also considered highly probable that legislation for improved manufacturing conditions will lead to a requirement for trade compliance with microbiological criteria related to codes of good manufacturing practice (GMP).

The panel was almost unanimous in deciding that there was no possibility that legislation would require that new products deteriorate in a pre-determined and safe manner giving organoleptic warning of deterioration. However, the panel was ambivalent as to the likelihood that safety and quality criteria would stifle innovation in process and product development. The upper and lower quartile ranges for this question changed only slightly from $P = 0.2$–0.55 for 1981 to 0.2–0.58 for 2001. The median response increased from $P = 0.2$ in 1981 to 0.4 in 1991 onwards. Because of the possible relevance of individual backgrounds it was again worthwhile to consider to what extent these forecasts may have been affected by individual experience. As before, there was no correlation between training, experience or current employment in the individual response, nor was the overall response affected by personal weightings.

There was a high degree of accord that by the end of the present decade consumer pressures will lead to even more stringent legislation on the use of all food additives; but a much lower degree of agreement concerning the implications of such legislation. A small proportion of the group considered that legislative restriction on the use of chemical preservatives would lead to increased public health problems of microbiological origin. Others felt that although not totally banned, the control of chemical preservatives would lead to increased use of safe, natural preservatives, to greater use of microbiological fermentation processes and of microbiological compositional analysis. The responses to this last question were completely at variance with those to the earlier suggestion that microbiological compositional analysis would play a greater part in future food control.

A proportion of the panel considered that biotechnology developments would lead to production of new, high quality, safe foods and feeds but many members of the panel were sceptical as to the opportunities in this area.

TABLE 3

Effects of technological change

	Year event most likely to occur (P = 0.80)			Panel response for year 2001 (probability × 100)		
	Upper quartile	Median	Lower quartile	Lower quartile	Median	Upper quartile
C1. The following changes in food technology practice will occur:						
(A) In processing systems:						
(i) UHT processes and aseptic packaging will totally replace traditional processes;	*	*	*	20	40	70
(ii) irradiation pasteurization of foods;	1996	*	*	40	59	80
(iii) new biocidal processes developed for foods;	*	*	*	40	44	76
(iv) greater use made of food fermentations;	1991	1999	*	60	89	100
(v) 'rapid' freezing processes will replace traditional processes.	1996	*	*	30	50	80
C1. (B) In packaging and distribution systems:						
(i) thermostable plastics will replace all cans and jars;	*	*	*	26	40	60
(ii) controlled atmosphere packaging will be extended to a wider range of foods;	1989	1996	2001	80	100	100
(iii) vacuum packaging will become commonplace for low temperature storage or retail cuts of meat;	1996	1996	*	40	80	80
(iv) controlled atmosphere packs will introduce new microbial problems by selection for 'new organisms';	1989	2001	*	40	80	100
(v) increasing use of domestic microwave ovens will result in an increase in 'unit packs' of foods;	1989	2001	*	40	80	100
(vi) liquid will be marketed increasingly in	1996	2001	*	40	80	90

concentrated form to reduce the bulk for transportation;	1991	1991	2001	80	100	100
(vii) improved cold chain distribution will maintain product at <2°C.	1995	*	*	40	60	100
C2. The need for energy conservation will lead to:						
(A) Increased use of						
(i) irradiation;	1996	*	*	25	60	90
(ii) HTST processes;	1991	1996	*	70	80	98
(iii) combined preservation methods;	1991	1996	*	60	80	100
(iv) processes based on microbial antagonism.	1996	*	*	40	50	80
(B) Reduced use of thermal processing of foods.	1999	*	*	40	40	100
(C) Development of new 'ambient' shelf-stable foods.	1991	1996	*	50	85	100
(D) A trend from unit packs towards bulk packs.	*	*	*	20	40	58
C3. Legislative changes affecting food manufacture and distribution:						
(A) Product liability legislation;	1987	1996	2001	80	90	100
(B) Legislation will require trade compliance with criteria related to GMP codes;	1991	1994	2001	80	98	100
(C) Statutory requirements for new products will require that foods deteriorate in a predetermined and safe manner;	*	*	*	19	30	55
(D) Insistence on criteria will stifle innovation;	*	*	*	20	45	64
(E) Consumer pressure will lead to stringent legislation on food activities;	1988	1991	1996	80	90	100
(F) Legislative restrictions on chemical preservatives will lead to increased public health problems.	1996	*	*	34	40	80
(G) Although rot totally banned, control of chemical preservatives will lead to increased use of:						
(i) natural preservatives;	1991	*	*	40	60	100
(ii) microbiological fermentation;	1991	1996	*	55	80	100
(iii) 'microbiological compositional analysis'.	1996	*	*	30	60	100
(H) Biotechnology developments will result in new high-quality, safe foods and feeds.	1996	*	*	60	70	100
(J) Legislation will be introduced to control production of seafoods consumed without cooking.	1991	1991	*	60	90	100

(continued)

TABLE 3 (*continued*)

	Year event most likely to occur (P = 0.80)			Panel response for year 2001 (probability × 100)		
	Upper quartile	Median	Lower quartile	Lower quartile	Median	Upper quartile
C3. (K) Because of their implication in the aetiology of cancer, traditional smoked foods will be banned.	*	*	*	20	30	54
C4. Nutrition						
(A) Consumer demand will lead to 'milder' processes and fewer additives.	1991	*	*	21	60	80
(B) Consumer awareness will lead increasingly to demands for 'natural, unrefined' foods.	1993	*	*	40	57	100
(C) Consumer demand will lead to 'nutritionally modified' foods.	1996	*	*	40	60	80
(D) New humectants will be developed to replace sugar in jams and conserves.	1991	*	*	40	60	85
(E) Nutritional claims will be made for 'new' confectionery products.	1991	*	*	40	45	100
(F) Nutritional legislation will control fat and carbohydrate content of foods.	1996	*	*	20	59	90
C5. Raw materials will be restricted in availability and more costly. This will lead to:						
(A) Widespread ingredient substitution.	1989	1996	*	60	80	100
(B) Greater use of 'low grade' meats and fish in manufactured food products.	1986	1991	2001	80	100	100
(C) Greater use of plant and microbial proteins as 'meat substitutes'.	1986	1991	2001	80	98	100
(D) Development of proteins with specific functional properties.	1991	1996	*	60	80	100
(E) Increased use of bulk 'pre-mixes' incorporating unconventional raw materials.	1991	2001	*	40	80	100

	1986	1989	*	76	98	100
C6. The use of 'novel' food ingredients will require legislative permission.	1986	1989	*	76	98	100
C7. Microbiological systems for control of quality and safety will be:						
(A) Confounded increasingly by problems of raw material supply.	1991	*	*	20	50	100
(B) Improved by better knowledge of pathogen dissemination from animal sources.	1991	1996	*	60	80	100
(C) Directed towards improved quality of 'potable' waters.	1996	*	*	40	60	98

Other aspects

C8. Manipulation of micro-organisms in non-sterile foods will ensure that 'fail-safe' situations result from abuse handling and storage.	1996	*	*	41	50	100
C9. Natural, native, ethnic and home-prepared foods will be identified as being more hazardous than commercially processed foods.	1986	1991	*	60	90	100
C10. Foods for thermal processing will be given a pre-treatment to ensure a consistently low spore heat resistance.	1996	*	*	24	40	100
C11. Greater consumption of food outside the home will lead to regulatory licensing control of food servicing establishments.	1989	1996	*	60	80	100

*Probability level of $P = 0.8$ not reached.

There was a reasonably high degree of consensus concerning the introduction of legislation for control of production methods, particularly for
seafoods likely to be consumed without cooking, but there was an even
greater degree of consensus that traditional smoked foods would not be
banned as a consequence of their implied involvement in the aetiology of
cancer.

B. Nutrition (Table 3, C4)

In all there were five questions on aspects of nutrition which might be seen
to have some relevance to food microbiology. A few panel members
declined to answer these questions on the grounds that they were not
microbiology. The panel was asked first to what extent consumer demand
for quality would lead to the use of milder processes and fewer additives. In
essence this question was the same as the earlier one relating to the likely
introduction of legislation on food additives but the response was very
different. The clear implication is that consumer demand for improved
nutrition, will lead to the use of yet milder processes. The panel was
ambivalent about the remaining questions in the area, less than 50%
believing that any of the changes suggested would occur before the end of the
century.

C. Raw materials (Table 3, C5–C11)

Certain raw materials for food manufacturing purposes will become increasingly restricted in availability and more costly. The majority of the panel
agreed that there would be extensive modification of food products with
widespread ingredient substitution and that greater use will be made of low
grade meats and fish and of plant and microbial proteins in manufactured
products. Such substitutions are already occurring and are under attack in
the UK in relation to meat products regulations (see, for instance, several
articles in *The Guardian* and *The Times* of June 1981). Nevertheless, the
panel has recognized that the probability for extending ingredient substitution is considerable. The panel recognized, too, that development of
proteins with specific functional properties from novel, surplus or otherwise
downgraded protein foods was highly probable within the next decade and
that the use of certain forms of novel food ingredients will require legislative
permission. The one area on which the panel was less in accord related to
increased use of bulk pre-mixes which might incorporate unconventional
raw materials so that the food manufacturer would not necessarily be aware
of the ingredients.

It was felt that microbiological systems for control of quality and safety
would not be confounded by the increasing problems of raw material supply

but that by better knowledge of the routes for pathogen dissemination from animal sources, quality systems would be enhanced. Once again it was felt that the manipulation of micro-organisms in non-sterile foods to ensure fail-safe conditions from abuse storage was unlikely to occur.

Considered equally unlikely was a suggestion that by the end of the century foods for thermal processing will be given a pre-treatment to ensure a consistently low spore-heat resistance. This I find somewhat surprising in view of the work which has been done in this area in recent years. There was considerable agreement that natural, native, ethnic and home-preserved foods will be identified within the present decade as being equally, or more, hazardous than commercially processed foods. There was also a reasonably high degree of consensus that greater consumption of food outside the home will lead to a requirement for regulatory licensing control of food servicing establishments.

6. Public Health Food Microbiology

A. Food-borne pathogens (Table 4, D1)

There was a high degree of consensus that foods will be more widely recognized as an important vector of organisms pathogenic to man. It was agreed also that within the present decade many lesser known bacterial pathogens will be identified as significant aetiological agents of human disease in developed countries; psychrotrophic pathogens are likely to be less important than food-borne viruses, the real significance of which will become apparent by the end of the century. However, in both areas members of the panel differed considerably in their responses. It was not considered that food-borne parasites would become of significance, in spite of increased environmental pollution; but at least 50% of the panel considered that bacterial endotoxins, mycotoxins and various chemical toxins, allergens and trace metals would become of greater significance. However, suggestions that the occurrence of carcinogenic microbial metabolites will lead to increased demand for the provision of sterile foods were completely refuted.

B. Public health control (Table 4, D2)

The majority of the panel did not believe that increasing problems with antibiotic resistant pathogens in human medicine would lead to changes in veterinary medicine but they did concur that there would be a much greater control of animal feeds and feed production processes in order to prevent the

TABLE 4

Public health food microbiology

	Year event most likely to occur (P = 0.80)			Panel response for year 2001 (probability × 100)		
	Upper quartile	Median	Lower quartile	Lower quartile	Median	Upper quartile
D1. *Food-borne pathogens*						
(A) Food will be widely recognized as an important vector of organisms pathogenic to man.	1991	1993	*	45	80	100
(B) The following will be identified as increasingly significant aetiological agents of human disease in the developed countries:						
(i) lesser known bacterial pathogens (e.g. *Campylobacter* and *Yersinia*);	1986	1991	2001	80	100	100
(ii) psychrotrophic pathogens in foods	1996	*	*	40	50	80
(iii) food-borne viruses;	1989	*	*	58	60	100
(iv) food-borne parasites;	2001	*	*	20	40	80
(v) bacterial enterotoxins;	1991	1996	*	50	80	100
(vi) fungal toxins (mycotoxins);	1991	2001	*	55	80	100
(vii) chemical toxins, allergens and trace metals.	1989	1991	*	49	86	100
D2. *Public health control*						
(A) Increasing problems with antibiotic resistant pathogens in human disease will lead to changes in veterinary medicine and agriculture.	1991	*	*	40	60	100
(B) There will be greater control of animal feeds to prevent the spread of zoonotic diseases.	1989	1991	*	60	80	100
(C) Control of foods of animal origin will move from the slaughterhouse and factory to the farm.	1996	*	*	20	60	90
(D) Veterinary certification will be a prerequisite for slaughter of food animals.	1998	*	*	40	59	81

(E)	The apparent incidence of food-borne disease will increase because of improvements in reporting.	1986	1986	*	60	98	100
(F)	The reported incidence of gastro-enteritis due to E. coli, Aeromonas and Yersinia will increase at least 50% over present levels.	1994	*	*	40	60	89
(G)	Investigation of the causes of food-borne disease will take precedence over documentation and description.	1991	*	*	40	76	98
(H)	Cost-benefit analysis of public health hazards will lead to improved co-operation between professional bodies.	1991	1996	*	60	90	100
(J)	Regulatory control of foods will increase significantly.	1991	1996	*	60	80	100
(K)	Regulatory agencies will install automatic monitoring devices in food processing plants.	1996	*	*	40	54	95
(L)	Regulatory agencies will recognize the limitation of 'zero tolerance' regulations.	1986	1996	*	50	80	100
(M)	New systems of health screening will be introduced for food industry personnel.	2001	*	*	39	60	80
(N)	. . . and arguments over the efficacy of health screens will cease.	*	*	*	20	40	70

D3. *Public health implication of changes in food technology practice*
(A) Processes will be developed to:

(i)	reduce surface contamination of carcasses and other food materials;	1990	1991	1996	80	98	100
(ii)	detoxify microbial toxins (especially mycotoxins);	1991	2001	*	56	80	100
(iii)	control the formation of hazardous chemicals in foods;	1991	1996	*	60	80	100
(iv)	apply terminal biocidal treatments to infant foods;	1989	*	*	40	74	100

(continued)

TABLE 4 (*continued*)

	Year event most likely to occur (P = 0.80)			Panel response for year 2001 (probability × 100)		
	Upper quartile	Median	Lower quartile	Lower quartile	Median	Upper quartile
D3. (v) improve process plant hygiene;	1991	1996	*	60	100	100
(vi) purify water to overcome problems from toxic chemicals derived from biocides.	1992	1996	*	40	76†	100
(B) There will be a 25% reduction in the real incidence of food poisoning as a result of improvements in food technology and education.	2001	*	*	25	60	80
(C) Greater use of raw materials and manufactured foods from 'developing countries' will lead to a significant increase in exotic diseases.	*	*	*	20	40	70
(D) Increasing use of poultry meats in 'further prepared' products will exacerbate the salmonella problem.	*	*	*	20	40	60
(E) Increased consumer demand for 'natural' foods will lead to increased health hazards.	*	*	*	20	40	70
(F) Before use is allowed novel food ingredients will require microbial hazard analysis.	1993	1996	*	60	90	100
(G) Increasing distances between producers and consumers will lead to increased health risks.	*	*	*	20	30	40
(H) New process plants will be required to install equipment which complies with hygienic design standards.	1989	1991	2001	80	80	100
(J) The identification of new food-borne pathogens will require revision of technological practices to ensure food safety.	1991	*	*	18	40	95

D4. Public health training and education

Education in food hygiene will become of paramount importance; in particular:

(A) All food handlers will be required to receive formal training.	1995	*	40	60	100
(B) A 'licence to practise' will be required for all personnel involved in:					
(i) management of food production plants;	1996	*	40	54	100
(ii) food inspection services;	1989	1996	60	98	100
(iii) food microbiology and QC.	1996	2001	60	80	100
(C) Granting of a 'licence to practise' will require both formal (academic) qualifications and experience.	1993	2001	40	80	100
D5. There will be international agreement on the routes to qualify for a 'licence to practise' food microbiology.	*	*	20	40	60
D6. Schools will be required to teach food hygiene and nutrition.	1996	*	40	60	90
D7. Food courses designed for the consumer will be presented via the mass media.	1991	*	40	70	80
D8. Academic teaching will increasingly use industrial and research institute personnel as support staff.	1991	*	40	60	100
D9. Application of existing knowledge will be found more cost effective than continued research in food science and technology.	*	*	20	34	60

*Probability level of $P = 0.8$ not reached.
†Probability level for 2001 < for 1996.

spread of zoonotic diseases. A small proportion of the panel considered that by the end of the century control of foods of animal origin would move from the slaughterhouse and the factory back to the farm and that veterinary certification of animal health would be a prerequisite for slaughter of food animals. However, the majority of the panel members did not consider this likely.

There was a high degree of consensus that within a few years the apparent incidence of food-borne disease would increase because of improvements in both national and international reporting systems. There was a lesser degree of concurrence that the reported incidence of gastro-enteritis due to *Escherichia coli*, *Aeromonas*, *Yersinia* and similar organisms would increase significantly; or that investigation of the causes of food-borne disease outbreaks would take precedence over present-day preoccupation with documentation and description. As with the veterinary health control, the differences in opinions of panellists on these questions were very wide.

There was a general consensus that increased regulatory control and surveillance of foods would occur but that regulatory agencies will rapidly come to realize the limitation of so-called 'zero-tolerance' regulations which are used frequently in relation to microbial pathogens and toxins. It was also agreed that cost-benefit analysis of public health hazards would lead to improved co-operation between the various professional bodies. However, the likelihood that regulatory agencies would install automatic devices in food processing plants was considered to be low. The panel considered that the introduction of new systems of health screening for food industry personnel is very unlikely and that arguments over the efficacy of health screens, and particularly of techniques such as stool-testing, will continue for many years to come.

C. Public health implications of changes in food technology practice (Table 4, D3)

The panel was almost unanimous that, by the end of the present decade, methodology will be introduced to reduce surface contamination of carcasses and other food materials; to control the formation of hazardous chemicals in foods; and to improve process plant hygiene. There was a lower degree of consensus that systems would be developed for detoxification of microbial toxins in foods especially mycotoxins and for terminal biocidal treatments for infant foods. Many panellists considered that new processes for water purification would be introduced to overcome current problems from toxic residues derived from biocides in water.

It was generally agreed that there was little likelihood of a significant reduction in the real incidence of food poisoning as a result of improvements in food technology practice and education. The panel also agreed that the greater use of raw materials imported from developing countries would not lead to any significant increase in exotic diseases; nor that the use of poultry meats in a wider range of further prepared products would exacerbate the salmonella problem. The panel felt that increased consumer demand for natural food products, which lack the safety margins of commercial foods, would lead to problems from the consumption of natural, unadulterated and ethnic food materials. The panel did not consider that increasing distances between producers and consumers will lead to any significant increase in health risk.

It was considered that by the end of the century, all new process plant will be required to conform with some hygienic design standards. Whilst this is undoubtedly a highly desirable objective, it is difficult to accept that it will occur within the time-scale indicated, because of the conservatism of much of the food manufacturing industry and the high cost of replacing process plant.

Panellists confirmed their earlier view that all new and novel food ingredients will require legislative clearance and that this will be related, at least in part, to microbial hazard analysis. In general they were less convinced that the identification of new food pathogens would lead to concern that existing technological practices are inadequate to ensure food safety.

D. Public health training and education (Table 4, D4–D9)

There was much divergence of opinion relating to the questions in this section. The suggestion that all food handlers would require formal training in food hygiene was considered by a minority of the panel to be highly probable; but the majority did not subscribe to this view, any more than they considered that a licence to practise would be required for people involved in the management of food production plant. However, they agreed that a licence to practise would be required for all persons involved in food inspection services and clearly this already is the case in many countries. There was strong support for the view that by the end of the century a licence to practise food microbiology would be required and that the granting of such a licence would require both academic qualifications and appropriate practical experience. However, there was no consensus for international agreement on the various routes to qualify for a licence to practise. Panellists were equally pessimistic concerning a suggestion that schools would be required formally to teach food hygiene and nutrition. A small proportion of the panellists believed that: food courses designed for the consumer would be presented via the mass media; and, academic teaching in food science will increasingly use research and industrial personnel as support staff. In view of the recent UK promulgations concerning financial support for universities I think it is highly probable that such an event will occur.

Finally, the panel was asked whether it will be realized that the application of existing knowledge and technology will be more cost-effective than continued research in food science and technology. Interestingly, there was a high degree of consensus that such an event is unlikely.

7. Biotechnology and the Food Industry (Table 5)

There was a high degree of consensus that by the end of the century we will have many microbial biotechnology products available for use in the food industry including gums, vitamins, colours, flavours, antimicrobial agents, food enzymes and proteins. The only area in which the panel did not believe that biotechnology has anything to offer was that of fats with specific dietary attributes and this may have arisen from a lack of knowledge of current developments in this area. Whilst there was a high degree of consensus on the use of microbial proteins for animal feeds and feed supplements, there was much less certainty amongst the panel concerning the possible use of microbial proteins for human foods. A majority of the panel members considered that by the end of the century microbial proteins with particular functional properties would be developed but few believed that specifically designed proteins would be available for dietary purposes for patients with genetic deficiency diseases. Again, there was a high degree of consensus that microbial proteins will be produced as by-products from other processes; for instance, gasohol production or fermentation of sewage, industrial trade waste and effluents. Such proteins would be primarily for use as animal feed, rather than for human foods.

It was agreed that microbial enzymes will be used increasingly in food processing applications and, of course, this is already established in some areas. Specific enzymes will be developed by genetic manipulation of appropriate organisms for specific processing requirements. The majority of the panel considered that by the end of the present decade, microbial enzymes will be used extensively in immobilized forms for continuous food processing and for effluent treatment.

There was a reasonable consensus that microbial fermentation processes will be used to upgrade waste materials and to modify cellulose. The prospects for energy production at factories by fermentation and for detoxification of food materials were considered less likely. However, at least one UK factory is already using a microbial fermentation process to produce gasohol and upgrade its trade effluent. The low degree of probability attached to this type of process may reflect a lack of awareness of current developments. There was a very high consensus that before the end of the present decade micro-organisms and their enzymes will be used increasingly for assay systems, especially for assays of carcinogens and mutagens.

It was considered that before the end of the century genetic manipulation of micro-organisms and plants will have yielded strains capable of producing high levels of specific metabolites for food use and that public concern over the safety and acceptability of genetic manipulation will be overcome. The panel did not agree that genetic manipulation will improve plant or animal resistance to adverse conditions nor lead to improved biological control systems based on the Nurmi concept.

A. Fermented foods (Table 5, E7)

The panel generally considered that in Western society there is little likelihood of an expansion of oriental-type vegetable fermentation processes but that there might be some increase in international trade in various ethnic fermented foods. The end of the decade is considered to be the most likely time for the introduction of improved technology for continuous and solid state fermentation processes. This will tie in with a wider availability of reliable starter cultures, for industrial use, many of which will have been produced by genetic manipulation to develop specific attributes.

8. Trends in Food Consumption (Table 6)

Suggestions for these trends came from one member of the panel. The panellists considered that, with one exception, few of the trends would be likely to occur before the end of the present century. The exception related to an increase in the consumption of natural or unadulterated foods, for which there was a high consensus amongst panel members.

In some cases panellists suggested that the most likely date or the latest date when such trends would be likely to occur would be in the year 3000+. Outliers of that nature were all recorded as the year 3000 in the analysis. In summary the implication of these trends is that in the early part of the next century consumption of fully thermal processed foods will fall, as also will consumption of frozen foods, largely because of the massive increase in energy costs. Consumption of intermediate moisture foods will have increased significantly by the end of the century and microbial protein will account for some 10% or more of all food proteins consumed by the early

TABLE 5

Biotechnology and the food industry

	Year event most likely to occur (P = 0·80)			Panel response for year 2001 (probability × 100)		
	Upper quartile	Median	Lower quartile	Lower quartile	Median	Upper quartile
E1. Specific biotechnology products of importance to the food industry will include:						
(A) Microbial polysaccharides, e.g. gums.	1991	1996	*	74	80	100
(B) Vitamins.	1991	1996	*	40	80	100
(C) Colours and flavours.	1991	1996	*	60	80	100
(D) Antimicrobial agents.	1991	*	*	40	70	100
(E) Fertilizers, insecticides, herbicides and other agrochemicals.	1991	2001	*	60	80	100
(F) Fats having specific dietary attributes.	*	*	*	34	49	74
(G) Enzymes.	1986	1991	2001	80	100	100
(H) Peptides and proteins.	1986	1996	*	60	80	100
E2. Microbial proteins will be:						
(A) Used increasingly for human foods.	1996	*	*	40	69	100
(B) Developed with particular functional properties.	1991	2001	*	50	80	90
(C) Developed for use in dietary foods for patients with genetic deficiency diseases.	2000	*	*	39	59	84
(D) Used as animal feeds and supplements.	1986	1991	1991	89	100	100
(E) Recovered as by-products from other microbial processes.	1989	1994	*	79	100	100
E3. Microbial enzymes will be:						
(A) Used increasingly in food processing applications.	1988	1991	2001	80	100	100
(B) Developed by genetic manipulation for specific processing requirements.	1988	1991	2001	80	100	100
(C) Used in immobilized form for						
(i) continuous food processes; and	1989	1993	*	60	80	100
(ii) effluent treatment.	1991	2001	*	60	80	100

E4. Microbial fermentation processes will be introduced for:						
(A) Modification/hydrolysis of cellulose.	1991	1996	*	60	96	100
(B) Detoxification of food materials.	1998	*	*	41	66	95
(C) Upgrading by-products and wastes.	1991	1996	*	79	90	100
(D) Production of 'energy', especially at food factories.	1996	*	*	40	60	80
E6. Microbes and their enzymes will be used for the assay of chemicals.	1989	1991	2001	80	100	100
E7. The market for fermented foods in Western society will be increased by:						
(A) Expansion of oriental-type vegetable fermentation processes.	2001	*	*	40	60	80
(B) Increased international trade in 'ethnic' fermented foods.	1994	*	*	40	58	95
(C) Improved technology for continuous and solid state fermentation processes.	1991	1996	*	60	90	100
(D) Wider availability of reliable starter cultures for industrial and domestic fermentations.	1991	1996	*	60	80	100
E8. Biological control systems based on the Nurmi concept (positive application of organisms) will be introduced for foods.	1996	*	*	40	59	80
E9. Genetic manipulation of micro-organisms and plants will be used increasingly:						
(A) To develop 'starter cultures' having specific attributes.	1986	1991	1996	80	98	100
(B) For production of high yields of metabolic products for food use.	1986	1996	*	74	99	100
(C) For improvement of plant or animal resistance to adverse conditions by transfer of microbial genetic material carrying 'markers' for, e.g. psychrotrophism.	1996	*	*	40	59	100
E10. Public concern over 'safety' of genetic manipulation will be overcome.	1989	1996	*	74	88	100

*Probability level of $P = 0.8$ not reached.

TABLE 6

Trends in food consumption

	Year event likely to occur (median responses)		
	Earliest	Most likely	Latest
F1. Consumption of fully thermally processed foods will fall to <20% of the present volume.	1995	2010	2040
F2. Consumption of frozen foods will fall to <10% of present levels because of massive increases in energy costs.	2000	2010	2030
F3. Consumption of intermediate moisture foods will constitute >25% of all manufactured foods.	1995	2000	2020
F4. Microbial protein will account for >10% of all food proteins consumed.	2000	2005	2030
F5. Consumption of 'natural', 'unadulterated' foods will increase to >5% of all manufactured foods.	1988	1995	2000
F6. At least 10% of consumers will become self-sufficient and home food preservation will increase significantly.	2000	2020	2055

part of the twenty-first century. Suggestions that at least 10% of all consumers will become self-sufficient and that home food-preservation will increase significantly each yielded a 'most likely date' of 2020; even that, in my view, is wildly optimistic.

9. Conclusions

From the vast amount of data collated from the panellists it is clearly very difficult to draw any specific conclusions. In general the panel considered that many of the suggestions made by their colleagues would occur by the end of the present century or in the early part of the next century.

Is the assessment of a small panel of people realistic? What is the view of the majority of food microbiologists? The objective of doing this work was not to say that these events will occur; it is to say that in the view of a small proportion of experienced food microbiologists there is a high probability that certain events will or will not occur by the end of the century. The objective of using forecasting techniques is firstly to sharpen the views of the individual panellist and, hopefully, then to sharpen the views of those who read the forecast. As a profession, we have a duty not only to look back to what has happened before but also to look to the future. Forecasting techniques may be imprecise but they stimulate people to consider the future. If the outcome is merely to make people aware of what might happen then the exercise will have been worthwhile.

I am indebted to the following: the Panel, without whom the work could not have been done; Mrs M. Barnfield, for extensive and invaluable secretarial assistance; Mr A. Hines and his colleagues, for assistance with graphics; Messrs I. S. Jarvis and D. Garner, for data processing; Mrs E. A. Nash, for assistance with the bibliography and in sorting and classifying the round-one responses; and the Director of the Leatherhead Food R.A. for permission to undertake the study.

10. References

DALKEY, N. C. 1967 Delphi. In *Long Range Forecasting Methodology,* pp. 1–11. Arlington, Va: Office of Aerospace Research.

DALKEY, N. C. & HELMER, O. 1963 An experimental application of the Delphi method to the use of experts. *Management Sciences* **9**, 458–467.

EARL, V. 1969 *Technological forecasting.* Brief No. 1. London: The Economist.

GRIFFIN, M. 1968. Technological forecasting in project planning. In *Innovation for Profit*, pp. 43–52. (SIRA Conference, Eastbourne). London: Adam Hilger.

HUDSON, J. E. 1972 New protein foods in the U.K.—A delphi exercise. *Chemistry and Industry, London* 18th March, 251–254.

JANTSCH, E. 1967*a* Forecasting the future. *Science* **3** (10th October), 40–45.

JANTSCH, E. 1967*b* Technological Forecasting in Perspective. Paris: O.E.C.D.

KATZENSTEIN, A. W. 1975 The food update survey—forecasting the food industry ten years from now. *Food Product Development* **9**, 20–24.

KOPLAN, J. P. & FARER, L. S. 1980 Choice of preventive treatment for isoniazid-resistant tuberculosis infection. *Journal of the American Medical Association* **264**, 2736–2740.

OZBEKHAN, H. 1967 Automation. *Science* **3**, 67–72.

PARKER, E. F. 1969 Some experience with the application of the Delphi method. *Chemistry and Industry, London* 20th September, 1317–1319.

PARKER, E. F. 1970 British chemical industry in the 1980s—a Delphi method profile. *Chemistry and Industry, London* 31st January, 138–145.

QUINN, J. B. 1967 Technological forecasting. *Harvard Business Review* March/April, 89–106.

STEINER, E. H. 1975 Delphi exercises in forecasting technology for the food industry in the U.K. *Food Product Development* **9**, 46–54.

ZWICKY, F. 1962 Morphology of propulsive power. *Monographs on Morphological Research No. 1.* Pasadena, Ca.: Society for Morphological Research.

Selected Abstracts of Papers and Posters
Presented at the Summer Conference

**The Development of the Microflora on Cured Meat Products Stored in
Different Gaseous Atmospheres**

E. BLICKSTAD
Swedish Meat Research Institute, Fack, 244 00 Kävlinge, Sweden

The development of the microflora on two cured meat products, smoked pork loin and a frankfurter-type of sausage, was followed during storage in vacuum, N_2 or CO_2 atmosphere at 4°C. The predominant organisms on the fresh products, when taken directly from the processing plant, were *Bacillus* spp., coryneform bacteria, *Flavobacterium* spp. and *Pseudomonas* spp.

The total aerobic count on the smoked pork loin reached 7 log units/g after 38 days in vacuum, 44 days in N_2 and 50 days in CO_2. The corresponding value for the sausage was 73 days in vacuum, while the growth declined at 5 log units/g after 98 days in N_2, and no growth at all was noticed in CO_2 after 140 days.

At the end of the storage time *Lactobacillus* spp. and coryneform bacteria constituted the entire flora on both products in the three different gas atmospheres tested. However, the numbers of these organisms in the different gas atmospheres varied and the number of coryneform bacteria decreased in order: vacuum $> N^2 > CO_2$. Large groups of unidentifiable homofermentative *Lactobacillus* spp. and coryneform bacteria were found.

**Sulphite: The Elective Agent for the Microbial Association in
British Fresh Sausages**

J. G. BANKS AND R. G. BOARD
*Department of Microbiology, School of Biological Sciences, University of Bath,
Bath BA2 7AY, Avon, UK*

We have confirmed that sulphite in sausages elects a microbial association dominated by Gram positive bacteria and yeasts. The latter are the only ones that do not show any response to the preservative, whereas it suppresses the growth of the principal bacterial contaminant, *Brochothrix thermosphacta*, this effect being accentuated by chill storage. Organisms isolated on violet red bile glucose agar at 30°C flourished to such an extent in unsulphited sausages that they dominated the microbial association of sausages stored at 10 or 22°C.

FOOD MICROBIOLOGY
ISBN 0 12 589670 0

Inhibition of Clostridia by Sodium Nitrite

L. F. J. WOODS AND J. M. WOOD

Leatherhead Food RA, Randalls Road, Leatherhead, UK

Sodium nitrite inhibits the growth of *Clostridium sporogenes* by inhibition of the phosphoroclastic system, which converts pyruvate to acetate with production of ATP. The active form of nitrite is nitric oxide which appears to react with the non-haem iron of pyruvate : ferredoxin oxido-reductase. This inhibition results in a marked fall in the intracellular ATP level and accumulation of pyruvate in the medium. Pyruvate accumulation in the presence of nitrite has also been shown in six strains of *Cl. botulinum*. The nitrite concentration which gave the maximum rate of pyruvate accumulation with *Cl. sporogenes* was 0·7 mmol/l at pH 6·0 and 0·3 mmol/l at pH 5·0. Above these nitrite concentrations pyruvate accumulation decreased presumably due to additional inhibition at a site in the glycolytic system.

A Rapid Identification System for *Bacillus* Species Isolated from Food and Other Sources

N. A. LOGAN AND R. C. W. BERKELEY

Department of Bacteriology, The Medical School, University Walk, Bristol BS8 1TD, UK

Bacillus species are important in the food industry, especially as contaminants and as spoilage agents; three familiar examples are 'bitty cream defect' caused by members of the *B. cereus* group, 'rope' in bread usually caused by *B. subtilis* and related organisms, and 'flat sours' in canned food due to strains of the thermophilic species *B. coagulans* and *B. stearothermophilus*. These spoilage problems are chiefly due to the ubiquity and the heat resistance of the *Bacillus* spores.

The identification of these organisms is often of importance, but it is frequently neglected in many laboratories. This is largely due to the inconvenience of using the classical methods of *Bacillus* identification, methods which are expensive, slow, inconvenient and poorly reproducible, even in the hands of experts (Logan, N. A. & Berkeley, R. C. W. 1981 Classification and identification of members of the genus *Bacillus* using API tests. In *The Aerobic Endospore-forming Bacteria* ed. Berkeley, R. C. W. & Goodfellow, M. pp. 105–140. London & New York: Academic Press).

Tests in the API system have been shown to give better reproducibility than the classical tests (Logan, N. A., Berkeley, R. C. W. & Norris, J. R. 1979 Results of an international reproducibility trial using the API system applied to the genus *Bacillus*. *Journal of Applied Bacteriology*, **45**, xxviii–xxix) and a taxonomy based upon API tests is in good agreement with those of other authors. A system based upon tests in the API 20E and API 50CH strips together with supplementary tests may be used with diagnostic tables and/or a computer data base (generated from nearly 1000 strains) for the rapid (48 h) and accurate identification of *Bacillus* isolates from food and other sources.

The Taxonomy of Psychrotrophic *Pseudomonas* spp. and their Distribution on Meat and Fish
G. MOLIN
Meat Research Institute, P.O. Box 504, S-244 00 Kävlinge, Sweden

A numerical study of the taxonomy of 193 *Pseudomonas* isolates from meat and of 18 reference strains has earlier been performed using 174 tests. The isolates were divided into five major groups designated as *Pseudomonas fragi* (57%); *Ps. 'bovis'* (11%); *Ps. fluorescens* (13%) and *Ps.* (or *Alteromonas*) *putrefaciens* (38%). A simplified classification scheme for psychrotrophic *Pseudomonas* spp. was proposed.

The scheme was used for the classification of a total of 265 new isolates originating from pork (205), herring (38) and beef (22). *Pseudomonas fragi* was the most frequently occurring species (67%), followed by *Ps. 'bovis'* (23%). The others were only sparsely represented, *Ps. fluorescens* (3%), *Ps. putrefaciens* (2%), *Ps. putida* (1%; only from herring) and 12 isolates were incompatible with the classification scheme.

Occurrence and some Characteristics of *Lactobacillus* spp. Isolated from Fish
S. KNØCHEL
Technological Laboratory, Ministry of Fisheries, Technical University, Lyngby, Denmark

It has been indicated that lactobacilli may be used in the preservation of fish. Very few data have been published concerning the natural occurrence of these organisms in live fish. In this study, 70 fish were investigated—40 from the North Sea and 30 from the Baltic Sea, respectively. Lactobacilli were found in all fish, both on skin and in intestinal content. The size of the population varied from 10/g up to 2×10^3/g. The *Lactobacillus* flora proved to be somewhat heterogeneous in terms of cell morphology. Eighteen isolates were subsequently examined and classified as 'atypical strepto-bacteria'. In general, they were psychrotropic and moderately salt-tolerant. None was able to reduce TMAO or produce H_2S.

Some Factors Affecting the Survival of *Salmonella typhimurium* in Systems of Low Water Activity at 10°C
R. ISON AND W. F. HARRIGAN
Department of Food Science, Reading University, Reading RG1 5AQ, UK

The paucity of information concerning the mechanisms of survival of micro-organisms in combined stress systems has been highlighted by the increasing interest in 'hurdle' preservation techniques employed by the food industry.

The interactive effect of reduced a_w (controlled by glycerol), potassium sorbate, nutritional complexity of media, O_2 tension and pH on the survival of *Salmonella typhimurium* held at 10°C was studied over a 28 day period. Counts were performed by membrane filtration employing 0·1% peptone diluent at a_w 0·94.

After 28 days a_ws 0·94 and 0·86 gave 1 and 4 log cycle reduction in survivors, respectively. Sorbate improved survival, but this was dependent on a_w and time. The difference between counts for systems with 400 μg/ml undissociated sorbic acid and no sorbic acid were +0·30, +0·93 log cycles, day 7, and +0·04, −0·45 log cycles, day 28, for a_ws 0·94 and 0·86, respectively.

Improved survival in minimal medium over complex medium was enhanced by a reduction in a_w, the difference being +0·32 and +0·49 log cycles for a_ws 0·94 and 0·86, respectively.

The effect of oxygen tension was enhanced by increasing undissociated sorbic acid concentration and dependent on a_w. At a_w 0·94, high O_2 tension decreased survival, but increased survival in a_w 0·86 systems.

Susceptibility of *Salmonella typhimurium* to Freeze-death in Chicken Exudate

A. OBAFEMI AND R. DAVIES
National College of Food Technology, University of Reading, St. Georges Avenue, Weybridge, Surrey KT13 0DE, UK

Experiments were conducted to determine the contribution that the freezing process could make towards reducing the level of salmonella contamination in processed chicken. The exudate obtained by a repeated freeze-thawing process was used as a model system. Mid-exponential phase, TSB-grown cells of *Salmonella typhimurium* LT$_2$ at 37°C were resuspended in sterile chicken exudate then subjected to different freezing/storage/thawing régimes.

At least 1 log cycle reduction was achieved by freezing in a liquid Freon system at −30°C for 44 min followed by immediate thawing at 40°C for 13 min. Similar results (90% reduction) were achieved by freezing in a dry-ice/acetone bath at −78°C for 8 min. However, thawing at 4°C for 70 min immediately after freezing in liquid freon was found to achieve 2$\frac{1}{2}$ log cycles reduction. Up to 3 log cycles death were achieved by each of the above freezing methods when followed by storage in a domestic chest freezer for at least 24 h. The presence or absence of Puron 604 polyphosphate at 0·4% w/v conc, did not affect the susceptibility of cells to freeze-death.

A Comparison of Media and Methods for the Enumeration of Yeasts and Moulds in Refined Sugar Products

L. JANES AND R. H. TILBURY
Tate & Lyle Ltd., Group R & D, Philip Lyle Memorial Research Laboratory, P.O. Box 68, Reading RG6 2BX, UK

Six media, two incubation temperatures and three incubation times were compared for the routine enumeration of yeasts and moulds in refined sugar products. The membrane filtration technique was used and both agar and broth forms of each medium were tested. Also studied were pure cultures of organisms commonly found in sugar products. Bacteria were not completely suppressed in any of the media. Agars were easier to handle than broths. Yeast morphology agar modified by the addition of chloramphenicol prior to sterilization performed best. It gave the highest yeast counts, inhibited mould spread and performed well in pure culture studies. Three days incubation at 30°C is recommended.

A Chemical Preservative for Milk Samples

G. L. PETTIPHER AND UBALDINA M. RODRIGUES
National Institute for Research in Dairying, Shinfield, Reading, Berks. RG2 9AT, UK

Of the bacteriostats we tested, that comprising boric acid, glycerol, potassium sorbate and nystatin was the most suitable in preventing multiplication of bacteria and yeasts whilst causing least changes in the direct epifluorescent filter technique (DEFT) count of bacteria and somatic cells. The relationship between the plate count of fresh milk samples and the DEFT count of the same samples after preservation and storage for 3 days at room temperature (20–22°C) was good, with a correlation coefficient (r) of 0·80. Good agreement was obtained between the Coulter count of somatic cells in fresh milks and the DEFT count of somatic cells in stored preserved milks ($r = 0·90$). The addition of the preservative did not affect the determination of fat by infra-red analysis but did affect the determination of protein and lactose.

The Potential for Impedimetry in Brewing Microbiology

H. A. V. EVANS
Bass Ltd., High Street, Burton-on-Trent DE14 1JZ, UK

Microbial populations can be monitored by measuring impedance changes of the medium in which they are growing. A series of preliminary experiments have been performed to explore the potential for impedimetry in brewing microbiology.

Eight species and strains of bacteria (five *Lactobacillus* spp., two *Pediococcus* spp. and one unidentified Gram negative rod), and 10 species and strains of yeast (six brewing yeasts and four non-brewing or 'wild' yeasts) were grown in both brewery wort and defined media and their impedance responses recorded using the 'Bactometer 32' instrument (Bactomatic Inc., Marlow, UK).

Bacteria gave different impedance responses to yeasts whilst different species and strains of the latter were distinguishable in terms of their growth rates and generation times. Wort was superior to defined media either because it promoted faster growth and hence faster impedance changes (in the case of bacteria), or because it minimized or prevented carbon dioxide gas break-out in the Bactometer sample chambers (in the case of yeasts).

The beer spoilage potential of bacteria and yeasts at concentrations less than 10 cells/ml was determined using a new protocol. This involves membrane filtration of samples of infected beer before and after a period of incubation followed by incubation of the membranes in wort. Measuring impedance responses in the wort gave a true indication of the beer spoilage capacity of the organisms. The entire test took between 2 to 4 days, compared to traditional methods of 'forcing' the infected beer at 27°C to yield a visually detectable haze which took between one and two weeks.

Heat Resistance of *Salmonella senftenberg* as a Function of Previous Growth Temperature

N. MOHAMED AND R. DAVIES
National College of Food Technology, University of Reading, St. Georges Avenue, Weybridge, Surrey KT13 0DE, UK

The growth temperature of vegetative microbial cells is considered to have a profound influence on their subsequent heat resistance. This concept has been re-examined for cultures of *Salmonella senftenberg* grown at various temperatures between 10 and 44°C. Growth-phase dependent changes in heat resistance were observed for all growth temperatures used and were characterized by periods of maximum sensitivity for mid-exponential phase cells shifting to relative resistance in stationary phase. These shifts were of greater magnitude for 775 W than for non-heat-resistant strains. Maximum cell heat sensitivities were similar for all growth temperatures except 44°C which produced more resistant cells. Maximum heat resistance, however, was achieved at a common level independent of growth temperature by cells harvested from extended stationary phase. The fact that incubation for prolonged periods at low temperatures (e.g. 456 h at 10°C) produces resistant cell populations may have practical implications for refrigerated foods.

How do Bacteria get into Cream Cakes?
SUSAN C. MORGAN-JONES
The East of Scotland College of Agriculture, West Mains Road, Edinburgh EH9 3JG, UK

Cream cakes are a luxury commodity with a relatively short shelf-life and so would be expected to have a high bacterial standard. In a survey carried out in the Edinburgh area only 8% were classified as satisfactory (10^4/g), 18% acceptable (10^4–10^5/g) and 74% unsatisfactory (10^4/g) according to the classification of Goldberg & Elliot (1973 The value of agreed non-legal specifications. In *The Microbiological Safety of Food* ed. Hobbs, B. C. & Christian, J. H. P. pp. 359–368. London & New York: Academic Press).

Sampling during production showed that the whipping and dispensing of the cream was the main source of contamination. Storage at 4°C was effective to reduce the bacterial count for 5 h but not 24 h.

Recovery of Bacterial Spores after Heating
J. H. HANLIN, M. J. CLOUTIER AND R. A. SLEPECKY
Biology Department, Syracuse University, Syracuse NY 13210, USA

That DNA damage occurs in bacterial spores following severe heating is established (Northrop, J. & Slepecky, R. A. 1967 *Science* **155**, 839–840; Kadota, H. *et al.* 1978 *Spores* **VII**, 27–30; Amin, I. R. & Grecz, N. 1981 *Abstracts of the Annual Meeting of the American Society of Microbiology*, p. 202). We are currently using several approaches aimed at assessing the degree of damage to, and the ability to repair, DNA in heated spores. When spores of *Bacillus subtilis* NCTC 3610 and *B. megaterium* ATCC 19213 are exposed to 5 μg/ml ethidium bromide (EB) for 20 min, no killing is observed; however, survivability to various heat treatments was reduced when compared with untreated spores. The ability to form a colony is the criterion for survival in these experiments; however, there are a number of important steps in the conversion of a spore to many vegetative cells and we are currently attempting to identify the inhibited stage of outgrowth in the EB exposed and heated spores. When spores are heated then plated on complex and minimal media, the recoverability of heated but not unheated control spores is reduced on the minimal medium. This suggests that repair of damaged DNA following heating may be a prerequisite for normal colony formation and that a nutritional factor may be important. Experiments are currently in progress to determine the heat-resistance patterns of spores with an impaired ability to repair damaged DNA and to assess the nature of DNA damage following spore heating.

Carriage of *Clostridium perfringens* in Different Groups of the Population

M. F. STRINGER AND R. J. GILBERT
Food Hygiene Laboratory, Central Public Health Laboratory, Colindale Avenue, London NW9 5HT
J. E. HASSALL AND J. G. WALLACE
Public Health Laboratory, St. Anne's Road, Lincoln LN2 5RF, UK

Two of the criteria important in the confirmation of an outbreak of *Clostridium perfringens* food poisoning are (i) the presence of relatively high numbers of *Cl. perfringens* ($>10^5$ organisms/g) in the faeces of ill persons and (ii) the isolation of the same serotype from the majority of faecal specimens collected from those with symptoms. Repeated outbreaks of food poisoning and cases of diarrhoea in a number of hospitals have suggested the possibility that in some institutions certain serotypes of *Cl. perfringens* may be unusually common. A survey in one hospital showed that geriatric 'long-stay' patients carried high numbers (often $>10^6$ organisms/g of faeces) of the same serotypes of *Cl perfringens* and yet remained free from diarrhoeal illness. Young 'long-stay' patients in the same hospital carried low numbers of *Cl. perfringens* (10^3–10^4 organisms/g of faeces) and a variety of serotypes.

The implications of these findings and the frequency of isolation of enterotoxigenic strains of *Cl. perfringens* from different sources were discussed in relation to the investigation of outbreaks of *Cl. perfringens* food poisoning.

A Survey of some Hundred Outbreaks of Food-borne Disease Mainly Microbiological Aetiology in Haarlem and Surrounding Towns

SIMONE A. BOUWER-HERTZBERGER
Keuringsdienst van Waren, Nieuwe Gracht 3, 2011 NB Haarlem, The Netherlands

In Haarlem a system of reporting cases of acute gastro-enteritis to the Food Inspection Service without delay is in operation. It attempts to trace the causes of alimentary infections and intoxications, particularly by examining epidemiologically valid left-overs of suspect food. These are first of all studied by quantitative differential microscopy. The results are combined with a food anamnesis of the patients and controls, and conclude with a direct bacteriological examination.

In addition some 100 outbreaks of gastro-enteritis of acute onset, in which the suspect left-overs of food were epidemiologically valid because of the rapid reporting, and their causes were discussed. The main pathogens involved were *Bacillus cereus*, followed by *Clostridium perfringens*, *Staphylococcus aureus* and Enterobacteriaceae. These results do not indicate the real incidence of occurrence of these pathogens. They are determined by (i) the limitations of methods available for the isolation of pathogens, and (ii) the fact that patients are more likely to complain about diseases contracted in restaurants where oriental foods, like rice dishes and noodles, are served, than about similar events due to Dutch food served in restaurants.

A striking result of this study was also that food at home is never suspected by patients. Nevertheless they report incidents contracted at home because they assume that the ingredients they brought are contaminated with 'pathogens'.

Contribution by Ingredients to the Contamination of British Fresh Sausages with *Salmonella*
J. G. BANKS, JANICE PAIN, G. PRICE AND R. G. BOARD
Department of Microbiology, School of Biological Sciences, University of Bath, Bath BA2 7AY, Avon, UK

We have used a most probable number technique (five tubes; pre-enrichment at 37°C followed by enrichment in tetrathionate broth with subculture on to three differential media after incubation for 24 and 48 h at 43°C) to assess the contribution of ingredients to the contamination of sausages with *Salmonella*. Lean meat was the principal source of contamination of the 36 batches of sausages examined.

Development and Application of a Bacteriocin-typing Scheme for *Clostridium perfringens*
G. N. WATSON, M. F. STRINGER AND R. J. GILBERT
Food Hygiene Laboratory, Central Public Health Laboratory, Colindale Avenue. London NW9 5HT, UK
D. E. MAHONY
Department of Microbiology, Faculty of Medicine, Dalhousie University, Halifax, Nova Scotia, Canada

Studies on the production of bacteriocins by *Clostridium perfringens* and their potential value in typing strains from outbreaks of food poisoning have been reviewed by Mahony (1980 *Methods in Microbiology* **13**, ed. Bergan, T. & Norris, J. R. London & New York: Academic Press).

A set of 49 bacteriocins were used to type 204 strains of known serotype isolated from food poisoning outbreaks. Preliminary results indicate that strains of the same serotype from a single outbreak give the same pattern of susceptibility to bacteriocins, whereas strains of different serotype or of the same serotype but from different sources produce many variations in susceptibility patterns.

The 204 strains, along with isolates from a wide range of sources were also screened for their ability to produce bacteriocins. A much greater proportion of the strains from food poisoning outbreaks were bacteriocinogenic than were isolates from human and animal infections, various foods and the environment.

Bacteriocin typing may be a valuable complementary technique to serotyping in the investigation of food poisoning outbreaks and the epidemiological study of human and animal infection.

The Microbiology of Liquid Egg
SUSAN C. MORGAN-JONES
The East of Scotland College of Agriculture, West Mains Road, Edinburgh
I. C. MARTIN
H. D. Hardie & Co. Ltd., Edgefield Industrial Estate, Loanhead, Midlothian, UK

Raw whole egg, pasteurized whole egg, raw egg white and 11% salted egg yolk were examined for total viable bacteria, pseudomads, thermoduric bacteria, salmonellas, yeasts and moulds. The total viable count of the raw whole egg showed a seasonal variation, there being a decrease in numbers during the winter months. Pseudomonads were present in greater numbers in the raw pasteurized product while thermoduric bacteria were the dominant bacterial flora of the pasteurized whole egg. The bacterial load of the egg white was similar to that of whole egg. Salt added to the egg yolk decreased the total viable count of bacteria within 20 h.

Numerical Classification of some Psychrotrophic Bacilli Isolated from Frozen Foods
M. A. SHEARD
School of Hospitality Management and Home Economics, Leeds Polytechnic, Leeds
F. G. PRIEST
Department of Brewing and Biological Sciences, Heriot-Watt University, Edinburgh EH1 1HX, UK

Psychrotrophic, aerobic, endospore-forming bacteria were isolated from a variety of home-frozen and commercially frozen vegetables and other frozen food by incubating heat-treated (70°C, 15 min) samples in tryptone soy broth at 0–4°C for 14 days. Seventy-five strains were purified by plating from the broth on to tryptone soy agar and combined with five named *Bacillus* psychrotrophs and 14 mesophilic *Bacillus* strains representing the principal species. All bacteria were examined for 107 unit characters and data were analysed using the simple matching (Ssm) and Jaccard (S_J) similarity coefficients with average linkage (UPGMA) clustering. In the Ssm dendrogram, 68 of the food isolates were recovered in 11 clusters at 82·5% similarity. One phenon contained the type strain of *B. insolitus*, one may represent psychrotrophic variants of *B. cereus*. The remaining eight phenons could not be identified. Cluster representatives have been examined for ability to germinate, grow and sporulate at low temperature; the implication of these results for food spoilage were outlined.

Incidence of Food Poisoning in the Sudan with Special Reference to *Staphylococcus aureus* in Milk and White Cheese (Jibna Baida)

AMNA S. KHALID AND W. F. HARRIGAN
*Department of Food Science, University of Reading, London Road,
Reading RG1 5AQ, UK*

Records of food-poisoning cases involving milk and cheese were compiled from records in hospitals and the Ministry of Health, Khartoum, Sudan. It was found that presumptive *Staphylococcus aureus* food poisoning from 1977 to March 1979 accounted for 25% of all incidents. Six out of 25 incidents investigated personally by the author were found to be staphylococcal food poisoning. Of these six incidents, three were children and milk was implicated; in all the other three, cheese proved to be the vehicle.

Jibna Baida was mostly made from raw milk in farmhouse factories using very simple equipment. The basic processing steps are: (i) addition of salt (6–20%) to the raw milk; (ii) coagulation of the salted milk by rennet; (iii) transfer of curd to moulds to drain; (iv) collection of whey for packing the cheese, although brine may also be used. The consumer steeps the cheese in tap-water (if it is highly salted) before consumption.

Milk and cheese samples were collected during manufacture and storage of cheese in the Sudan. Presence of enterotoxigenic *Staph. aureus* isolates was proved by the optimum sensitivity plate method. The effect of manufacture and storage on growth of, and toxin production by, enterotoxigenic *Staph. aureus* was investigated. It was found that if toxin was not produced before cheese making and the milk is highly salted toxin may not be produced; but the steeping process encourages growth of surviving *Staph. aureus* and therefore constituted a threat to public health.

Comparison of Calcium Alginate Swab-rinse Technique and a Commercially Available Agar Contact Slide for Monitoring Bacteria on Work Surfaces and Equipment

J. A. PINEGAR
*Leeds Regional Public Health Laboratory, Bridle Path, York Road,
Leeds LS15 7TR, UK*

This survey, carried out in a hospital kitchen, compared the calcium alginate swab-rinse technique and a commercially available agar contact slide for the enumeration of bacteria on defined sites before and after cleaning. Results obtained with agar contact slides showed good agreement with those obtained by the more traditional swab-rinse method. The commercial kit was found to have certain advantages over the swab-rinse technique, being easy to use, convenient to transport and requiring on arrival at the laboratory nothing more than incubation at an appropriate temperature.

The agar contact slide was found to be a rapid and reliable means of bacteriologically monitoring work surfaces and equipment to assess the effectiveness of cleaning procedures. It has also been found of particular value in hygiene education by providing, to food handlers, a visual demonstration of the presence of bacteria in their working environment.

The agar contact slides used, Hygicult-TPC, were supplied by Gibco Europe Ltd.

Subject Index

THE SOCIETY FOR APPLIED BACTERIOLOGY SYMPOSIUM SERIES

General Editor: F. A. Skinner